Chemische Technik / Verfahrenstechnik

Henner Schmidt-Traub · Andrzej Górak

Integrated Reaction and Separation Operations

Modelling and experimental validation

With 148 Illustrations

Professor Dr.-Ing. Henner Schmidt-Traub
Universität Dortmund
FB Bio- und Chemieingenieurwesen
Emil-Figge-Str. 70
44227 Dortmund
schmtr@bci.uni-dortmund.de

Professor Dr.-Ing. Andrzej Górak
Universität Dortmund
FB Bio- und Chemieingenieurwesen
Emil-Figge-Str. 70
44227 Dortmund
andrzej.gorak@bci.uni-dortmund.de

Library of Congresss Control Number: 2006926661

ISBN 978-3-540-30148-6 Springer Berlin Heidelberg New York

Springer is a part of Springer Science+Business Media

springer.com

Cover design: medionet AG, Berlin
Typesetting: Digital data supplied by editors

Printed on acid-free paper   68/3020/m-5 4 3 2 1 0

Preface

Economic needs as well as ecological demands are major driving forces in improving chemical processes and plants. To meet these goals processes have to be intensified in order to get products of higher quality, to increase yield by reducing or even suppressing by-products and to minimise energy consumption. A preferred principle for such intensifications is process integration, especially integration of reaction and separation operations. Scientific research in this field has been boosted by certain extremely successful examples like the Eastman-Kodak process for methyl acetate or the MTBE process which are milestones for this method. In 2002 the German Research Foundation defined process integration as one of the major research topics for the next decade.

In 1998 the Department of Biochemical- and Chemical Engineering at the University of Dortmund decided to pool its activities for concerted efforts in process integration and to form a joint research cluster. Our interest was to find out the general challenges as well as obstacles of integrated processes and to work out methods for their design and valuation. Soon it became clear that theoretical work only cannot give reasonable answers. Corresponding experiments are essential to validate process modelling and to verify process simulation. But experiments also confront research and development with process reality as well as conditions of plant design and operation. This necessitates an open-minded debate if theoretical assumption is to suit real process behaviour and plant design. In order to cover a broad variety of chemical processes, reactive distillation and absorption, reactive gas adsorption, chromatographic reactors and reactive extraction were chosen as references. But the aim to develop general design methods cannot be met only by a collection of example processes. Own research added by information from open literature have to be the source for generalised conclusions implemented in a system for process synthesis. From the point of view of process operation, integrated processes are getting more and more complicated as their degree of freedom is reduced with increasing integration. Therefore research in process control was another essential.

The linkage and interrelation of our research cluster is illustrated by Figure 1.1.

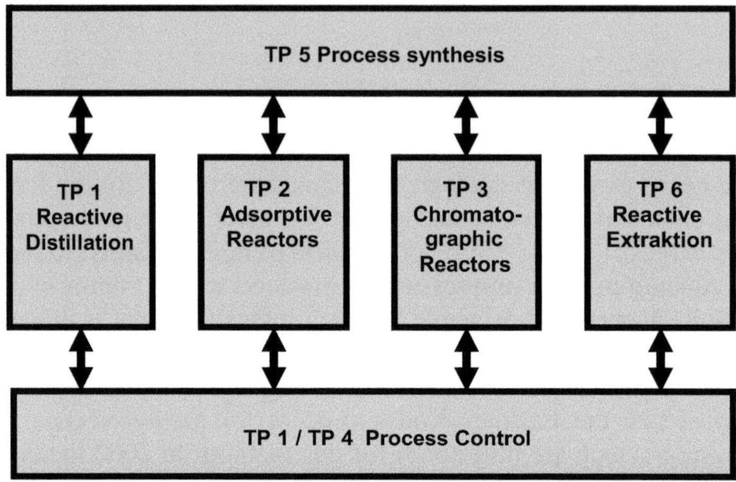

Fig. 1.1: Research cluster: Integrated Reaction and Separation Operations

Our research cluster was sponsored by the German Research Foundation (DFG) during two phases each of three years. The members in phase I are: Prof. D. Agar, Reaction Engineering; Prof. S. Engell, Process Control; Prof. A. Gorak, Thermal Separations, Dr. G. Schembecker, Process Design Centre GmbH; Prof. H. Schmidt-Traub, Plant Design;

During phase I it became clear that the degree of freedom for process design and operation is an important criterion for synthesis of efficient and effective processes. This means that the distribution of the functionalities reaction and separation within the plant may vary between homogeneous integration and a non-integrated sequential process. Therefore the research area was expanded to add other processes but also to investigate the degree of process integration in more detail. The supplemented research topics in phase II were headed by: Prof. A. Behr, Technical Chemistry; Dr. M. Grünewald, Reaction Engineering; Prof. O. Hinrichsen, Technical Chemistry.

A group of PhD students formed the basis of our research as they worked in a really integrated manner. We have to thank S. Barkmann, T. Borren, C. Dittrich, M. P. Elsner, J. Fricke, S. Geisler, A. Hoffmann, E. Y.

Kenig, P., M. Kloeker, P. S. Lawrence, C. Noeres, S. Reßler, J. Richter, G. Sand, J. Seuster, C. Sonntag, M. Völker, A. Toumi, M. Tylko for their hard and ambitious work.

Special thanks also to all technicians of our laboratories. Dirk Lieftucht, Queens University Belfast, supported us during major changes at the reactive distillation plant. Special thanks go to Christian Dusny for his technical assistance.

Several companies supported our work. In particular we would like to thank BASF AG, Bayer AG, Grace Deutschland GmbH, Degussa AG, Protocatalyse SA, Sasol Germany GmbH, Siemens-Axiva GmbH, Süd-Chemie AG and Umicore AG & Co. KG.

The support by Deutsche Forschungsgemeinschaft (DFG) by funding the research cluster FOR 344 "Integration von Reaktions- and Separationsoperationen" is gratefully acknowledged.

The results and conclusions of our research are summarised in this book where each chapter stands for a different topic of research. The overall findings and conclusions can be summarised as follows: Process integration offers interesting challenges for process intensification. But it is not a general recipe for successful improvements. The distribution of the functionalities has to be worked out thoroughly and validated by experiments. Process simulation is a must in order to find out optimal areas for process operation.

Corresponding Authors

Prof. rer. nat. David Agar
University of Dortmund
Department of Biochemical and
Chemical Engineering
Chair of Chemical Reaction Engi-
neering
44221 Dortmund
Germany

Prof. Dr. rer. nat. Arno Behr
University of Dortmund
Department of Biochemical and
Chemical Engineering
Chair of Chemical Process Devel-
opment
44221 Dortmund
Germany

Prof. Dr. Ing. Sebastian Engell
University of Dortmund
Department of Biochemical and
Chemical Engineering
Chair of Process Control
44221 Dortmund
Germany

Prof. Dr. Ing. Andrzej Gorak
University of Dortmund
Department of Biochemical and
Chemical Engineering
Chair of Fluid Separations
44221 Dortmund
Germany

PD. Dr. Ing. Markus Grünewald
Bayer Technology Services GmbH
PT-RPT-REC
D-51368 Leverkusen
Germany

Prof. Dr. rer. nat. O. Hinrichsen
University of Leipzig
Department of Chemistry and Min-
eralogy
Institute of Technical Chemistry
Linnéstr. 3-4
04103 Leipzig
Germany

Prof. Dr. Ing. Gerhard Schembecker
University of Dortmund
Department of Biochemical and
Chemical Engineering
Chair of Plant and Process Design
44221 Dortmund
Germany

Prof. Dr. Ing. Henner Schmidt-
Traub
University of Dortmund
Department of Biochemical and
Chemical Engineering
Chair of Plant and Process Design
44221 Dortmund
Germany

Table of contents

1 Introduction

Chemical manufacturing companies produce materials based on chemical reactions between selected feed stocks. In many cases the completion of the chemical reactions is limited by the equilibrium between feed and product. The process must then include the separation of this equilibrium mixture and recycling of the reactants (Figure 1.1).

$$A + B \rightleftarrows C + D$$

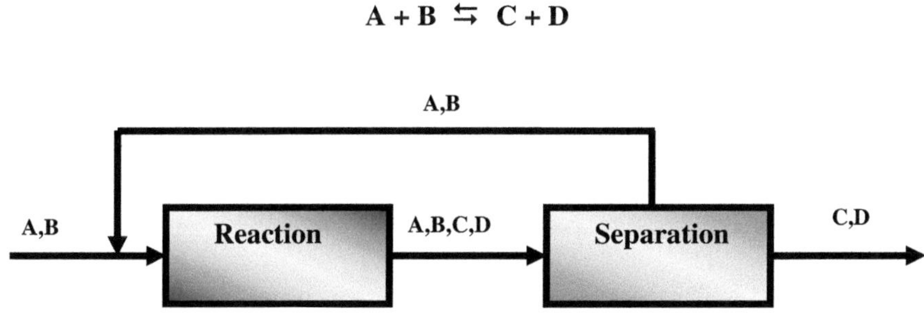

Fig. 1.1: Standard process scheme for reversible reactions in which the conversion is limited by the chemical equilibrium

The number of the separation steps depends on the number of products, catalysts, solvents as well as reactants which are not converted. The main objective functions to increase process economics are selectivity as well as reaction yield what influences the reactor design. For example the type of main and side reactions determines the selection of a continuous stirred tank reactor or plug flow respectively batch reactor. In other cases it is advantageous to achieve a 100% conversion of one reactant like in case of hydrogenations or chlorination.

Conventionally, each unit operation – whether mixing or absorption, distillation, evaporation, crystallisation, in fact, any of the heat-, mass- and momentum-transfer operations so familiar to chemical engineers – is typically performed in individual items of equipment, which, when arranged together in sequence, make up the complete process plant. As reaction and separation stages are carried out in discrete equipment units, equipment and energy costs are added up from these major steps. However, this historical view of plant design is now being challenged by seeking for combination of two or more unit operations into the one plant unit. The potential for capital cost savings is obvious, but there are often many other process advantages that accrue from such combinations. This combination has been recognised by the chemical process industries for having favourable economics of carrying out reaction simultaneously with separation for certain classes of reacting systems, and many new integrated processes have been invented based on this technology. These new units may be also treated as a kind of multifunctional reactors in which the functionalities of several processes are combined to generate the new reactor concept.

The most obvious way to improve the reaction yield in an integrated unit is a continuous separation of one product out of the reaction zone (Figure 1.2). This allows for getting a 100% conversion in case of reversible reactions.

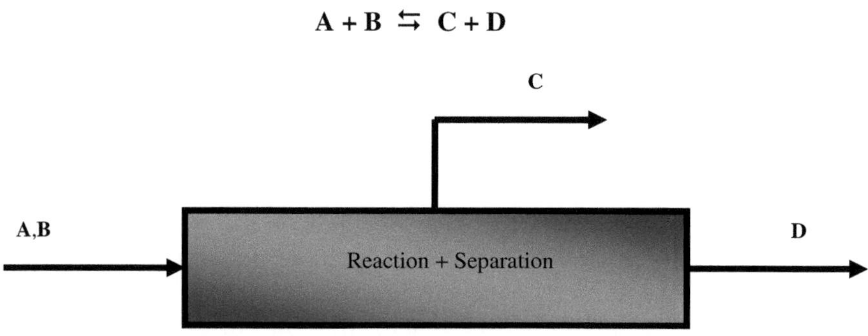

Fig. 1.2: Complete conversion of reactants in case of equilibrium reaction

But the global optimum of a process is not the sum of individually optimised process steps. Since the component separation very often is a major cost block further improvements have to focus on simplification of the separation units by changing the composition of the reactor outlet. The total conversion of one reactant e.g. to avoid of harmful or corrosive components in the reactor outlet have been mentioned already. Another important problem is e.g. separation of azeotropic mixtures which may require an additional feed stream as well as further separation units. An elegant solution of this problem is to avoid an azeotropic mixture by changing the composition of the reactor outlet as illustrated in Figure 1.3.

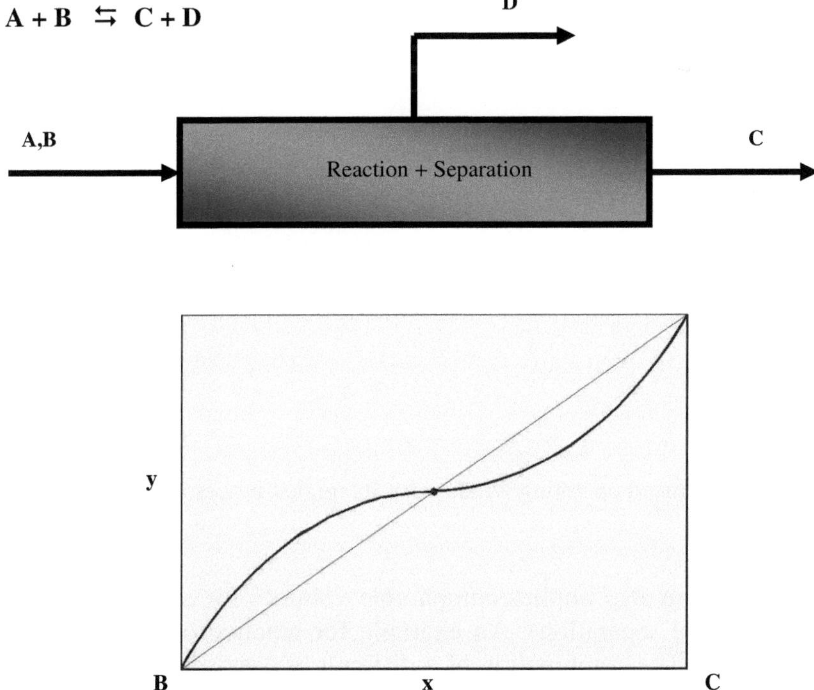

Fig. 1.3: Breaking the azeotrop B/C by process integration

The Eastman-Kodak process is one of the most striking examples for the integration of reaction and separation. But such integration may also lead

4

to some disadvantages. One of them is the necessity to operate the reaction and separation at the same pressure and temperature what reduces the degree of freedom. Also equipment design influences the operating window of an integrated process (Figure 1.4).

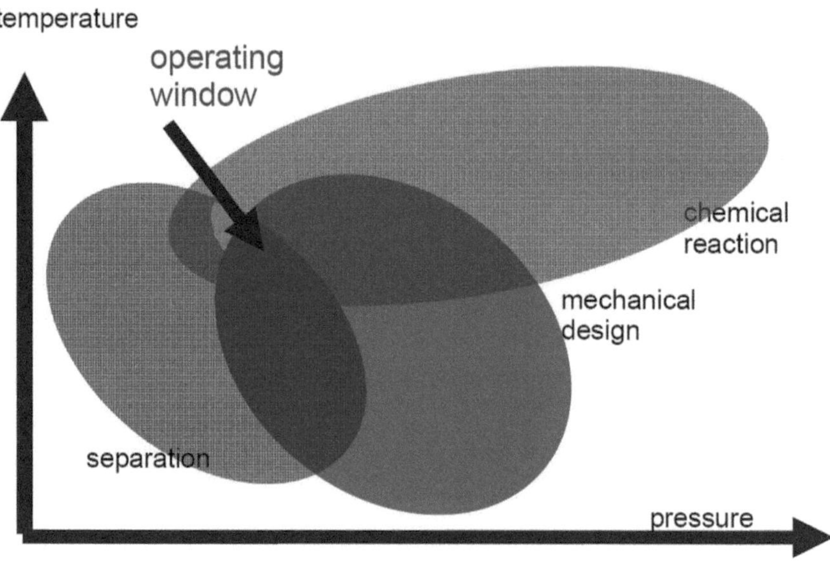

Fig. 1.4: Reduced operating window for integrated processes

Integration also implies comparable volumes for each functionality (i.e. reaction and separation). An example for reactive distillation is given in Figure 1.5. The combination of a difficult or easy separation with a fast or slow heterogeneous reaction results into a portfolio of different equipment designs. In case of a low separation factor and a high reaction rate a distillation column with quite high number of plates is needed. If only a short residence time is necessary for the reaction both functionalities can be integrated in one column. But in case of a low reaction rate more residence time respectively volume is needed which cannot be economically integrated within the column. Therefore reactors are placed outside the column as shown in Figure 1.5 down left. Quite a different design is appropriate if the separation is simple and can be achieved for instance in a small column or even by a flash drum. In this case integration is reduced to a recycle flow between two completely mixed units or a reactor and a small scale separation column.

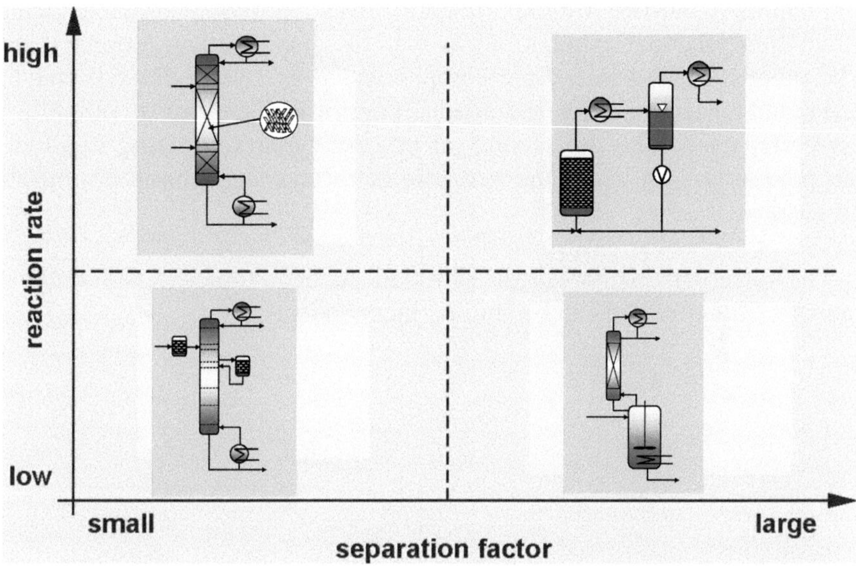

Fig. 1.5: Influence of reaction rate and separation factor on equipment design

All examples described in Figure 1.5 clearly demonstrate that the degree of integration of both functionalities, reaction and separation, is another parameter for process optimisation. Therefore, it has to be checked in each individual case whether integration is advantageous or not.

Process intensification comprises novel equipment, processing technologies and process development methods that, compared to conventional ones, offer substantial improvements in (bio-)chemical manufacturing and processing. Integration of reaction and separation into a single unit is one of the most important methods of process intensification. The integration of reaction and separation into one unit (i.e. in a single piece of equipment) or the integration of several separations leads to reactive separations or hybrid separations, respectively. This book is devoted to the first class of the integrated processes, i.e. to integrated reaction and separations.

The following chapters deal with process integration for gas/liquid, gas/solid, liquid/solid as well as liquid/liquid systems which covers a broad span of applications in chemical engineering. Theoretical modelling and simulation is verified by experimental investigations in order to demonstrate their practical applications as well as to evaluate an optimal design. The process related issues are complemented by two additional chapters: one on advanced model-based process control which ensures optimal process performance and the second on process synthesis, giving a guide for the process design using the heuristic rules and algorithmic optimisation procedures.

The different chapters deal with independent topics. They quote the relevant literature and can be studied independently.

2 Synthesis of reactive separation processes

Markus Tylko, Sabine Barkmann, Guido Sand, Gerhard Schembecker, Sebastian Engell

2.1 Introduction

The synthesis of an integrated reactive separation process is a task which can be addressed by different methods: knowledge or heuristic rule-based methods and optimization-based methods as e.g. mathematical programming. As knowledge-based methods often use graphic representations, their applicability to complex systems is limited. Moreover, it is not possible to find an optimal design by heuristic methods. As Daichendt and Grossmann pointed out, mathematical optimization techniques can be used fruitfully in the conceptual design, since important structural decisions are made at this stage of process design which can be tackled by mathematical programming techniques (Daichendt and Grossmann 1998). Typically, mathematical optimization in conceptual design is based on a superstructure representation of the process. Setting up an appropriate superstructure, which is assumed to comprise the optimal structure, is always based on process insights. Thus, a combination of mathematical and heuristic methods in the conceptual process design seems to be an obvious approach.

Ismail et al. use multifunctional process modules to describe the superstructure of the process (Ismail et al. 1999). Every operation is considered either as material or energy transfer operation. Binary variables denote the existence of certain modules in the optimal structure. The modules do not represent trays of a distillation column but column sections.

Hostrup et al. propose a three-step procedure for process synthesis (Hostrup et al. 2001). The first step is the problem formulation step which results in a superstructure of all possible flowsheets and a good initial flowsheet. Subsequently, the superstructure is reduced based on process knowledge and then optimized. Finally, the flowsheet is verified using a process simulator.

Bedenik et al. describe a strategy for the synthesis of process flowsheets which combines optimization and an analysis of the optimal solution (Bedenik et al. 2004). The optimization of the superstructure is performed in three steps. First, a reactor network is optimized, which is then extended by separation processes. Finally, heat integration is considered. The optimization is followed by a rigorous simulation. The simulation results are interpreted in a so-called economic attainable region. If promising process extensions can be gathered from this analysis, the superstructure is extended and the model is optimized again.

Similar to the literature mentioned, we propose a hierarchical approach to process synthesis. In this work, the emphasis is on the synthesis of an integrated reaction and separation process by means of heuristic rules rather than on a reduction of a large variety of process alternatives. Firstly, the aim of the process is defined. As a result the reaction paths as well as the occurring components are given. The next step is the acquisition of thermodynamic as well as chemical properties of the system. With the help of graphic methods a qualitative design suggestion is made. A superstructure model which contains this suggestion is then optimized by means of mathematical programming techniques. Heuristic information about feasible process structures is employed to generate good initial values for the optimization.

2.2 Fundamental process synthesis concepts

In this book several different integrated reaction and separation processes are investigated and discussed. Although they differ strongly from each other, from a general point of view there are also many characteristics they have in common. In sequential steps we shall develop a generic approach to the design of reactive separation processes. The starting point will be a definition of basic concepts.

Functionalities

In the remainder of this chapter, the term functionality is defined as a part of equipment element that performs a specific task inside an integrated reaction and separation (IRS) process. The four major types of functionalities considered here are:
– reaction,
– separation,

– reactive separation, and
– mixing.

They are distinguished according to the classification of Harmsen (Harmsen 2004).

A reaction functionality represents all kinds of single phase reactions while a separation functionality represents all kinds of separation operations. In contrast to a simple splitter, a separation operation always involves at least two different phases (as, for example distillation, extraction etc.).

The purpose of a reaction functionality is to enable chemical conversion. Here enhancing the rate of the main reaction is aimed at while simultaneously suppressing undesired side reactions. This can be achieved by choosing favorable operating conditions and a suitable residence time distribution as well as by using an effective catalyst. Another task of a reaction functionality is to overcome thermodynamic limitations like solidus curves, binodal curves, and distillation boundaries.

A separation functionality is introduced to purify the products. It may also help to increase chemical conversion by recycling unconverted reactants back to the reactive functionality. Thus chemical equilibrium-limited reactions may be driven to higher rates of conversion.

A reactive separation functionality is not only a combination of reaction and separation but represents a specific functionality with specific potentials and limitations. On the one hand, it combines some of the advantages of reaction and separation, on the other hand adverse effects may occur when using this functionality improperly like, for example, the occurrence of reactive azeotropes (see section 2.3.7).

The functionality of mixing may include, besides the conventional blending of separate streams, operations as, for example, recycling or feeding to support one of the other three functionalities.

Each of the four functionalities serves different purposes and is subject to specific constraints. A particular functionality can only be realized if all of its specific constraints can be satisfied simultaneously. Otherwise it will not be active. For the process designer, this provides several options to activate or to deactivate a specific functionality without influencing the other functionalities.

Some of the constraints mentioned are valid for all functionalities. These are
– feasible operating conditions (for a separation functionality, for example, the choice of pressure and temperature has to ensure phase split-

ting; for a mixing functionality streams to be mixed must usually have the same phase condition),
– sufficient residence time, and,
– sufficient driving forces (concentrations away from the chemical or thermodynamic equilibrium).

Other constraints are only valid for some functionalities. For a reaction functionality or a reactive separation functionality, for example, all reactants have to be present in the reaction phase simultaneously in space and time, and, in addition, a catalyst that remains active for a sufficiently long time under theses conditions is often necessary.

For a separation functionality or a reactive separation functionality the existence of (at least) two phases is essential.

The classification of different functionalities and the knowledge of the corresponding specific constraints are crucial for the development of IRS-processes and also assist in the application of the operation window method (explained below in this section), especially in terms of the spatial distribution of functionalities.

Phase model

The major characteristic of an integrated reaction and separation operation is the existence of at least two phases, a reaction phase and a transport phase. While the reaction phase is primarily used to provide residence time for the reaction, the transport phase is used to supply or to withdraw material (reactants, products, by-products or inerts) or energy (heating or cooling) to or from the reaction phase. The interfacial mass transport is caused by the differences in the chemical potentials of the components. Different separation properties cause a material flux through the interfacial area. Some examples are summarized in Table 2.1.

The removal of reaction products from the reaction phase is the most common application of a transport phase in integrated reaction and separation processes (see, for example, the Claus process in chapter 4). The main benefit here is to push conversion of chemical equilibrium-limited reactions to higher values (sometimes up to 100 %) according to the law of mass action. In case of side reactions (consecutive reactions) the transport phase can also be used to obtain favorable concentrations in the reaction phase due to the withdrawal of reactants of the side reaction.

Another possibility to suppress side reactions is to use the transport phase as a supplier of reactants. If higher selectivities are expected from low reactant concentrations, a slow diffusion-controlled phase transition of the reactants can be advantageous (see example 4 below).

Table 2.1: Phase combinations in reactive separation functionalities

Separation property	Reaction phase	Transport phase	Measure to generate interfacial area	Process example
Permeability	Gas	Gas	Membrane (sweep gas)	Reactive vapor permeation
Permeability	Liquid	Gas	Membrane (sweep gas)	Reactive pervaporation
Volatility	Liquid	Gas	Gas	Reactive stripping
Volatility	Liquid	Gas	-	Reactive distillation
Volatility	Liquid	Gas	-	Reactive condensation
Permeability	Liquid	Liquid	Membrane	Reactive membrane process
Solubility	Liquid	Liquid	-	Reactive extraction
Solubility	Liquid	Liquid	Solvent	Reactive extraction
Solubility or melting point	Liquid	Solid	-	Reactive crystallization
Adsorbability	Liquid	Solid	Adsorbent	Reactive chromatography
Diameter	Liquid	Solid	Filter	Reactive filtration
Adsorbability	Gas	Solid	Adsorbent	Reactive adsorption

Apart from these possibilities to improve the performance of the reaction functionality, integrated operations also offer the opportunity to enhance separation processes. An example is the chemical conversion of components that participate in azeotropes to simplify downstream processing. Industrial examples of this are the synthesis of methyl-acetate (MeAc) or methyl tertiary butyl ether (MTBE) via reactive distillation (Agreda et al. 1990, Beckmann et al. 2002, Schoenmakers and Bessling 2003).

Heuristic rules

Decisions of process engineers are based on expert knowledge. Sometimes the conditions under which this knowledge is applied can be quite complex. A common technique to capture expert knowledge is to formulate rules, termed heuristic rules. A heuristic rule is structured like an "if – then" clause. The first part defines a goal and a specific situation in which the rule is valid, the second part provides a potential solution to achieve the goal. Heuristic knowledge is applied in many areas of process engineering, for instance in the choice of unit operations, design of plant layouts, column sequencing, etc. Heuristic rules often have the character of rules of thumb but there is usually a scientific or monetary reason on which each rule is based. Two examples may help to explain this statement.

One major heuristic rule for plant design says that "if pumps are needed in the process then they should be positioned on the ground floor". While this rule appears to be merely a rule of thumb, there are several reasons for it:
– Cavitation can be avoided,
– Pumps often have to be placed on special foundations,
– Vibrations in the steel frame are avoided,
– As pumps need to be maintained frequently, it is easier to access and to replace them if they are put on the ground floor,

Another heuristic rule in a different context says that "if conversion of endothermic reactions has to be increased (higher equilibrium constant) then temperature should be increased". This rule is based only on thermodynamic considerations and can easily be proved by the van't Hoff equation.

Many different heuristic rules have been stated in the available literature for the design of reaction and of separation operations (Blass 1997, Douglas 1988, Levenspiel 1996, Seider et al. 2004). The structure of the process synthesis strategy introduced later in this chapter allows us to apply many of these rules, even in the design of integrated reaction and separation processes.

Tasks

In contrast to single-phase reactors, integrated operations offer the opportunity to influence the concentration profile inside the reaction phase without feeding additional components. To achieve this, the non-reactive transport-phase of an IRS-unit has to fulfill several tasks. As described above, these tasks are the separation of products as well as the supply of reactants by means of interfacial mass transfer.

Besides these two tasks, in counter-current operations a third one is possible which is called accumulation. In a counter-current process, the driving forces for interfacial mass transfer change along the apparatus (for example the temperature profile in a distillation column). Components with intermediate separation factors (intermediate boilers) will first move with one phase (vapor phase) along its flow direction (towards the top of the column). When the driving force to leave this phase becomes strong enough (lower temperature) then these components will move to the other phase (liquid phase) and move into the opposite direction (towards the bottom of the column). Depending on the operation parameters (mass ratio of outlet streams, recycles, reflux ratio, pressure and temperature profiles), a beneficial accumulation effect can be achieved for specific components in some cases (Tlatlik and Schembecker 2005).

This accumulation effect can be observed, for example, in the synthesis of ethylene glycol via reactive distillation described by Ciric and Gu (Ciric and Gu 1994). The reaction system consists of one main reaction

$$\text{ethylene oxide (EO)} + \text{water (H}_2\text{O)} \rightarrow$$
$$\text{mono-ethylene glycol (EG)} \tag{2.1}$$

and several side reactions to higher glycols. Here only the first one will be considered:

$$\text{ethylene oxide (EO)} + \text{mono-ethylene glycol (EG)} \rightarrow$$
$$\text{di-ethylene glycol (DEG)} \tag{2.2}$$

The boiling points in this system increase from EO over H_2O and EG to DEG. According to conventional heuristic rules for reaction processes, the concentration of EG should be low in the reaction phase and the concentration of water should be rather high. The concentration profile of the reactive distillation column shows exactly this behavior (Ciric and Gu 1994). Besides the necessity of distributed feed streams (water above EO) and a non-reactive section at the bottom of the column (to purify the product, see section 2.3.9), the accumulation effect (excess of water by 200) regarded here is the reason for the successful (but on the other hand energy consuming) operation of this integrated column (Tlatlik 2004).

All tasks of the transport phase mentioned so far serve the purpose of influencing the concentration profile. In addition, IRS operations are often characterized by intensive heat-integration. In these cases, the transport phase has the additional function to supply or withdraw energy to or from the reaction phase. In some publications this task is described as being very useful, for example when the heat of reaction is used to support the reboiler of a reactive distillation column. Example 1 (sections 2.3.11 and 2.4.3), however, shows that this kind of direct heat integration is not always favorable. In some cases it should rather be avoided by decoupling reaction and separation from each other by distributing functionalities.

Level of integration

The structuring of IRS-processes is possible on the micro-scale and on the macro-scale. Micro-scale means distributing reaction functionality (catalyst) and separation functionality (e.g. adsorbens) within a single particle (see chapter 9), macro-scale means distributing functionalities in or between one or more unit operations. The process synthesis strategy described in this chapter is limited to the macro-scale distribution of functionalities.

Combining reaction and separation in one single unit operation always leads to dependencies on the operational parameters as well as to two-way interactions and influences. These dependencies are mathematically described by the phase rule (Perry et al. 1997).

$$N_F = N_C - N_P - N_R + 2 \tag{2.3}$$

With N_F = degrees of freedom
N_P = number of phases
N_C = number of components
N_R = number of independent equilibrium reactions

The term "2" represents the operation parameters pressure and temperature. The number of independent reactions N_R is equal to or lower than the number of reactions in the system. To determine the number of independent reactions see Bearns et al. (Bearns et al. 1999).

Distribution of the functionalities reaction, separation and reactive separation on a process influences the parameters N_C, N_P and N_R in eq. (2.3). A classification of the different possibilities to distribute functionalities is shown in Figure 2.1. On the left, a completely integrated apparatus is shown only performing the functionality reactive separation. This type of integration will be called a homogeneous distribution. It represents the highest level of integration where the number of degrees of freedom is the lowest. Typical examples of this type of processes are fully catalytic reactive distillation columns, reactive gas adsorptions, or simulated moving bed reactors in reactive chromatography as well as all single-phase reactors and all two-phase separation units. The second type of distribution of functionalities is called heterogeneous distribution Figure 2.1,b. Here at least two different functionalities are arranged within one single apparatus, for example a reactive distillation column with reactive and non-reactive stages. The number of degrees of freedom F in the non-reactive section (assuming only one reaction) increases by one ($N_R = 0$) compared to the reactive section ($N_R = 1$), according to eq. (2.3). As the separation functionality in such a structured apparatus performs a task that is different from the task of the reactive separation functionality, this additional degree of freedom may help to achieve the design objectives (see example 1, MTBE process).

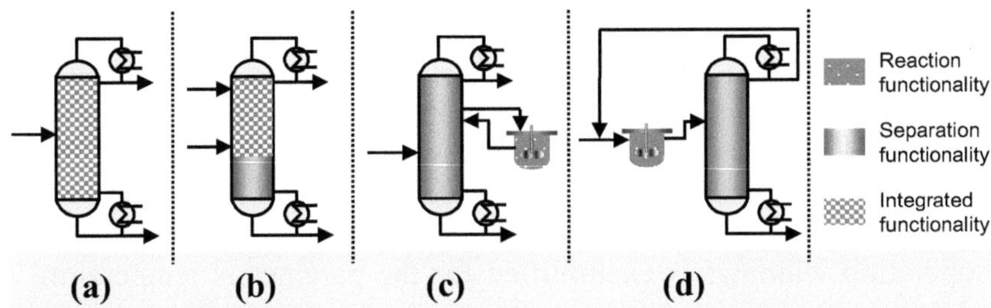

Fig. 2.1: Basic classification for different levels of integration (a: homogeneous distribution, b: heterogeneous distribution, c: partially integrated distribution, d: sequential arrangement)

In the case where an integrated reaction-separation process offers potential cost savings compared to a conventional process, but where the operating parameters for reaction and separation do not match (see the section on operation windows), a third type of distribution of functionalities called partially integrated distribution can be useful (Figure 2.1,c). In this case at least two pieces of equipment that realize different functionalities are needed (for example a distillation column ($N_P = 2$) with external reactors ($N_P = 1$) or a chromatographic Hashimoto process (Hashimoto et al. 1983). The overall process, however, is still an integrated one. Although some of the operation conditions such as temperature and pressure are decoupled, both the external reactor and the corresponding distillation stages are still interacting closely with each other. In Figure 2.1,d the fourth basic type of functionality distribution is the conventional sequential arrangement of reaction and separation.

Operation windows

The number of boundary conditions for integrated processes increases with the level of integration. This can be illustrated by the operation window method developed by Schembecker and Tlatlik (Schembecker and Tlatlik 2003) . This method helps to evaluate whether coupling of reaction and separation is possible in a single apparatus. For the design of inte-

grated unit operations, three different aspects have to be taken into account:
– chemical reaction,
– phase separation,
– apparatus design.

The influence of these aspects on the process design can be represented by multi-dimensional operation windows. The dimension of the operation windows is given by the number of operating parameters that have a limiting influence on the process, such as pressure, temperature, pH value, residence time, concentration level of a specific component (corrosive, prevention of explosive atmosphere) etc.. In Figure 2.2 the superposition of the operation windows is exemplified for the parameters temperature and residence time.

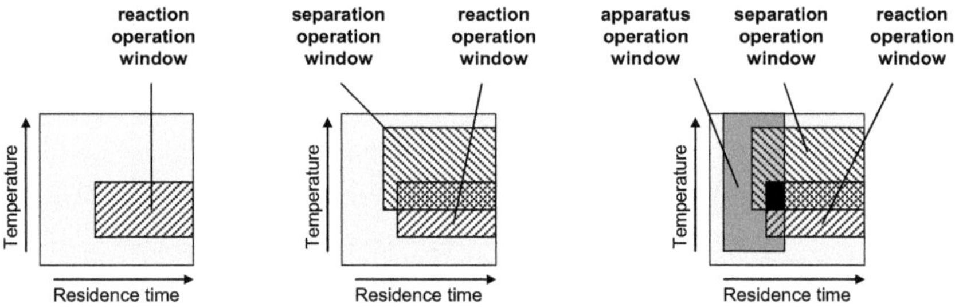

Fig. 2.2: Superposition of operation windows (schematic for two operation parameters only (Schembecker and Tlatlik 2003))

On the left side of Figure 2.2, the shaded area shows all feasible operation points for the reaction, resulting for example from a minimum temperature for catalyst activation and a maximum temperature for the disintegration of a component and a minimum residence time for sufficient conversion. By adding the separation step (Figure 2.2, middle) and choosing a specific type of apparatus (Figure 2.2, right) two more operation windows are defined. Each of them causes additional limitations to the designer, and only where all three operation windows overlap, an integrated process is feasible.

It is not very likely that the optimal operating conditions for reaction and separation as well as for the cost of the equipment lie in this overlap of the three operation windows. Hence choosing an operation point always means finding a compromise among all parameters.

Of course it is possible that the specific operation windows do not overlap at all. In this case it has to be checked whether it is possible to adjust or to change some of the limiting factors. In Figure 2.3 the example of a reactive distillation process is shown. Here the operation windows overlap for reaction and separation and also for separation and apparatus (distillation column). However, since the distillation column cannot provide enough residence time to achieve sufficient chemical conversion per stage no overlapping between reaction and distillation column

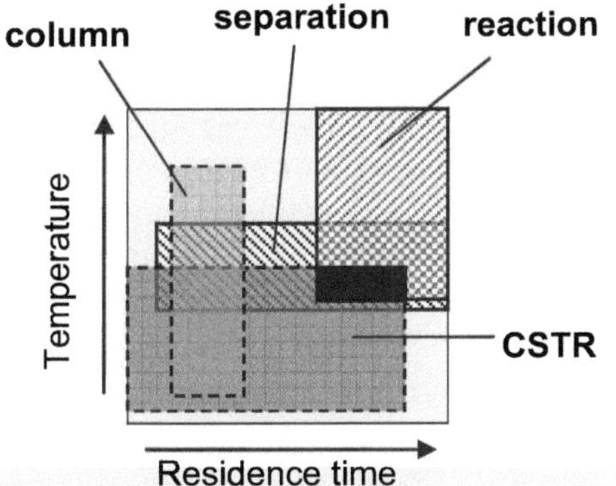

Fig. 2.3: Operational windows for partially integrated distribution of functionalities

can be found. Alternatively, a continuously stirred tank reactor (CSTR) can provide the residence time required. In this case combining a distillation column with one or more external reactors might be a useful option to accomplish an IRS process. The resulting structure will be partially integrated according to Figure 2.1. Combined structures like this have already been mentioned in several publications. The intentions here are not always to increase residence time but also to change other operation parameters like pressure and temperature or to simplify catalyst exchange (Baur and Krishna 2004, Citro and Lee 2004, Jakobsson et al. 2002).

2.3 Process synthesis strategy

To find not only feasible but also, economically seen, promising process alternatives in the enormous number of possible arrangements of unit operations is one of the most challenging tasks for process designers. Many decisions are made in the early phases of process development. The major problem is that these decisions have, on the one hand, a great impact on plant economics and, on the other hand, have to be made based upon limited information (Clark and Lorenzoni 1978). This is true for conventional processes, and even more so for integrated processes. Therefore, the process synthesis strategy proposed in the sequel will provide guidance to develop integrated reactive separation processes based on information that is easily accessible in the available literature or by experimental investigations. The strategy will first be explained in general and then be applied to four examples that have been investigated at Dortmund University.

2.3.1 Process goals

The plant or equipment design is strongly dependent on the specific goals for the process under consideration. For the same reaction, the equipment design may differ strongly if different goals are pursued, for example whether a product should be produced and purified to be sold or a large amount of a toxic component must be converted (see chapter 9).

Typical goals for the introduction of integrated reaction and separation processes are the increase of conversion or selectivity, the achievement of a better space-time yield than in conventional processes, the optimization of heat integration, and the simplification of downstream processing.

2.3.2 Data acquisition / thermodynamic analysis

For process development it is essential to have reliable and complete thermodynamic data for the component system under consideration. Furthermore, it is desirable to know as much as possible about the interactions of the species in the system with potential auxiliary components (solvents, adsorbents etc.). In reality, not all information required is easily assessible. For common chemicals the most important data as, for example, vapor-liquid(-liquid) phase behavior have already been investigated experimentally and are published in the accessible literature or available in data bases like DETHERM or Dortmunder Datenbank (DMD). These experimental data sets are necessary for the mathematical simulation of the phase behavior using, for example, activity-coefficient models like NRTL (Renon and Prausnitz 1968), UNIQUAC (Abrams and Prausnitz 1975) or WILSON (Wilson 1964). For components for which no data has been measured or published yet, the UNIFAC approach (Fredenslund et al. 1975, Gmehling et al. 1979) offers an alternative to experimental investigations. This approach, however, only provides semi-quantitative information on the thermodynamic behavior and is less useful for some types of molecules (for example large molecules with numerous functional groups) as it was developed primarily for short-chain hydrocarbons.

Special simulation tools offer the opportunity to generate information for process synthesis, for example topologies of distillation and reactive distillation areas, determination of reactive and kinetic azeotropes, shapes of the reaction space (Bessling et al. 1997, Frey 2001, Ung and Doherty 1995c), non-reactive and reactive miscibility gaps or entrainer effects.

In order to obtain information about reaction kinetics, interaction with auxiliary materials like solvents (solubility), adsorbents (isotherms), membranes (permeability) and also technically relevant information like temperature or pressure sensitivities of the participating components and catalysts, experimental investigations are often necessary.

2.3.3 Investigation of the reaction phase

The main goal in the design of reactive separation processes is to enhance reaction efficiency in terms of higher conversion and/or selectivity. Therefore the optimal conditions in the reaction phase have to be determined first. Here conventional heuristic rules for reaction engineering (Bühner 2001, Dröge 1996, Fried 1991, Levenspiel 1999, Westhaus 1995) may be applied to figure out advantageous concentration profiles (for example to suppress side reactions), optimal operation conditions (for example temperature, pressure and residence time) and also the optimal catalyst for the reaction phase.

2.3.4 Identification of incentives

Integrated processes offer potentials for saving investment or operational costs of chemical process plants; but on the other hand, due to their high level of integration, they are often not easy to operate and may also cause new difficulties that do not occur in conventional sequential arrangements of reaction and separation. Therefore it is quite important to check which advantages can be expected from a combination of reaction and separation in one piece of equipment.

2.3.5 Selection of the separation process

When the optimal conditions for the reaction phase have been determined and an additional transport phase proves useful to attain reaction conditions that cannot be achieved in a single phase reactor, a suitable separation process must be determined which fulfils the demands that result from the investigation of the reaction phase, for example "withdraw the main product".

The decision as to which kind of transport phase is most suitable or economical cannot be made by a general heuristic approach as there are too many factors that influence the decision. Schembecker provides a useful guideline for the choice of the transport phase (Schembecker 1998a):

1. Separate phases that already exist in the system (using for example a decanter, sedimentation).

2. For phase creation, prefer the input or the removal of energy as, for example, pressure and temperature change (condensation, crystallization) over the use of auxiliary components (for example extracting agent), to avoid additional separation steps and recycle streams.
3. Prefer single-stage processes instead of multi-stage processes (flash, membrane separations).
4. When additional components are necessary for phase separation (extraction, azeotropic distillation, adsorption), prefer components which are already present in the process over external components. Separation and recycle of these components is usually easier.

Table 2.1 may give assistance to find the suitable separation property and therefore a separation process for a given separation task.

Example: In case highly purified ethanol (> 99 wt.%) has to be extracted out of a liquid 50 wt.% mixture with water, the problem cannot be met with the first step of the selection order given above, because the mixture does not separate into two phases (a miscibility gap) at ambient conditions. Solubility (Table 2.1) is therefore not a suitable separation property. A certain difference in volatility of both components, however, can be found. Nevertheless, simple energy supply in a single-stage separation process (flash separator; steps 2 and 3 of the selection order) is not useful to gain pure ethanol. The difference in the volatilities is not large enough. A multi-stage counter-current operation like a distillation column is able to separate the components even when the difference is small. In this particular case, however, a low-boiling azeotrope occurs at 95.6 wt.% of ethanol. At this point the difference of the volatilities is zero, thus concentrations above 95.6 wt.% cannot be reached in a distillation column. Therefore, this separation property alone is not sufficient. Another separation property mentioned in Table 2.1 is permeability. Simple polyvinylalcohol/polyacrylnitril membranes offer the possibility to separate ethanol from water with purities above 99 wt.% by pervaporation (Rautenbach, 1997). This technique offers a feasible solution to the separation problem at hand. Phase creation is achieved by a heterogeneous auxiliary component, the membrane, together with a change of pressure (energy supply). In contrast to additional homogeneously distributed auxiliary components as strip gases or extracting agents, further separation steps and recycles of the auxiliary components are not necessary.

In reality, for example for the production of bio-alcohol, distillation (up to 95 wt.% ethanol) and membrane separation (over 99 wt.%) are coupled to reduce the cost of the expensive membranes.

The guideline for the selection of a separation process obviously intends to save energy as well as equipment cost and was originally proposed for conventional separations. For integrated reaction and separation processes it is nonetheless applicable, as the examples in section 2.3.11 show.

2.3.6 Knock-out criteria

Having figured out the type of IRS-unit operation, the next step is to check whether it is possible to integrate both functionalities into one single apparatus. This can be done by checking all specific constraints for each functionality, for example whether a suitable catalyst is available or residence time is sufficient. Here the operation window method may offer valuable assistance in checking different knock-out criteria. However, it has to be considered that even if the operation windows do not overlap (for example in case a distillation column does not provide enough residence time for the reaction), a partially integrated reaction-separation process (with an external CSTR) as shown in Figure 2.1,c might still be feasible. A knock-out criterion therefore is only valid when it prohibits the realization of an integrated process under all circumstances and none of the possible additional steps (section 2.3.9) allows the integration of reaction and separation.

2.3.7 Estimation of product regions for full integration

The estimation of product regions via graphic methods is very useful for the process designer to develop separation sequences. Because of the summation condition

$$\sum_{i=1}^{N_C} x_i = 1 \tag{2.4}$$

with N_C = number of components and
$\quad x_i$ = molar fraction of component i

an N_C-component system can always be displayed in an $N_C - 1$ dimensional simplex (Grassmann 1971), called component space, with each component lying in one vertex of the simplex. A three-component system, for example, can be displayed in a two-dimensional triangle diagram. In such a diagram the composition of the feed stream entering a separation apparatus can be marked and, according to the lever rule, feasible splits by separation can be deduced. However, for a special separation process and specific properties of the component system not all splits will be feasible (the intermediate boiler of an ideal three-component system can never be purified in a simple distillation column). The limiting factors depend on the separation process considered. For distillation, the product region is bordered by distillation curves, azeotropes and distillation boundaries (Doherty and Perkins 1978a, Doherty and Perkins 1978b, Stichlmair 1988). For crystallization processes, the corresponding limitations are the solidus and liquidus curves, melting points, eutectics and crystallization areas (Dye 1995, Myerson 2002). For extraction processes, tie-lines, binodals and solutropes have to be taken into account (Samant and Ng 1998a, Samant and Ng 1998b, Smith 1950).

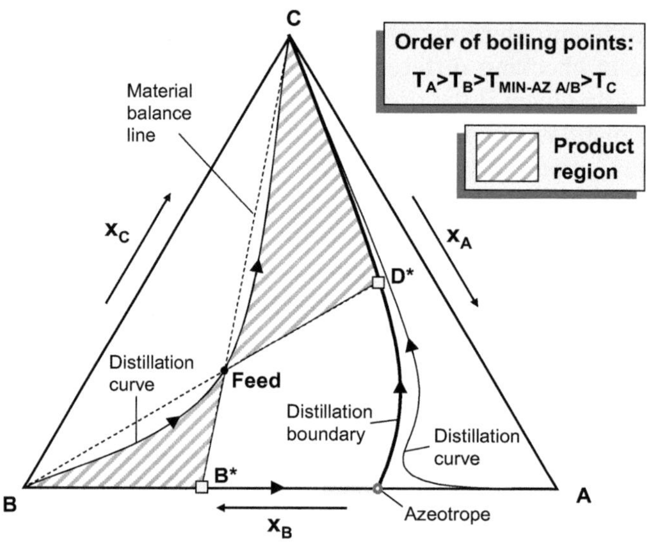

Fig. 2.4: Product areas for a ternary distillation with minimum azeotrope

A simple example will help to understand the meaning of the term product region.

In Figure 2.4 the ternary system ABC with the high boiler A, intermediate boiler B, low boiler C and a minimum boiling azeotrope between A and B is shown. For the estimation of product regions, distillation curves or residue curves are necessary. A distillation curve describes a column composition profile (from bottom to top) for a distillation column at total reflux ($x_{i,\ Stage\ n} = y_{i,\ Stage\ n+1}$ with n = 0, 1,...) and vapor-liquid equilibrium on each stage. A residue curve describes the change of the composition of the equilibrium liquid residue of a simple batch distillation still over time. A detailed description of residue and distillation curves can be found for example in Widagdo and Seider (Widagdo and Seider 1996).

The distillation boundary between the binary azeotrope and component C in Figure 2.4 divides the component space into two distillation areas.

The feed composition is located on the left side. Feasible splits in a distillation column without side-draws are marked by the dashed areas, the so called product regions. The shape of the product regions can be found by the application of two simple rules:
- The distillate, feed and bottom compositions are connected by a straight line,
- The distillate and the bottom composition must lie on the same distillation curve.

To satisfy overall and component balances, the feed composition must be an intermediate point of a straight line between distillate and bottom product. This line is called material balance line (Figure 2.4, dashed). The lever rule is valid for this line. The other constraint for the shape of the product regions is that both distillate and bottom composition must lie on the same distillation curve. Therefore, the product regions are bordered by the distillation curve through the feed composition and the one connecting the points B, B*, Azeotrope, D* and C (which includes the distillation boundary: Azeotrope, D*, C) in Figure 2.4.

Here either the lowest boiling component C can be purified towards the top of a distillation column (with bottom composition B*) or component B (the highest boiling component in left distillation area) can be purified towards the bottom (with distillate D*). Other splits inside the product region with the feed composition as an intermediate point of a straight line between distillate and bottom are also possible. A purification of the highest boiling component A is not possible because the distillation boundary cannot be crossed (top and bottom composition must lie on the same distillation curve). Crossing the distillation boundary is only possible when the feed composition lies on the concave side of a highly curved distillation boundary and is also located quite close to this boundary (Seader and Henley 1998). A detailed treatment of product regions in distillation processes is given by Wahnschafft et al. (Wahnschafft et al. 1992).

For other separation processes like extraction, crystallization, membrane separations or chromatography similar considerations lead to comparable results for the product regions (Schembecker 1998a). The thermodynamic non-idealities (like azeotropes) are referred to as solutropes (Smith 1950, Vriens and Medcalf 1953), eutectics (Myerson 2002), arheotropes (Huang et al. 2004) or adsorptivity reversal (Basmadjian and Coroyannakis 1987, Basmadjian et al. 1987a, Basmadjian et al. 1987b, Grüner 2004).

In processes that are not restricted to the component system of the chemical reaction but build at least one of the phases by adding an additional component (for example membrane separation, gas adsorption or chromatography), these effects are usually avoided by changing the phase-creating material, either the adsorbent, the membrane or the solvent. In a distillation process, the components themselves constitute both phases (liquid and vapor). Avoiding azeotropes is therefore not possible because the phase-creating material cannot be changed.

The product region concept is also applicable to reactive systems. For N_R independent equilibrium reactions, the number of degrees of freedom in the system is reduced by N_R according to eq. (2.3). The resulting component space can be projected on a simplex with N_C-1-N_R dimensions. For a simple esterification process of the type

$$A + B \rightleftharpoons C + D$$

the component space is a three dimensional tetrahedron while the equilibrium surface, called reaction space, has only two dimensions (see Figure 2.5). The transformation procedure for projecting this equilibrium surface on a plane is described in detail by Ung and Doherty (Ung and Doherty 1995a, Ung and Doherty 1995b). The major difference to the original component space is that the positions inside the transformed component space are no longer described by molar compositions (x_i) but by so called transformed compositions (X_i) which are only valid for chemical equilibrium.

The estimation of product regions for reaction separation processes using transformed composition diagrams is possible in the same way as in molar based diagrams. Distillate, feed and bottom compositions are connected by a straight line. Distillate and bottom product now lie on the same reactive distillation curve ($x_{i,\ Stage\ n} \rightarrow x^*_{i,\ Stage\ n} = y_{i,\ Stage\ n+1}$ with x^*_i mole fraction of component i in chemical equilibrium; n = 0, 1,…) or reactive residue curve. It has to be stressed that the resulting product regions are only valid for completely reactive apparatuses, assuming chemical equilibrium at each location (Figure 2.1,a). If non-reactive sections are distributed within the apparatus (Figure 2.1,b), transformed composition diagrams are no longer valid. Since the process synthesis strategy for integrated reaction and separation processes described here starts with the investigation of a single reactive separation functionality, these diagrams may be used to estimate feasible split fractions. It is important to state that their application is restricted to cases where the desired product is one of the vertices of the transformed diagram. If the desired product is one of the reference components for the transformation procedure, it is not one of the vertices and a transformed composition diagram is not useful.

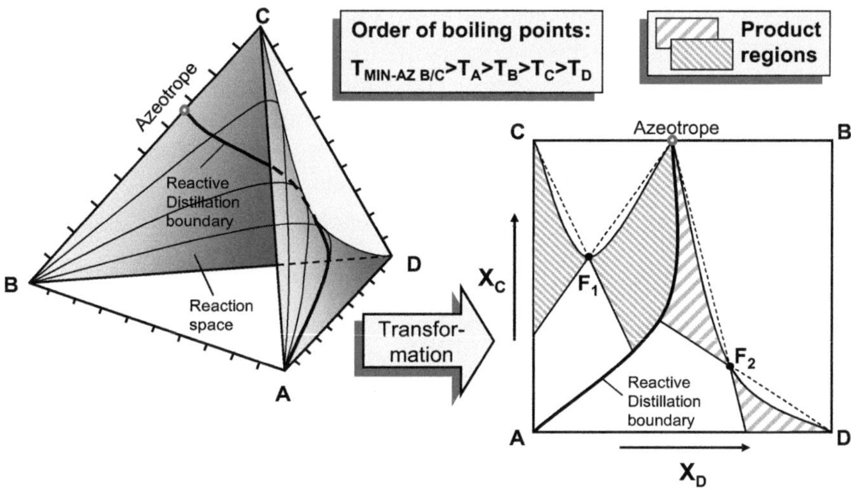

Fig. 2.5: Transformation of a reaction equilibrium surface in component space (left) to transformed coordinates (right). Here product areas can be determined in two dimensions.

In reactive systems, non-idealities like azeotropes can be broken by reaction if they are not part of the reaction space. However, due to the reaction new non-idealities called reactive azeotropes (Song et al. 1997), reactive arheotropes etc. can occur when reaction and separation compensate each other (Kienle et al. 2005). These limitations cannot be overcome by a simple reactive separation functionality. Therefore additional functionalities become necessary.

2.3.8 Measures to achieve the desired product quality

If the product regions that were determined in the molar composition diagram or in a transformed composition diagram do not include the vertex of the desired product, the composition of the feed stream is usually not suitable (for example, located in the wrong reactive distillation area). A different composition profile with one component in excess sometimes leads to better results. This excess can be reached in different ways:
- excess in the feed stream (separation of the excessive component will be necessary afterwards),
- recycle (requires an additional separation functionality),
- internal accumulation (only possible in counter-current processes).

If a change in the concentration profile is not possible or not advantageous, another option is the change of the operating conditions as, for example, temperature and pressure. The thermodynamic behavior of the component system (boiling and melting points, azeotropes, reactive azeotropes, chemical equilibrium, miscibility gaps, adsorbilities, permeabilities etc.) often depends strongly on these two parameters.

If the desired product region cannot be reached with any of these measures, additional functionalities have to be added to the flowsheet.

2.3.9 Necessity of additional steps

Additional steps can be necessary in the following cases:
- The operation windows for reaction, separation and apparatus do not overlap.
- A single reactive separation functionality is not sufficient to fulfil the goals defined at the beginning of the process development.

In both cases different functionalities have to be added to the reactive separation functionality. In case the operation windows do not overlap, a partially integrated structure might be useful; in case the goals (conversion, purity) are not reached, additional measures like recycle streams can be introduced by additional functionalities. These functionalities can either be distributed inside one equipment element together with the reactive separation functionality or they can be arranged sequentially. For each additional functionality the knock-out criteria (section 2.3.6) have to be checked again.

2.3.10 Simulation and optimization

Integrated reaction and separation processes influence both the upstream and the downstream units of a chemical plant. A comparison of the cost with conventional processes is therefore only valid when complete and optimized flowsheets are considered. Such optimizations can be rather time consuming. Therefore, in the process synthesis strategy discussed here, we first reduce the number of process alternatives by applying heuristic knowledge and then formulate and solve an optimization problem as described in section 2.4.

2.3.11 Examples

This section is dedicated to the detailed description of the synthesis strategy for integrated reaction and separation processes. The general methodology will be explained considering four different examples. In the course of this section, only the information that is relevant to the discussed reaction systems will be given; for example, the interaction with different adsorbent types during the thermodynamic analysis will not be shown, if the system favors a reactive distillation.

Example 1: Functional design

The first example deals with the production of methyl tertiary butyl ether (MTBE). This process is one of the few examples of reactive separations which have made their way to an industrial scale. The feed streams for this process are methanol (MeOH) and a C4-cut leaving a steam cracker. This C4-cut consists of isobutene (IB), isobutane, n-butene and n-butane. Butadiene is usually separated by an extractive distillation before entering the process (Onken and Behr 1996). Depending on the raw oil and the specific operating state of the steam cracker, the composition of the C4-cut may vary significantly (Qi 2002). For the problem at hand these varieties as well as feed impurities like water are neglected. It is assumed that the two feed streams consist of pure MeOH and of a mixture of isobutene and n-butane (molar ratio 65.2 : 34.8), respectively.

Process goals

Two major process goals can be specified. Firstly, MTBE has to be produced with a purity of more than 99 mole-%. Secondly n-butane has to be purified out of the C4-stream. As IB and n-butane show quite a narrow boiling behavior their separation is very difficult in unit operations which are based on volatility differences.

Data acquisition / thermodynamic analysis

The reaction system is assumed to consist of only one main reaction. Side reactions like the etherification of MeOH or the dimerization of IB are not considered. The exothermic main reaction is equilibrium limited and can be described by the following equation:

$$MeOH + IB \leftrightarrows MTBE \qquad\qquad \Delta h^0_{R\_298} = -37.7 \text{ kJ/mol} \qquad (2.5)$$

The fourth component in the system, n-butane, is inert. The temperature dependency of the chemical equilibrium constant as well as the kinetics and production rates of the reaction measured in liquid phase for the catalyst Amberlyst-15 are taken from (Rehfinger and Hoffmann 1990a). Equations are given in section 2.4.3 (eq. (2.55) – (2.59)). The net production rate is defined as the difference between the production rates for forward reaction and back reaction (eq. (2.6)).

$$P^{net} = P_{j=R1} - P_{j=R2} = H^{eff} k^{reac}_{j=R1} \left(\frac{a_{i=IB}}{a_{i=MeOH}} - \frac{1}{K^a} \frac{a_{i=MTBE}}{a^2_{i=MeOH}} \right) \qquad (2.6)$$

Its temperature dependency is shown in Figure 2.6. The graphs show a maximum since the chemical equilibrium constant decreases while the reaction rate increases with increasing temperature. Depending on the molar composition high net production rates will be achieved between 60 °C and 110 °C.

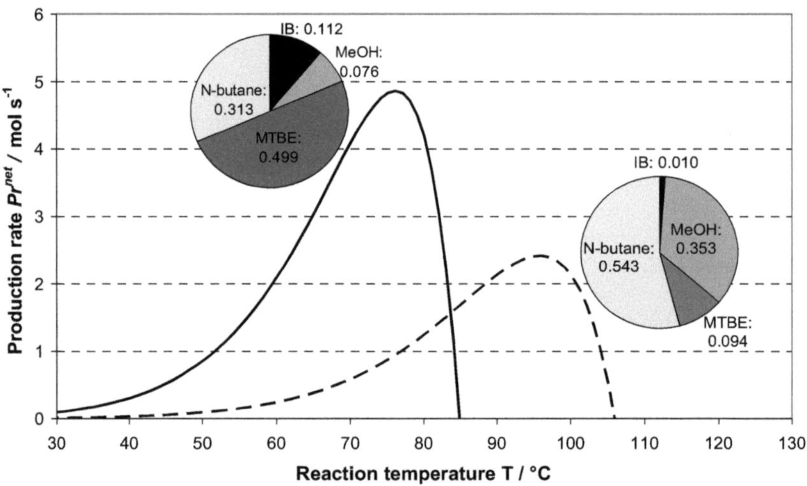

Fig. 2.6: Temperature dependency of production rate for different compositions

For the thermodynamic analysis the software tools PROSYN (Schembecker 1998b) and ASPEN PLUS® are used. Simulation studies applying the NRTL and Wilson approach show that the four-component system forms two small miscibility gaps: one between MeOH and IB and the other one between MeOH and n-butane.

Table 2.2: Distillation areas and boiling points of pure components and azeotropes (WILSON, 1 bar and 8 bar)

T_B/ °C (1 bar)	Distillation area (DA)	Components /Azeotropes	T_B/ °C (8 bar)	Distillation area (DA)	Components /Azeotropes
-7.5	DA1/DA2	MeOH-IB	59.5	DA1/DA2	MeOH-IB
-7.3	DA2	IB	61.4	DA2	IB
0.5	DA1/DA2	MeOH-n-butane	67.0	DA1/DA2	MeOH-n-butane
0.8	DA2	n-butane	69.4	DA2	n-butane
51.1	DA1/DA2	MeOH-MTBE	121.1	DA1/DA2	MeOH-MTBE
54.7	DA2	MTBE	128.0	DA1	MeOH
64.2	DA1	MeOH	136.2	DA2	MTBE

Miscibility gaps between IB and n-butane or between the product MTBE and any other component do not occur between 1 and 8 bar (Schubert et al. 2001). Boiling points of pure components and azeotropes are shown in Table 2.2 for 1 and 8 bar. At 1 bar the reactant MeOH is the highest boiling component of the system, at 8 bar it is MTBE. The boiling order alters around 4 bar.

The three azeotropes are connected by a distillation boundary which divides the composition space into two distillation areas. Under reactive conditions two of the binary azeotropes disappear and all reactive distillation curves lie inside the 2-dimensio-nal reaction equilibrium space (see Figure 2.7). It is divided into four reactive distillation areas bordered by two intersecting reactive distillation boundaries. The first one connects the vertices of

Table 2.3: Reactive distillation areas and boiling points of pure components and azeotropes (8 bar)

T_B / °C (8bar)	Reactive distillation area (RDA)	Components / Azeotropes
61.4	RDA2 / RDA4	IB
67.0	RDA1 / RDA3	MeOH-n-butane
69.37	RDA 1-4	Reactive azeotrope
69.4	RDA3 / RDA4	n-butane
128.0	RDA1 /	MeOH

MeOH and n-butane (see Figure 2.8), the second one connects the azeotrope between MeOH and n-butane with the IB vertex. At the intersection point of both reactive distillation borders a reactive saddle azeotrope is formed. As this reactive azeotrope is located very close to the n-butane vertex and the second reactive distillation border is located very close to the binary edges between MeOH/n-butane and n-butane/IB, only two of the reactive distillation areas (RDA 1 and RDA 2) are of technical interest. For the estimation of product regions the very small reactive distillation areas RDA 3 and RDA 4 can be neglected. Therefore in the following RDA 3 will be added to RDA 1, forming the "combined" reactive distillation area RDA 1* and RDA 4 will be added to RDA 2, forming the "combined" reactive distillation area RDA 2*. The reactive distillation border between RDA 1* and RDA 2* is shown in Figure 2.7 and Figure 2.8.

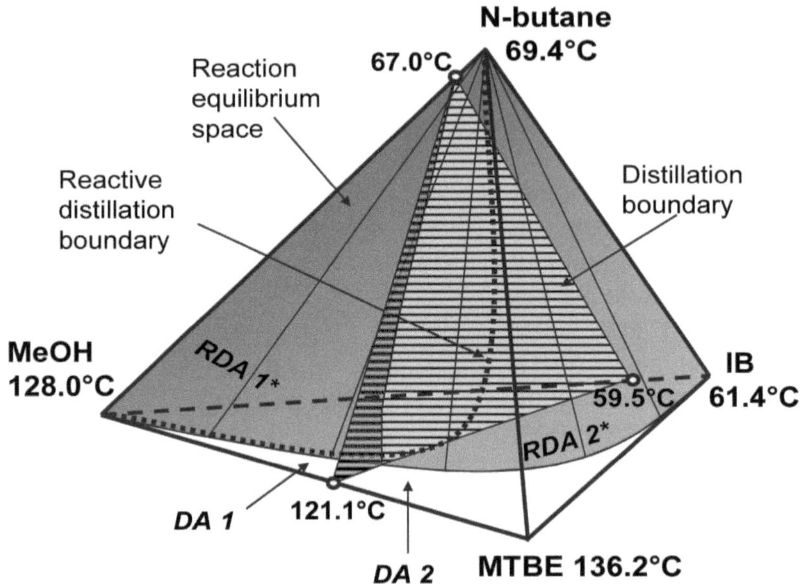

Fig. 2.7: Component space of the MTBE system with azeotropes, distillation boundary, reaction equilibrium and reactive distillation boundary (8 bar)

Investigation of reaction phase

As mentioned above, only one main reaction without any side reactions is considered. Since the reaction kinetics given in eq. (2.58) and (2.59) are activity based, conclusions concerning an advantageous concentration profile are hard to draw because of the mutual influence which all components exert on one another's activity coefficients. Here the activity coefficients will be neglected in the first step. As a result of this the concentration in the reaction phase (deduced from eq. (2.58) and (2.59)) for reactant IB should be as high as possible and for product MTBE as low as possible. As the inert component n-butane only dilutes the reaction mixture and therefore lowers the molar fractions of the other components, its concentration in the reaction phase should be rather low. The optimal temperature for the reaction phase is strongly dependant on the composition as can be deduced from Figure 2.6.

Identification of incentives

The optimal composition in the reaction phase as determined above cannot be adjusted in a simple single-phase reactor. Though excess of IB can be achieved in a tubular reactor with interstage feeding of MeOH, withdrawing the product MTBE or the inert n-butane is not practical. This can only be achieved in a multi-reactor network with intermediate separation of MTBE and n-butane which leads to high equipment demands like in the OXENO-Process (3 reactors, 4 columns (Peters et al. 2003)).

Contrary to this option, integrated reaction and separation technology can offer the opportunity to reduce equipment. Therefore an additional transport phase has to be added to the reaction phase. The transport phase must be able to separate IB and MTBE from the reaction phase.

Selection of separation process

The first step in selecting a separation process is to find a suitable transport phase which fulfills the needs defined during the Identification of incentives. As mentioned before, the reaction phase is in liquid state. Therefore another liquid phase is not useful to form the transport phase because the miscibility gap of the 4 component system is rather small and does not have an advantageous shape (direction of the tie-lines). For an extractive process a miscibility gap between the narrow boiling components IB and n-butane or between the product MTBE and at least one reactant is of interest only.

At first sight a vapor transport phase does not seem to be suitable either. The product MTBE is quite high-boiling compared to the other components so it cannot be withdrawn from the reaction phase. The inert n-butane has a lower volatility than IB so separating the inert also means removing one reactant from the reaction phase. In case of an integrated reaction and separation process consisting of a liquid reaction phase and a vapor transport phase, some other thermodynamic aspects have to be taken into consideration. The shape of the "combined" reactive distillation areas differs significantly from the shape of the distillation areas. Figure 2.8 shows the reactive distillation areas in transformed coordinates for a pressure of 8 bar. According to Ung (Ung and Doherty 1995b), the transformed compositions are calculated as follows, with MTBE as the reference component:

$$X_{IB} = \frac{x_{IB} + x_{MTBE}}{1 + x_{MTBE}} \tag{2.7}$$

$$X_{MeOH} = \frac{x_{MeOH} + x_{MTBE}}{1 + x_{MTBE}} \tag{2.8}$$

$$X_{N\text{-}butane} = \frac{x_{N\text{-}butane}}{1 + x_{MTBE}} \tag{2.9}$$

N-butane is the lightest boiling pure component in RDA 1* and relatively close to the unstable node (azeotrope between MeOH and n-butane). It can be purified towards the top of a reactive distillation column, which means that it is withdrawn by the vapor transport phase as demanded above. This is only possible when the concentration profile inside the column does not leave RDA 1*, otherwise IB is the unstable node. This demand will be checked during the Measures to achive the desired product quality.

At this point a reactive distillation seems to be a promising option to separate n-butane from IB which is one of the requirements from the Process goals. Following Schembecker (Schembecker 1998a) integrated processes with different phase combinations (for example reactive chromatography or reactive membrane separations) will not be taken into account at this moment since they require additional components to form the second phase. In general such process alternatives lead to extra separating costs.

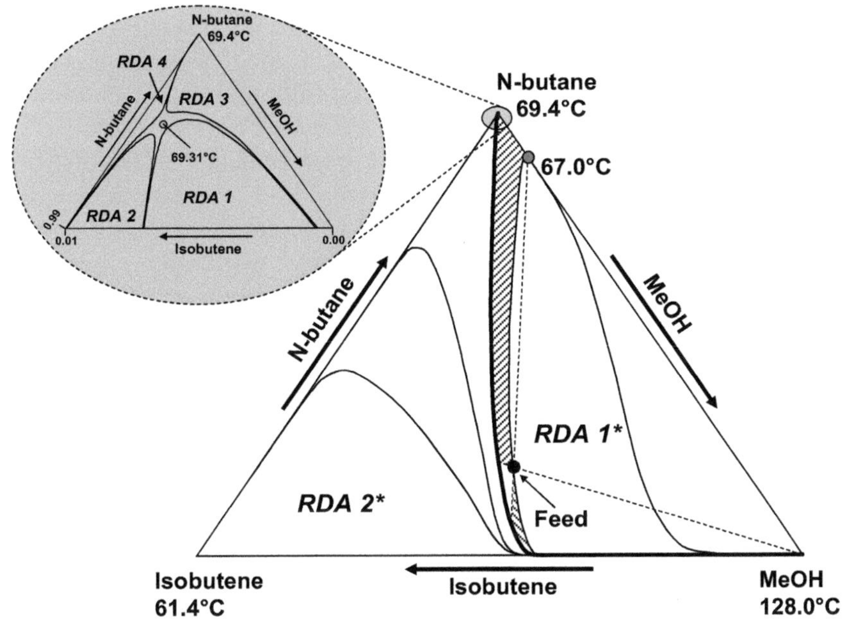

Fig. 2.8: MTBE system in transformed coordinates showing reactive distillation areas (RDA 1-4), "combined" reactive distillation areas (RDA 1* and RDA 2*) and product regions (hatched) at 8 bar

Knock-out criteria

As already mentioned, a catalyst for this reaction is available. It is the ion exchange resin Amberlyst-15. By using this catalyst the production rate is high enough for a reactive distillation column.

The operation windows for the reactive distillation process are shown in Figure 2.9 for the dimensions pressure and temperature. The operation window for reaction is not limited by pressure but limited by temperature. The temperature range is deduced from Figure 2.6 for production rates which are

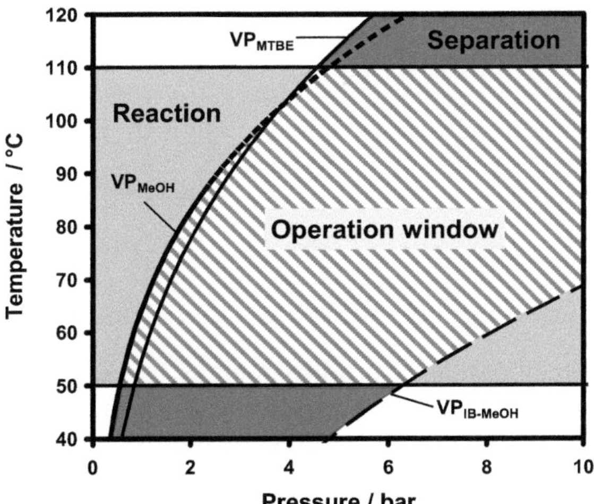

Fig. 2.9: Operation windows for the MTBE process for the dimensions pressure and temperature. The operation window for reaction is not limited by pressure but limited by temperature. The temperature range is deduced from Figure 2.6 for production rates which are

sufficient concerning the limited residence time on each stage of a reactive distillation column. The separation window is limited to the boiling conditions of the mixture. To guarantee boiling conditions inside the column at all times, the separation window is bordered by the vapor pressure curves (VP) of the azeotrope MeOH/IB and the high boilers MeOH or MTBE (depending on the pressure). For the pressure and temperature range regarded the apparatus window does not add any additional limitations concerning pressure and temperature. Thus a reactive distillation is feasible in the overlapping area of the reaction window and the separation window.

Estimation of product regions for full integration

The reactive separation functionality (reactive distillation column) has to fulfill two tasks. First, the production and purification of MTBE, second, the separation of n-butane. From thermodynamic analysis it is known that the quaternary system forms two "combined" reactive distillation areas of technical relevance. Figure 2.8 shows these two areas and also feasible splits for a single reactive separation functionality (see Figure 2.1,a). The diagram displays the four-component system in transformed coordinates. As MTBE is not part of the reaction equilibrium space it does not lie in a vertex of the triangle. Therefore it cannot be reached by any product region and so purification is not possible. MeOH is the stable node of both "combined" reactive distillation areas and will always be the bottom product. As already supposed during the Selection of separation process, the second task, i.e. separation of n-butane from IB, can be fulfilled in a reactive distillation functionality when starting from a feed point in RDA 1*. In this case the product region reaches the n-butane vertex (considering all 4 RDAs, only a point very close to the n-butane vertex can be reached).

Measures to achieve the desired product quality

The estimation of product regions for full integration shows that a single reactive distillation functionality gives the opportunity to separate n-butane from IB. Therefore the column profile must be located in RDA 1*. As the reactive distillation boundary is close to the stoichiometric composition of MeOH and IB, a slight excess of MeOH is recommended in the overall feed.

Necessity of additional steps

For the purification of *n*-butane RDA 1* must be reached at the top of the column only. A local excess of MeOH at the low boiling end of the reactive distillation functionality (upper part of the reactive distillation column) could therefore be sufficient. Local excess might either be reached by internal accumulation or by distributed feed streams. For the latter option the MeOH feed should be positioned in the upper part of the column to create the local excess. Having a look at the boiling points of the feed stream components (Table 2.2) it becomes obvious that this is possible. As MeOH is the higher boiling feed stream compared to the C4-cut consisting of IB and n-butane, feeding MeOH in the upper part of the column and C4 in the lower part results in a counter-current flow of the reactants. This does not object the specific constraint for a reactive separation functionality which requires all reactants to be present in the reaction phase simultaneously in space and time. Feeding MeOH into the low-boiling zone of the reactive distillation functionality should therefore allow the purification of n-butane in the distillate.

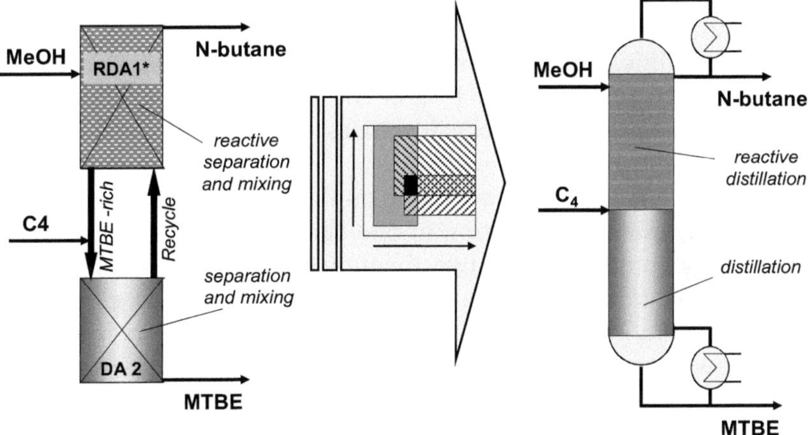

Fig. 2.10: Transfer from distributed functionalities (left) to recommended process design of a structured reactive distillation column (right)

The purification of the other product is more difficult. Since MTBE is not part of the reaction equilibrium space, a single reactive separation functionality cannot be used to produce and purify MTBE. The product stream (bottom of RD column) will always contain the other three components, too. Therefore, another functionality is necessary to separate MTBE out of this stream and recycle the other components. Since MTBE is the highest boiling component of the whole system (above 4 bar), a feasible separation functionality for this task might be a distillation. Here only the topology of the non-reactive distillation areas (DAs) is important. Figure 2.7 shows that MTBE is the stable node of DA 2 but it is not part of DA 1. The given feed composition, however, is located in DA 1. Therefore it has to be ensured that the composition of the component stream entering the distillation functionality is located in DA 2.

Two measures may help here. The first one is a local excess of IB achieved by the above-mentioned local distribution of feed streams. The C4-stream will not be fed into the reactive distillation functionality but into the distillation functionality (see Figure 2.10, left). Due to the recycle to the reactive distillation functionality IB will still be available for reaction.

The second measure is the reaction itself. Large parts of the reaction equilibrium space are located in DA 2 (see Figure 2.7). If chemical equilibrium is reached in the reactive distillation functionality, the distillation boundary between DA 1 and DA 2 will be crossed by reaction. In case the residence time for reaction is not sufficient to reach DA 2, external reactors combined with the reactive section of the column (as shown in Figure 2.1,c) can help to achieve the conversion required.

The process developed so far consists of two main functionalities, reactive separation and separation. The feed streams are distributed over these two functionalities so as to create a local excess of MeOH and C4 to achieve compositions in RDA 1* and DA 2, respectively. Therefore these two locations may also be considered as mixing functionalities (Figure 2.10, left).

Since both functionalities are based on the same separation process (distillation), a combined realization in one single apparatus should be evaluated. For the combination of two functionalities into one apparatus again the knock-out criteria for both functionalities have to be checked, especially the operation windows. Furthermore, the question has to be answered how the reaction can be deactivated in the separation functionality. The former constraint has already been answered by Figure 2.9 because the apparatus and the separation window for the separation functionality are the same as for the reactive separation functionality. The latter constraint can be held since the catalyst is heterogeneous so a spatial distribution on one apparatus is possible.

The process design recommended is a structured reactive distillation column with two separated feed streams as shown in Figure 2.10 (right). The pressure recommended for this process is 8 bar because at this pressure MTBE is the highest boiling component of the system and the distillate stream can be cooled with water instead of an expensive cooling medium.

Optimization

Detailed cost optimization using mathematical programming is described in section 2.4.3.

Example 2: External reactors

The following example shows that general statements for a liquid-vapor system like the formerly considered reactive distillation are still valid when the phase system is totally changed. Therefore, a solid-liquid system will be investigated.

In carbohydrate chemistry the general term sugar is used for all nutritive mono- and disaccharides such as glucose, fructose, galactose, maltose, and sucrose. Out of these, normally fructose containing syrups are the sweetest on the relative sweetness scale for natural sugars (Schenck 2003). The irreversible hydrolysis of saccharose leads to the isomers fructose and glucose. The chemical equilibrium constant for the isomerization between these two is around $K_{eq} = 1$ (Camacho-Rubio et al. 1996, Converti and Del Borghi 1998). The chemical reaction is shown below.

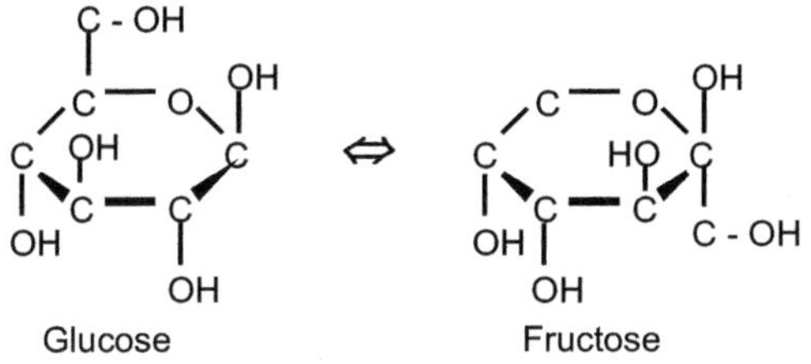

Glucose   Fructose

Since fructose is around 50% sweeter than glucose (with the same amount of calories) and can be metabolized much better by diabetics, it is widely used for diabetic sweetening of beverages, canning and baking cereals (Vuilleumier 1996).

Process goals

The feed for this process consists of a mixture of glucose and fructose with a content of 42 wt.-% (dry base) fructose. The major goals of this process are, on the one hand, to increase conversion of glucose higher than the chemical equilibrium which results in an higher yield of fructose and, on the other hand, to purify fructose from glucose.

Data acquisition / thermodynamic analysis

Both isomers are well soluble in water. Experimental investigations by Jupke et al. show a stronger adsorption of fructose on the cation exchanger resin Amberlite® CR1320 Ca, produced by Rohm & Haas (Jupke et al. 2002). The equilibrium reaction between glucose and fructose is catalyzed by the enzyme Sweetzyme® IT (Novozymes). The reaction kinetics for this isomerization are described by Borren and Schmidt-Traub based on a quasi-homogeneous approach (Borren and Schmidt-Traub 2004).

$$r_i = v_i \cdot 1.427 \cdot 10^{-3} \left[\frac{K_{eq}}{1 + K_{eq}} \left(c_{\text{Glucose}} + \frac{c_{\text{Fructose}}}{K_{eq}} \right) \right] \qquad (2.10)$$

Investigation of reaction phase

The catalyst for this reaction is an immobilized enzyme. The feed components glucose and fructose are solved in water. In analogy to reactive distillation processes with heterogeneous catalysts, the liquid phase will be considered as the reaction phase.

For a successful enhancement in conversion of glucose and therefore a higher yield of fructose the concentration of glucose in the reaction phase should be as high as possible. At the same time the concentration of fructose in the reaction phase should be as low as possible, thus shifting chemical equilibrium to the product side.

Identification of incentives

The requirements for the reaction phase determined in the previous section cannot be met in a simple single phase reactor because chemical equilibrium cannot be overcome. Additional separation functionalities become necessary which, firstly, separate fructose from the feed stream and, secondly, send the remaining glucose to the single phase reactor to reach chemical equilibrium again. This equilibrium stream must then be recycled to the separation functionality as presented in Figure 2.11.

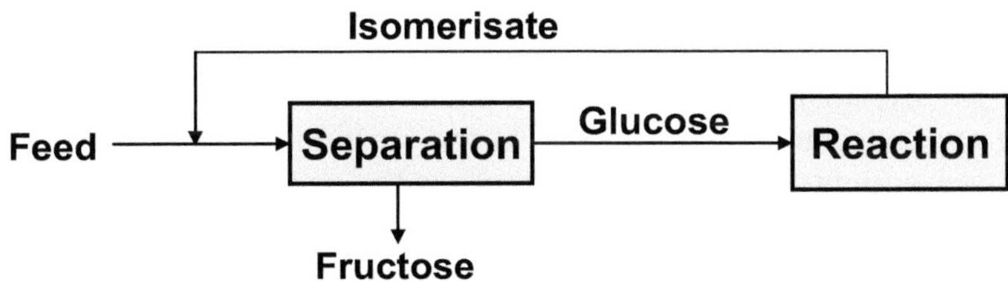

Fig. 2.11: Conventional process flow sheet for fructose production

In an integrated reaction and separation process an additional transport phase can help to meet the requirements from above and achieve an in-situ separation of the product fructose. Compared to a conventional arrangement, the additional separation section and the recycle stream may be avoided.

Selection of separation process

The reaction phase is a water solution of glucose and fructose with a solid catalyst and will therefore be assumed as (pseudo-) homogeneous in the liquid phase. The additional transport phase has to support the reaction phase in such a way that the requirements for the investigation of reaction phase are fulfilled. These are high concentration of glucose and low concentration of fructose in the reaction phase. The transport phase can either be in solid, liquid or vapor state.

As both sugars do not boil, a vapor phase is no option for becoming the transport phase. A liquid transport phase will be useful in case the two isomers have different solubilities in different liquid phases (one polar and one apolar). As the chemical properties of glucose and fructose are quite similar (same functional groups), their polarities differ insignificantly. A separation by different liquid phases is therefore not possible.

The only feasible transport phase therefore is solid. Here different options are possible:

Precipitation of fructose out of the liquid reaction phase which means reactive crystallization. However, as fructose has the highest water solubility (3.76 g per gram H_2O at 20 °C) of any known sugar, glucose will precipitate first (Cronewitz et al. 2000). This is exactly the wrong effect. Now glucose will leave the reaction phase and the conversion of fructose increases.

The second possibility to create a solid transport phase is to add it to the process. As known from the data acquisition, a suitable adsorbent is available which affects fructose more than glucose so it can extract the product out of the liquid reaction phase. This matches with one of the requirements for the investigation of the reaction phase. The other requirement, increasing the concentration of glucose in the reaction phase, can be met by a counter-current flow, which accumulates the reactant glucose in the reaction phase.

Such a process with a liquid reaction phase and a solid transport phase is called reactive moving bed chromatography. A schematic process flow sheet is given in Figure 2.12. An equimolar feed mixture of glucose and fructose enters the apparatus. The stronger adsorbed fructose will move to the extract port together with the solid phase. An extensive overview of chromatographic processes is given in (Schmidt-Traub 2005).

The operation modes as well as differences between several types of counter current chromatographic reactors are described in detail in chapter 5.

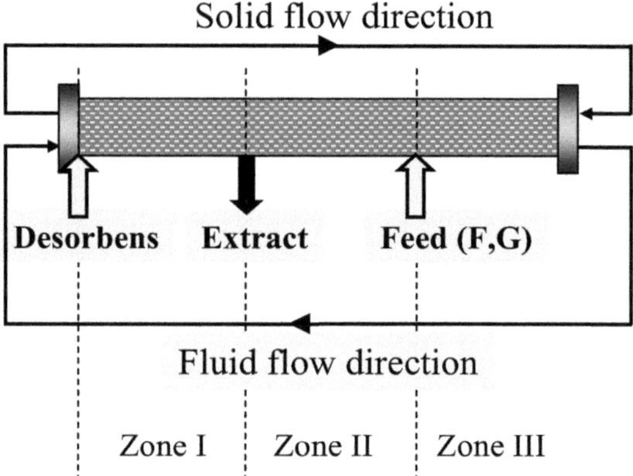

Fig. 2.12: Schematic process flow sheet of a counter current chromatographic reactor.

It should be pointed out that the process selection performed here intends to demonstrate how a suitable transport phase should be selected only. In this example membranes were not considered and only an aqueous reaction phase was taken into account. If the solubility of glucose in a solvent different from water is higher than that of fructose, a reactive crystallization process is still of interest.

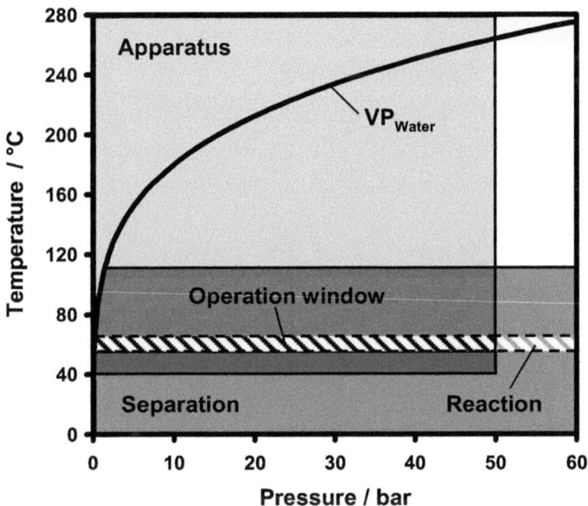

Fig. 2.13: Operation window for reactive chromatography

Knock-out criteria

Some important criteria for the counter-current reactive chromatography process have already been met. These are, for example, a suitable catalyst, adsorbent and solvent. Another design parameter is the flow pattern. A real counter-current flow of a liquid and a solid phase (true moving bed) is not very practicable. This problem is usually avoided by applying a so-called simulated moving bed with a stationary solid phase. Both process types are explained in detail in chapter 5.

The next task is to check whether the operation windows for reaction, separation and apparatus do overlap, so a feasible operation point for reactive chromatography can be found. Figure 2.13 shows such a diagram for the dimensions temperature and pressure. To ensure the liquid state of the aqueous phase, the temperature window for separation is bordered by the solidus curve of water (depending on pressure and sugar concentration approximately 0 °C) and the vapor pressure curve of water. The upper border for temperature is given by the degeneration temperature of fructose (110 °C). The optimal temperature range for reaction is between 55 and 65 °C. Below 55 °C the reaction rate is too low; above 65 °C the catalytic enzyme will be destroyed. The apparatus window (lab scale plant made of glass) is bordered by a maximum pressure of around 50 bar. Since chromatographic columns have quite a high pressure drop, the viscosity of the liquid phase is of importance, too. It will increase with lower temperatures. At temperatures below 40 °C the viscosity of the liquid phase is not acceptable.

Another important dimension of the operation windows not shown in Figure 2.13 is the pH-value. The enzyme shows the highest activity in a pH range between 7.7 and 7.9. The recommended operation point for this process is a temperature of 60 °C, a pH-value of 7.8 and a pressure of 40 bar.

Estimation of product regions for full integration

As this reaction is of the type A $\leftrightarrows$ B the component space (inert water neglected) is a one-dimensional line and the reaction space is a single point on this line, tagging the chemical equilibrium concentration (see Figure 2.14).

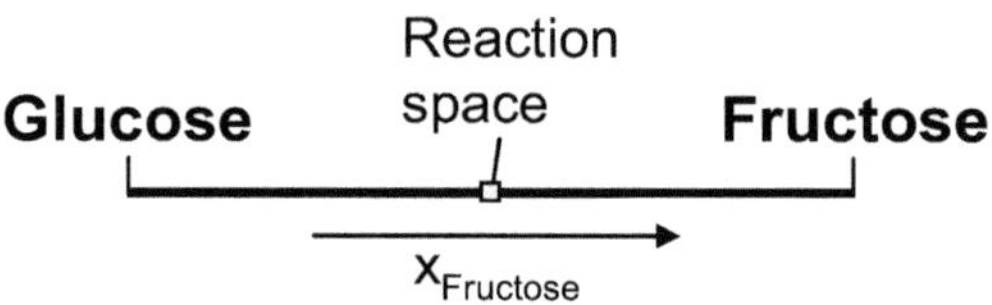

Fig. 2.14: Reaction space for the glucose-fructose isomerization

Assuming chemical equilibrium at all times, the concentration in a single reactive separation functionality (Figure 2.1a) will not leave this equilibrium point. Even the counter-current process as suggested in Figure 2.12 cannot be used to separate the desired product fructose. Whenever fructose is enriched towards the extract port, according to the adsorbent flow direction, back reaction will occur instantly. In real life systems the reaction rate is not infinite, so even in this unstructured process (according to Figure 2.1,a) a certain enrichment of fructose will be possible and is also reported by some authors (Borren and Schmidt-Traub 2004, Fricke 2005, Toumi and Engell 2004).

Measures to achieve the desired product quality

In section 2.3.8 three different measures (change of concentration, temperature or pressure) are suggested to achieve the desired product quality.

For the problem at hand none of these measures can be applied to meet the process goals defined above.

Necessity of additional steps

In a single reactive separation functionality it is not possible to leave the reaction space which in this case is a single point (the chemical equilibrium). An additional separation functionality is therefore necessary to increase the product concentration on the adsorbent without allowing back reaction. This can be achieved by a simple chromatographic separation using the same solvent and adsorbent as in the chromatographic reactor. Therefore, the same apparatus can be used with the new constraint to suppress chemical reaction in the section before the extract port. According to the specific constraints mentioned in section 2.2 this can easily be achieved by a spatial distribution of the catalyst. The schematic process design is shown in Figure 2.15. In real-life systems the Hashimoto process is usually preferred to realize such a process (Hashimoto et al. 1983), see chapter 5.

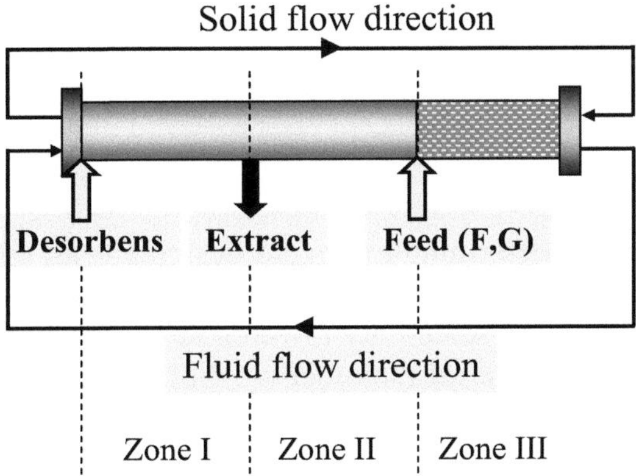

Fig. 2.15: Schematic process design for a reactive chromatography with distributed functionalities

Simulation

A comparison between a chromatographic separation (SMB) of glucose and fructose with a chromatographic reactor (SMBR) and a Hashimoto reactor can be found in Figure 5.15. It shows that for high purities the suggested process design (Hashimoto reactor) achieves significantly higher production rates than the SMBR. Nevertheless, it has to be stated that the process design made here is based on the assumption of chemical equilibrium. The simulation results shown in Figure 5.15 are based on the reaction kinetics of eq. (2.10).

A comparison of costs could not be found in the literature. As industrial standards require fructose concentrations of about 55 %, it is questionable whether the Hashimoto concept with its difficult structure and higher demand in process control is of advantage to the problem at hand.

Example 3: Impact on selectivity

In general, the selectivity of chemical reactions can be influenced via the choice of catalyst. In addition to the impact of this measure, a proper functional design of an integrated reaction and separation process will help to achieve a significant increase in selectivity. This idea can be illustrated by the transesterification of the ester ethyl acetate (EtAc) with a mixed stream of two alcohols, methanol (MeOH) and n-butanol (BuOH). Depending on the reaction step favored by the process design, either methyl acetate (MeAc) or n-butyl acetate (BuAc) is the main product in this reaction system.

Process goals

The objective is to determine a process design for an integrated reaction-separation process that allows both increasing the conversion of one alcohol and influencing the selectivity towards the desired main product. Consequently, the conversion of the other alcohol should be suppressed to obtain high selectivity. Furthermore, switching the selectivity between both products by changing the operating parameters only (and not the configuration of the column) was chosen as an additional design goal.

Data acquisition / thermodynamic analysis

The chemical system EtAc-MeOH-BuOH is described in detail in chapter 3.5.3. Therefore only a brief summary is given here. For the thermodynamic analysis the software tool PROSYN was used.

The feed components react in three competing transesterification reactions producing methyl acetate (MeAc), butyl acetate (BuAc) and ethanol (EtOH). The reaction system can be described as follows:

$$EtAc + MeOH \leftrightarrows MeAc + EtOH \qquad K^a \approx 0.8 \qquad (2.11)$$

$$EtAc + BuOH \leftrightarrows BuAc + EtOH \qquad K^a \approx 1.0 \qquad (2.12)$$

$$MeAc + BuOH \leftrightharpoons BuAc + MeOH \quad K^a \approx 1.25 \tag{2.13}$$

Catalyzed by acidic ion exchanger resins like Amberlyst-15 or Amberlyst-46 each reaction leads to conversions close to the equilibrium after short residence time (Beßling et al. 1997, Pöpken 2001, Steinigeweg 2003). Differences in reaction velocity between the three reactions are negligible. In addition, etherifications of the alcohols are neglected in this theoretical investigation.

By thermodynamic analysis, two distillation areas (DA) and two reactive distillation areas (RDA) in a five-dimensional component space are identified for the six-component system.

Table 2.4 schematically shows all components and azeotropes as well as their distribution between the two distillation areas. It becomes obvious that BuOH is the stable node in DA 1 and BuAc is the stable node in DA 2. All other nodes are part of the distillation border, so they belong to both distillation areas. The MeAc/MeOH azeotrope is the unstable node of the whole system, being present in DA 1 and DA 2. Under reactive equilibrium conditions two reactive distillation areas are present (RDA 1, RDA 2) and the component list is reduced to five pure components and three azeotropes only. The azeotropes between MeAc and EtOH and between EtAc and MeOH are not part of the reaction space because they are formed by a reactant and a product (Bessling et al. 1997). Reactive azeotropes have not been detected for the conditions under consideration. A further analysis reveals that the components EtOH and EtAc are saddle nodes in both distillation areas (see Table 2.4) but considering reactive distillation areas, EtAc and EtOH are only saddle nodes in RDA 2 and RDA 1, respectively.

Table 2.4: Distillation areas and boiling points of pure components and azeotropes (UNIFAC-DMD, 1 bar)

Temp (°C)	Distillation area (DA)	Components / Azeotropes
53.73	DA 1 / DA 2	MeAc-MeOH
57.05	DA 1 / DA 2	MeAc
57.05	DA 1 / DA 2	MeAc-EtOH
62.51	DA 1 / DA 2	EtAc-MeOH
64.53	DA 1 / DA 2	MeOH
72.65	DA 1 / DA 2	EtAc-EtOH
77.20	DA 1 / DA 2	EtAc
78.31	DA 1 / DA 2	EtOH
117.24	DA 1 / DA 2	BuAc-BuOH
117.68	DA 1	BuOH
126.01	DA 2	BuAc

Contrary to the five-dimensional distillation areas, the reactive distillation areas can be plotted graphically. In the system at hand two of the three reactions are independent which, according to the phase rule given in eq. (2.3), reduces the number of degrees of freedom by two. Thus a three dimensional plot of the reactive distillation areas is possible in transformed coordinates (section 2.3.7). Figure 2.16 shows the reactive component system in these coordinates with MeOH and BuOH as reference components. The transformed mole fractions are assigned as follows:

$$X_{EtAc} = x_{EtAc} - x_{MeOH} - x_{BuOH} \tag{2.14}$$

$$X_{BuAc} = x_{BuAc} + x_{BuOH} \tag{2.15}$$

$$X_{EtOH} = x_{EtOH} + x_{MeOH} + x_{BuOH} \tag{2.16}$$

$$X_{MeAc} = x_{MeAc} + x_{MeOH} \tag{2.17}$$

As a consequence of the summation condition, which is valid also for transformed molar fractions, X_{MeAc} is neglected for the three dimensional plot. The remaining transformed mole fractions are sufficient to specify reactive equilibrium compositions in Cartesian coordinates with origin MeAc. The resulting component space is a prism.

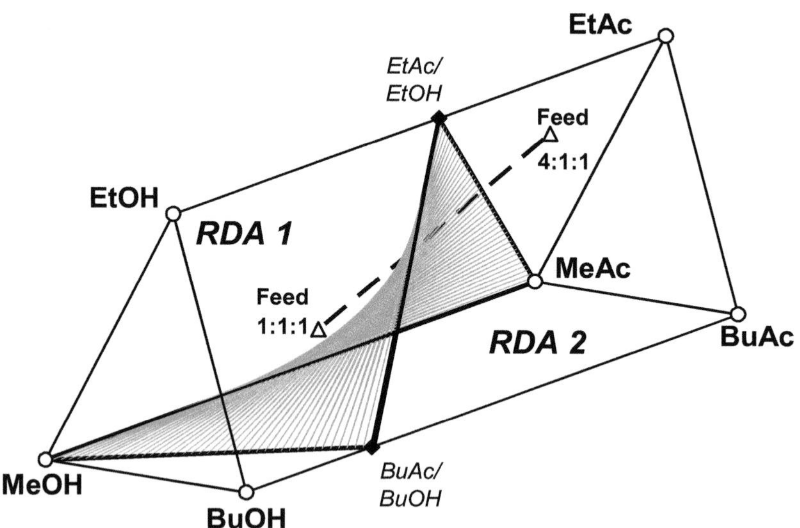

Fig. 2.16: Component System with schematic reactive distillation border and two different feed compositions in transformed coordinates

The thermodynamic analysis reveals that several nodes are located on the reactive distillation boundary separating RDA 1 and RDA 2. These nodes are the pure components MeAc and MeOH as well as the binary azeotropes MeAc/MeOH, EtAc/EtOH and BuAc/BuOH. In Figure 2.16 these nodes are connected and the resulting area is shown to give a qualitative impression of the position of the reactive distillation boundary.

As shown in Figure 2.16, the reactive distillation border separates the two reactive distillation areas from each other with EtOH and BuOH in RDA 1 and EtAc and BuAc in RDA 2. During the investigation of the liquid-liquid behavior of the system no miscibility gaps have been detected.

Investigation of reaction phase

As the kinetics of all three reactions are of the same type (1st order for each component) and all reaction rates and equilibrium constants are approximately in the same range, the only possibility to influence selectivity is to raise or lower concentrations by varying the stoichiometry of the reaction. Following general heuristic rules of reaction engineering, the concentrations of MeAc and BuOH should be low whereas the concentrations of MeOH and BuAc should be high when MeAc is the desired product. If BuAc is the desired product, the concentrations of BuAc and MeOH should be low and the concentrations of BuOH and MeAc should be high, respectively.

Identification of incentives

Increasing the concentration of one alcohol while decreasing the concentration of the other alcohol, as postulated above, is not possible with a mixed feed stream and an ordinary continuously operated stirred tank reactor (CSTR). It could be achieved by applying a recycle stream of the alcohol that is needed in excess. However, at least one separation unit is required in this case. An integrated apparatus can offer a promising alternative by accumulating the two alcohols in different phases. At the same time the main product can be withdrawn from the reaction phase, thus fulfilling the requirements above without (with less) expensive downstream processing equipment.

Selection of separation process

Considering the order of the boiling point temperatures, it becomes obvious that the two possible main products MeAc and BuAc are the lowest and the highest boiling pure components in this system, respectively. Regarding the reaction kinetics and the catalyst used, a liquid phase is chosen to be the reaction phase. Utilizing the difference in vapor pressure a gaseous transport phase offers the opportunity to fulfill the requirements formulated above and the introduction of an additional component to establish a transport phase can be avoided. For MeAc production the product has to be withdrawn from the liquid phase. For BuAc production MeOH - reactant for the side reactions (2.11) and (2.12) - has to be withdrawn from the liquid phase while BuOH has to be accumulated there.

Considering all this, the recommended integrated apparatus is a reactive distillation column. In case of the MeAc production it has to be admitted that there is still one azeotrope with a boiling point lower than pure MeAc. This azeotrope (MeAc/MeOH) cannot be overcome by reaction.

Other reactive separation processes like reactive extraction, reactive chromatography, reactive membrane separations or reactive adsorptions are excluded from further investigation since in all these processes additional components have to be added to the reaction system to generate the transport phase.

Knock-out criteria

As often quoted in literature chemical reactions taking place inside a distillation column are limited by one major constraint: sufficient reactive conversion must be reached under boiling conditions of the reaction system considered. To ensure this, heterogeneous catalysts are widely used for catalytic distillation proc-

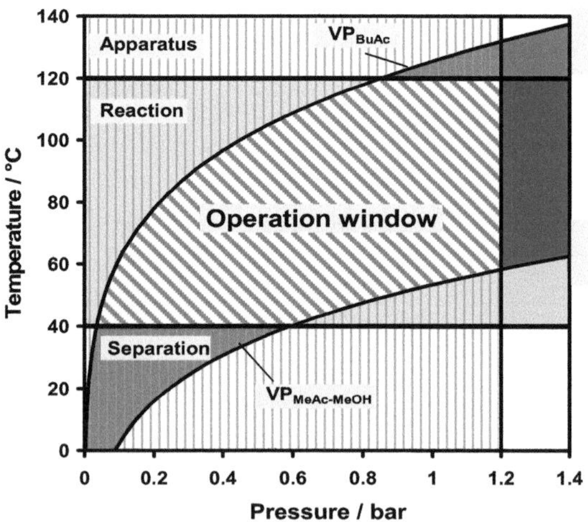

Fig. 2.17: Operation window for transesterification via reactive distillation

esses. In case of the transesterification of EtAc Amberlyst-15 is a suitable catalyst (Steinigeweg 2003).

In addition to the residence time, other parameters of the operation windows are pressure and temperature. For the problem at hand the pressure range is defined by the apparatus window. Since the utilized pilot plant column is made of glass, operating pressures between 0 and 1.2 bar can be realized.

The separation window is limited to the boiling conditions of the mixture. To ensure boiling conditions inside the column the separation window is bordered by the vapor pressure curves of MeAc/MeOH and BuAc.

An additional limitation occurs adding the reaction window. Here the lower bound over the whole pressure range is set at 40 °C because the reaction rate becomes too slow below this temperature. The upper boundary of 120 °C is given by the maximum operating temperature of the catalyst (Rohm and Haas 2005). The overlapping of the operation windows for reaction, separation and apparatus leads to the reactive distillation operation window shown in Figure 2.17. Inside this overlapping area all criteria for the operation of a reactive distillation process are fulfilled.

Estimation of product regions for full integration

At first we consider the production of methyl acetate in a single feed column. As the azeotrope between MeAc and MeOH is the unstable node in both reactive distillation areas, it will be gained at the top of the column independent from the feed composition. The content of 67 mole-% of MeAc in this azeotrope is equivalent to the conversion of MeOH fed into the column. For the same reason (33 mole-% of MeOH in azeotrope) full conversion of MeOH with EtAc is not possible in a homogeneously catalyzed single feed column.

A stoichiometric feed composition is located in

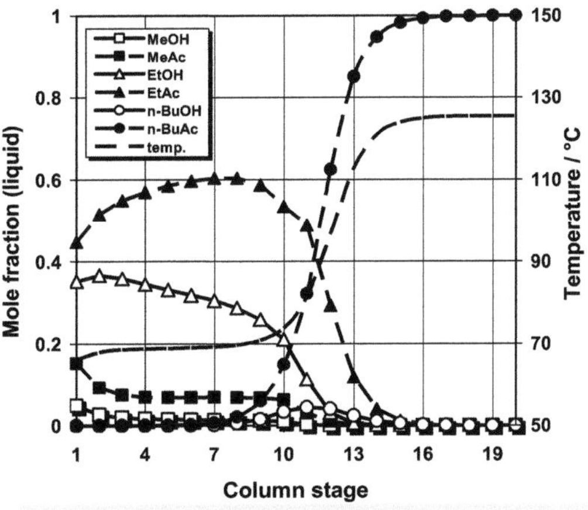

Fig. 2.18: Concentration profile for EtAc excess in feed (UNIFAC-DMD, 1 bar)

RDA 1 with BuOH as the stable node; therefore, the production of BuAc, which is only part of RDA 2, is not possible in a single feed column. To overcome this limitation additional measures have to be taken into account.

Measures to achieve the desired product quality

Having a look at the reactive distillation areas known from the thermodynamic analysis, the production of BuAc can be achieved by changing the feed composition. As displayed in Figure 2.16 the feed component EtAc is only part of RDA 2 with BuAc as stable node. Increasing the EtAc concentration in the feed stream should lead to a feed composition in RDA 2 and therefore to the opportunity to produce and purify BuAc at the bottom of a homogeneously catalyzed reactive distillation column.

As shown in Figure 2.18, a molar ratio between EtAc, MeOH and BuOH of 4:1:1 in the feed allows the production of pure BuAc. However, as a consequence of the high excess of EtAc, also high amounts of MeOH are converted to MeAc, which is not favorable.

Necessity of additional steps

As seen before, the production of MeAc and also of BuAc in a reactive distillation column is possible. However, the goal of high conversion with at the same time high selectivity has not been reached yet. MeAc cannot be purified because of the MeAc/MeOH azeotrope and for the production of BuAc the necessary excess of EtAc is very high. In addition also high amounts of MeOH are converted. For both processes additional functionalities are necessary to achieve the goals defined above.

In case of the MeAc

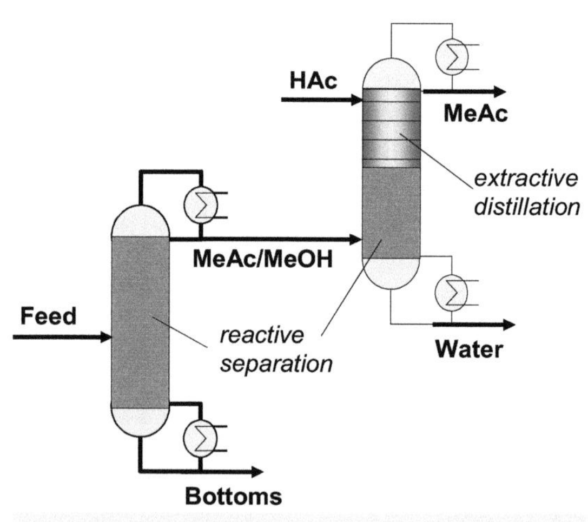

Fig. 2.19: Process design for MeAc production

production a feasible solution is already described in literature: The MeAc/MeOH azeotrope gained at the top of the reactive distillation column can be fed into another reactive distillation column. Here the remaining conversion of MeOH to MeAc is achieved by reaction of MeOH with acetic acid (HAc). The design of this column has already been described by different authors (Agreda et al. 1990, Bessling et al. 1998, Doherty and Buzad 1992). It consists of a non-reactive separation functionality at the column top to purify MeAc due to the entrainer effect of HAc and of a reactive distillation functionality to convert the remaining MeOH with HAc and purify the by-product water towards the column bottom. The column sequence suggested is shown in Figure 2.19.

For the production of BuAc it is necessary to switch the column profile to RDA 2. This is only possible with an excess of EtAc. Since BuAc has to be purified towards the column bottom an additional functionality can be utilized to switch the profile in the lower column section into RDA 2. This functionality is a mixing functionality. It can be introduced by first splitting the feed stream locations into an EtAc feed and an alcohol feed and second lowering the reflux ratio. The EtAc feed stream leads to a local excess of EtAc while the low reflux ratio avoids dilution of this local excess due to high internal fluxes. The column design is shown in Figure 2.20. This column design also offers the opportunity of switching the main product from BuAc to MeAc without any reconstructions. To achieve high selectivities towards MeAc, it has to be ensured that BuOH will not be converted according to eq. (2.12) but purified to the column bottom. This is possible when the column profile is switched back to RDA 1 only. Therefore, the mixing functionality around the EtAc feed has to be deactivated. This can be achieved by increasing the reflux ratio to overcome the local excess of EtAc. The azeotrope between MeAc and MeOH still can only be broken by an additional step as shown in Figure 2.19.

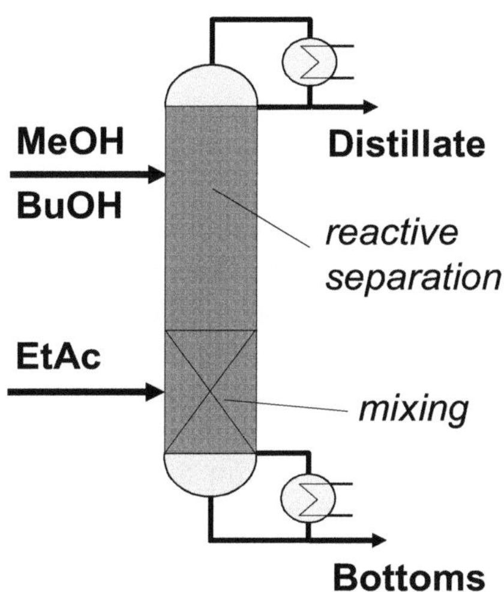

Fig. 2.20: Flexible process design for BuAc and MeAc production

Simulation

For the column design proposed two different operation ranges are possible, one for the production of BuAc and one for the production of MeAc. Switching between these two is possible when the column profile changes between the two reactive distillation areas. Although the overall feed composition lies in RDA 1, for the production of BuAc RDA 2 has to be reached. In Figure 2.21 the liquid column profiles are shown for both cases, simulated in ASPEN Plus. The calculations are based on the Unifac approach, assuming physical and chemical equilibrium on each of the 20 column stages. MeOH and BuOH are fed to stage 5, EtAc is fed to stage 15. The ratio between the three feed components is 1:1:1 (RDA 1).

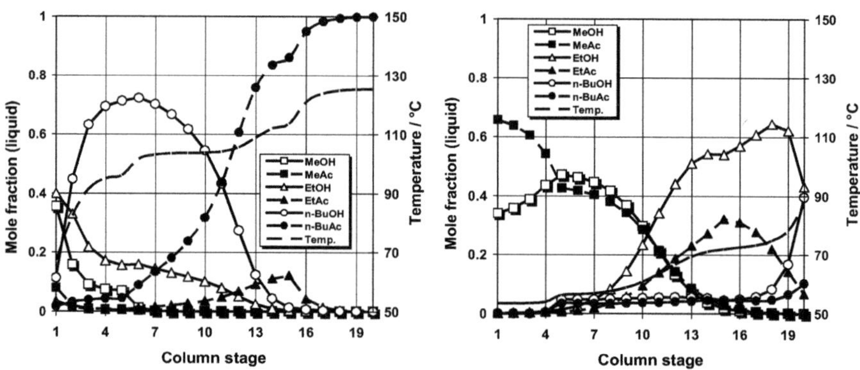

Fig. 2.21: Liquid column profiles for transesterification via reactive distillation. Left: Production of BuAc, right: Production of MeAc.

For the production of BuAc the distillate to feed ratio is set to D/F = 0.767 and the reflux ratio is RR = 0.7. The mixing functionality is located around the EtAc feed-port. This can be deduced from the high local excess of EtAc at this position compared to the other components (except for BuAc which is enriched towards the column bottom). Since BuAc is purified in the bottom stream it can be concluded that RDA 2 (with EtAc and BuAc) is reached in the lower column section. For the selective production of MeAc using the same column design, the mixing functionality has to be deactivated to change the column profile back to RDA 1. This is achieved by an increase in reflux ratio to RR = 12 and D/F = 0.333. Now EtOH dominates the concentration profile in the lower column section and therefore only a small amount of BuAc is produced.

Assuming full conversion of EtAc with ideal selectivity towards one of the esters (yield = 100 %), the distillate to feed ratio to purify this ester must be 0.333 (in case pure MeAc leaves the column at the top) or 0.667 (in case pure BuAc leaves the column at the bottom). Although a yield of 100 % is not reached in both cases, the reaction system illustrates very well that selectivity control can be achieved by applying integrated reaction and separation technology. Experimental results supporting the results of the theoretical design are given in chapter 3.

Example 4: Catalyst recycling

This example focuses on the production of mono-telomere (MT) from the reactants butadiene (But) and ethylene glycol (EG). As it will be shown during the investigation of reaction phase, a liquid phase process is useful only for this kind of problem. Therefore only relevant (and often qualitative) thermodynamic data will be given in the following, based on UNIFAC-DMD as well as our own experimental investigations.

Process goals

There are three major goals determined for this process. These goals are
- high conversion of ethylene glycol and butadiene with
- high selectivity towards mono-telomere (no adducts, no polymerization)
- and low losses of the expensive palladium catalyst.

Data acquisition / thermodynamic analysis

In Table 2.5 the boiling points of the main components are given, calculated with UNIFAC-DMD for 1 bar and for 12 bar. This data shows that the order of boiling points does not change over the pressure range. According to UNIFAC-DMD the component system butadiene, ethylene glycol and telomere does not form a miscibility gap. When adding small amounts of wa-

Table 2.5: Boiling points of pure components for two different pressures (UNIFAC-DMD)

Temp / K (1 bar)	Temp / K (12 bar)	Components
-4.9	81.93	Butadiene
99.65	188.06	Water
196.64	304.37	Ethylene Glycol
258.73	378.88	Telomere

ter, however, experimental investigations showed a large miscibility gap for the four-component system (chapter 6).

Investigation of reaction phase

The main reaction is irreversible according to the following equation:

$$\text{ethylene glycol} + 2 \text{ butadiene} \rightarrow \text{mono telomere} \tag{2.18}$$

In order to facilitate conversion of the reactants, first a reaction functionality is needed. There are two different catalysts available for this reaction, triphenylphosphine (TPP) and triphenylphosphinetrisulfonate (TPPTS), see chapter 6. Both are homogeneous liquid phase catalysts. The former can be dissolved in the organic reaction mixture, the latter can only be dissolved in an aqueous phase, which leads to an additional water demand in the process.

The reaction rate depends on the temperature showing a typical Arrhenius behavior. Pressure changes do not influence the reaction as long as no phase splitting (evaporation) occurs. As both catalysts for this reaction are homogeneous liquid phase catalysts, the reaction phase has to be in liquid state.

Besides the main reaction (eq. (2.18)) several side reactions may occur. These are one consecutive reaction and two parallel reactions as shown in the following:

Consecutive reaction:

$$\text{mono telomere} + 2 \text{ butadiene} \rightarrow \text{ditelomere} \tag{2.19}$$

Parallel reaction 1:

$$2 \text{ butadiene} \rightarrow \text{butadiene dimere} \tag{2.20}$$

Parallel reaction 2 (occurs only with TPPTS):

$$\text{butadiene} + \text{water} \rightarrow \text{water adducts} \tag{2.21}$$

All reactions are irreversible liquid phase reactions. Parallel reaction 1 is faster than the main reaction. The other two are slower (Behr and Urschey 2003). Parallel reaction 1 also occurs in vapor phase, especially at higher pressures (Le Chatelier, principle of least constraint). Parallel reaction 2 occurs with water present in the system only.

In order to increase selectivity several requirements can be postulated for the liquid reaction phase. Low butadiene concentration, for instance, hinders all three undesired side reactions. Low concentration of mono telomere hinders the consecutive reaction, and low concentration of water hinders parallel reaction 2. High concentration of ethylene glycol supports the main reaction without influencing any other reaction.

Identification of incentives

Out of a stoichiometric feed the favorable conditions for the reaction functionality, as postulated above, may not be achieved in a simple reactor using TPP as catalyst. This catalyst is dissolved in the organic phase only leading to poor selectivities towards mono telomere and none of the constraints above can be held for the reaction phase as both the consecutive reaction and parallel reaction 1 will occur (Behr and Urschey 2003). This problem can be avoided using special reactor designs like shown in Figure 2.22. Here local excess of ethylene glycol can be achieved.

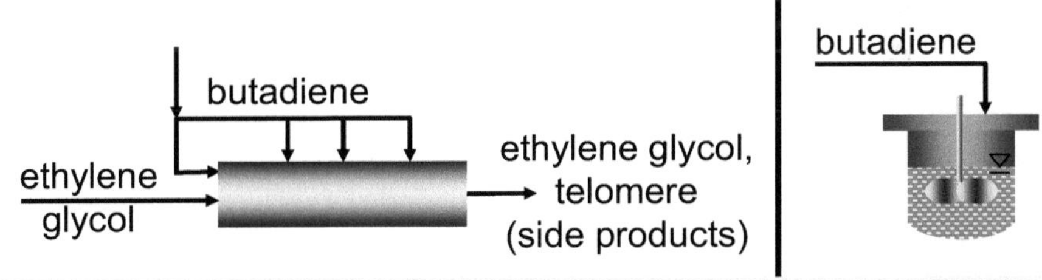

Fig. 2.22: Tubular reactor with side feed (left) and semi-batch reactor

Even with these designs the catalyst TPP has to be separated in an additional unit operation and recycled back to the reactor which often leads to catalyst deactivation (Urschey 2004).

Another possibility to increase selectivity is to exchange the reaction functionality by a reactive separation functionality thus creating a two-phase system. A reaction phase consisting mainly of ethylene glycol is desired whereas the second phase (transport phase) has to supply butadiene continuously in low amounts over inter-phase mass transfer and simultaneously withdraw the product. Thus all constraints for the reaction phase are fulfilled.

For the catalyst recycling it is favorable if the catalyst is only soluble in the reaction phase. If no material is withdrawn from this phase, catalyst losses may be prevented completely.

Selection of separation process

According to the constraints above the reaction phase is in liquid state. It consists of ethylene glycol and catalyst. The transport phase should contain butadiene and telomere preferably.

As butadiene shows the highest and telomere the lowest volatility in the system (Table 2.5) the transport phase cannot be gaseous with a liquid reaction phase consisting of ethylene glycol. So a liquid/vapor system will not work for the process at hand.

Contrary to this, a liquid/liquid system may be formed by adding some water to the reaction system. According to the components´ polarities (water and ethylene glycol rather polar, butadiene and telomere rather inpolar) the LL-envelope formed will have exactly the characteristics recommended above with water and ethylene glycol in the reaction phase and butadiene and telomere in the transport phase.

As the catalyst TPPTS is dissolved in water and water is hardly soluble in the organic transport phase, low catalyst losses due to leaching can be predicted. However, the water concentration in the reaction phase should be as low as possible because high water concentrations support the undesired parallel reaction 2.

According to these insights the recommended reactive separation functionality for this reaction system will be a reactive extraction.

Knock-out criteria

Having decided on a reactive extraction functionality the next important step is to check whether the operation windows of reaction and extraction do overlap and also to find an apparatus for the operation conditions required.

In Figure 2.23 a p-T operation window for this process is shown for reaction and for separation.

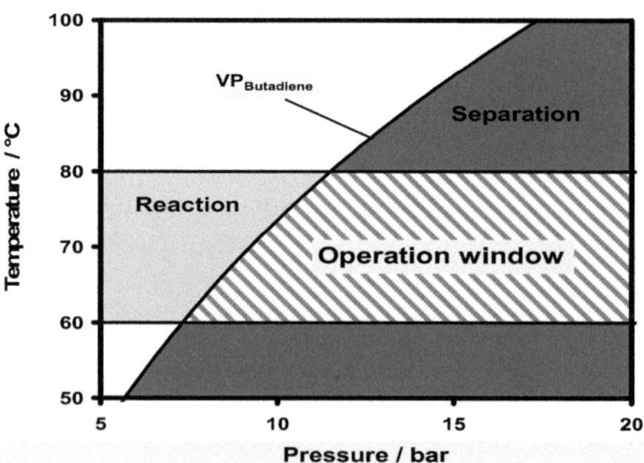

Fig. 2.23: Operation window for reactive extraction

The reaction is hardly dependent on pressure but only reasonable in a small temperature range between 60 °C and 80 °C. At lower temperatures the reaction rate of the main reaction is too small and at higher temperatures the reaction rates of the side reactions increase strongly. The catalyst is stable up to over 100 °C so this effect does not have to be taken into account here.

The two major constraints for the separation window are the formation of a miscibility gap and the inhibition of evaporation. The former leads to a water demand of 5-10 wt.%. The latter causes the vapor pressure curve of butadiene (highest volatility) being the upper border of the separation window.

The overlapping of the reaction and the extraction windows encloses all feasible operation conditions for the integrated process (hatched area in Figure 2.23). The apparatus window is not shown here because it does not limit the operation window. The experimental results according to this process design are given in chapter 6. The mini plant described there is made of stainless steel. Its operating pressure is limited to 12 bar in order to prevent parallel reaction 1. Furthermore, the mini plant is inertised with argon gas to lower the partial pressure of butadiene when operating close to its boiling point at around 80 °C as well as to prevent catalyst deactivation by oxygen.

Estimation of product regions for full integration

For the system with three reacting components and one inert, transformed coordinates are not a useful option for plotting the product regions. Telomere has to be the reference component for the transformation procedure and so it cannot be located inside the diagram (see section 2.3.7).

Keeping in mind that the amount of the inert water needed in this reactive extraction is rather low (only added to dissolve the catalyst), it is permitted to plot the product regions in a triangle diagram showing the three main reaction components containing a constant water fraction only. This plot is given in Figure 2.24. The miscibility gap is of type 2 according to Sørensen and Arlt (Sørensen and Arlt 1979). Only butadiene and telomere are completely soluble with each other. As the diagram is mass-based the pole Π approaches infinity and the stoichiometric lines (dashed) become parallel (Hugo 1965).

The reactive extractive functionality proposed is of a special kind. As only one phase (aqueous phase) is reactive and the reaction is irreversible, two different regimes inside the reactive two-phase region are possible. They are separated from each other by the stoichiometric line through the intersection of the binary edge between ethylene glycol and telomere and the upper binodal curve (P1 in Figure 2.24).

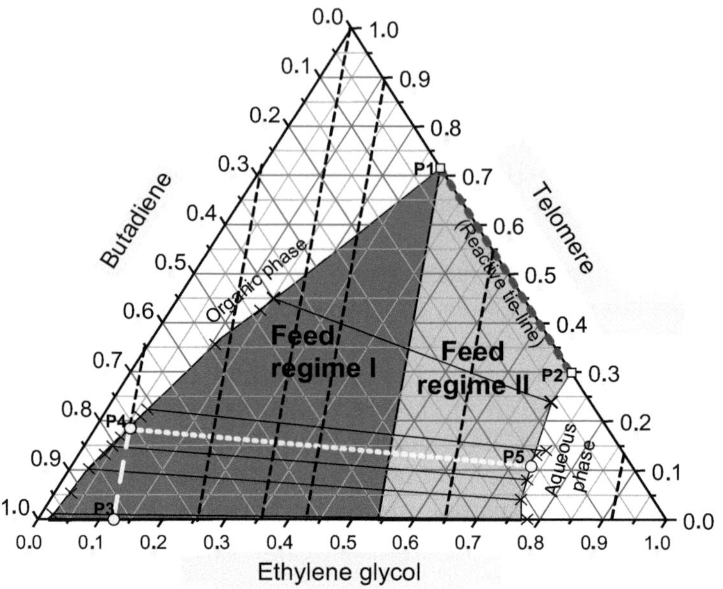

Fig. 2.24: Measured ternary LLE diagram for reactive extraction, mass based, 12 bar, 80 °C, ~10 wt.% water

For process synthesis this means that starting with a feed-composition in regime 1 or 2 will lead to completely different results after reaction and phase splitting. Feed-compositions in regime 1 always end up in two particular compositions for both phases, namely the end-points of the reactive tie-line (Samant and Ng 1998a), P1 and P2 in Figure 2.24. Contrary to this, starting in feed-regime 2 (P3) will lead to different phase-concentrations (here P4 and P5) depending on the feed-composition. The intersection of the stoichiometric line through the feed and the upper binodale (P1) gives the maximum concentration of telomere after infinite residence time, hence the reaction phase is not existent any more and the catalyst precipitates. Applied to Figure 2.24, this means that at least a slight excess of ethylene glycol compared to butadiene is necessary in the reactor to ensure the highest concentration of telomere in the transport phase after reaction. This corresponds with the requirements from the investigation of the reaction phase.

Measures to achieve the desired product quality

As mentioned before, an excess of ethylene glycol is necessary to allow highest yield of telomere. This leads to unconverted amounts of ethylene glycol which is not mandatory. Working with a stoichiometric feed is also possible in case ethylene glycol will be accumulated in the apparatus. This can be achieved either by a counter-current flow or by a recycle of the polar phase (ethylene glycol, water, catalyst) to the reactor.

Necessity of additional steps

As deduced from Figure 2.24, the transport phase can never consist of pure product, even if an excess of ethylene glycol is applied, all side reactions are neglected and infinite residence time is provided for the irreversible reaction.

An additional separation functionality is necessary to purify telomere from the other components in the transport phase (mainly butadiene). Regarding the differences in boiling points a simple flash separator in the downstream section might be sufficient. As telomere behaves like a surfactant in this system (more polar than butadiene), some amount of water and therefore catalyst will be dissolved in the transport phase after reaction (high telomere concentration). This catalyst cannot be separated from the telomere by flashing so it will be lost when leaving stage 2 together with the transport phase. The problem can be avoided by forcing water (catalyst) back to the aqueous phase before the transport phase is sent to the flash. This can only be achieved by an additional functionality that increases the difference in polarity of both phases, which may be accomplished by either increasing concentrations of butadiene (lowest polarity) and water (highest polarity) or decreasing concentrations of ethylene glycol and telomere (intermediate polarities).

As the amount of telomere after reaction is recommended to be as high as possible and the concentration of water should be as low as possible (because of parallel reaction 2), changing concentration of these two components will not be advantageous. Adding butadiene (for example from the gas phase of the flash) with an additional feed into the separation functionality (extraction stage 3 in Figure 2.25) provides an option. The difference in polarity increases and water together with the catalyst will be forced back to the reaction phase. Afterwards this catalyst is needed in the reactive stage to support reaction there which leads to a counter-current flow between the stages 2 and 3 (see Figure 2.25). To ensure the existence of two phases in stage 3, some amount of EG and water have to be added to stage 3 continuously. To achieve this either EG or water can be fed to stage 3. Adding EG (as a surfactant) will force back the catalyst to the organic phase in stage 3. Here it will be lost. The other option, adding water continuously, leads to an accumulation of water in the process because it is not meant to leave the process. The best option therefore is to use process-immanent streams. Sending one recycle stream from stage 2 to stage 3 will be a feasible solution. When recycling the aqueous phase from stage 2 to stage 3 also ethylene glycol will be recycled, which increases its concentration in stage 3. In order to avoid this, ethylene glycol has to be separated from the aqueous phase in a third functionality before the recycle. Separation of a specific component can generally be achieved in two different ways, physically or chemically. As ethylene glycol is a reactant of the main reaction, in this case chemical conversion as a mean of separation offers a good opportunity. Therefore stage 1 is a reactive separation functionality again, but now fulfilling a separation task due to reaction. It separates ethylene glycol out of the recycle (which then consists mainly of water) to stage 3 by converting it to telomere.

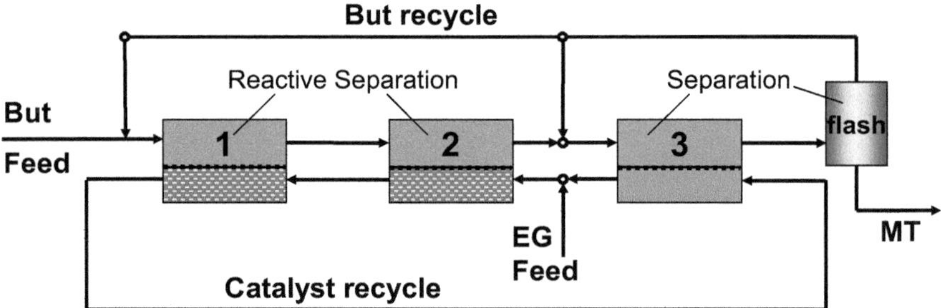

Fig. 2.25: Recommended partially integrated cross-counter-current process design for the reactive extraction

Considering all this the process design shown in Figure 2.25 meets all constraints defined during the investigation of reaction phase (ethylene glycol in excess in reaction phase, telomere not in reaction phase, only small concentrations of butadiene in reaction phase, catalyst not discharged by the transport phase but recycled).

Experimental results

For experimental results and a comparison to a simple counter-current flow see chapter 6.

2.4 Optimization of the process

The development of an integrated process based on thermodynamic insights provides a first structure of a possible process that is likely to attain the design goals. The parameters of the equipment and the operating parameters then have to be determined either using heuristics and approximations again or by optimization. If the optimization is restricted to the structure that resulted from the heuristics-based design, a potential for improvement is left unused. Often different feasible structures with considerably different degrees of integration of reaction and separation functionality exist. It is therefore recommended to perform a rigorous search in the space of feasible structures in order to determine the economically optimal plant structure or to quantify the effect of additional restrictions on the performance of the process.

Optimality here refers to the value of an objective function f, which usually consists of the sum of the operating costs and the annualized investment costs reduced by the revenues from the sales of the products. The variables of the design problem in general comprise both continuous variables (dimensions of the equipment, operating parameters) as well as discrete variables that describe the existence of certain functionalities. Generally speaking, the values of the continuous variables z and of the discrete variables have to be determined so that the objective function f is minimized subject to equality constraints g and inequality constraints h. Most variables are restricted by lower and upper bounds z^{lo} and z^{up}. The discrete decisions can be modeled by binary variables δ, which can assume the values 0 or 1 only. The optimization problem in its general form thus reads

$$\begin{aligned}
\text{Minimize} \quad & f(z,\delta) \\
\text{subject to} \quad & g(z,\delta) = 0 \\
& h(z,\delta) \leq 0 \\
& z^{lo} \leq z \leq z^{up} \\
& \delta \in \{0,1\}
\end{aligned} \tag{2.22}$$

The constraints g and h reflect material balances, reaction kinetics or restrictions on the operating range or on the dimensions of the equipment. Bounds on variables may result, for example, from purity requirements for products or physical limitations as positive material streams. If a subset of the constraints is non-linear - as it is usually the case in chemical process design - the above formulation defines a so-called Mixed-Integer Non-Linear Program or MINLP. Mathematical optimization as a tool in the synthesis of integrated processes has gained increasing attention during the last decade.

The focus in this section is on the optimization of (reactive) distillation processes. A common feature of the different approaches that have been proposed in the literature is the use of a superstructure model which is supposed to include the optimal process structure. Usually the choice of the superstructure is based on problem-specific knowledge (see section 2.3).

The first authors who considered mathematical optimization in the design of a reactive distillation (RD) column were Ciric and Gu (Ciric and Gu 1994). Similar to many superstructure formulations, they modeled the existence of an optional column tray in the optimal solution by a binary variable. These variables appear in bilinear terms which can cause numerical difficulties. Moreover, non-existing trays cause zero flow rates of vapor and liquid streams which are numerically unfavorable (Grossmann et al. 2005). The model was solved by a generalized Benders decomposition (GBD) algorithm. Cardoso et al. extended this model by a non-ideal description of the vapor-liquid equilibrium (VLE), and solved it by a simulated annealing (SA) algorithm and an adaptive random search algorithm (Cardoso et al. 2000). In many models, the existence of the chemical reaction on a certain tray is a continuous degree of freedom. In contrast, Frey and Stichlmair used a binary variable to model the existence of the reactive functionality (Frey and Stichlmair 2000). In their model, reaction and separation are performed in consecutive blocks. The existence of column trays is modeled by a variable position of the reboiler (Frey 2001). To avoid numerical difficulties with zero flow rates, Jackson and Grossmann applied generalized disjunctive programming (GDP) to formulate the process superstructure (Jackson and Grossmann 2001). They modeled the existence of column trays by Boolean variables which deactivate a subset of model constraints for non-existing trays. The model was solved by a modified outer approximation (OA) algorithm.

In this chapter a different modeling approach is described. The existence of column trays is modeled by a switched separation efficiency, thus avoiding zero flow rates on non-existing trays as described in section 2.4.1. The solution method is described in section 2.4.2. A branch-and-bound algorithm is used for the optimization of the discrete decisions. Different initialization strategies based on problem-specific knowledge or on stochastic algorithms are compared.

2.4.1 The optimization model

The optimization model of a RD column is composed of a total number of N uniform elements that describe the column trays as well as the condenser and the reboiler. The model is described in different subsections below: the model of the elements, the substance system, the superstructure, constraints on the geometry and operating parameters of the process, and the cost function.

Equations of the elements

The model equations which describe the uniform elements indexed by k (from the bottom upwards) are based on the so-called MESH equations (material balance, phase equilibrium, summation condition, enthalpy balance) extended by reaction terms.

$$F_{k,i} + \begin{pmatrix} L_{k+1,i} - L_{k,i} & \text{for } k < N \\ -L_{k,i} & \text{for } k = N \end{pmatrix} - D_{k,i}$$
$$+ \begin{pmatrix} V_{k-1,i} - V_{k,i} & \text{for } k > 1 \\ -V_{k,i} & \text{for } k = 1 \end{pmatrix} + \sum_j v_{i,j} P_{k,j} = 0 \quad \forall k,i \tag{2.23}$$

$$y^{\text{eq}}_{k,i} \varphi_{k,i} p = x_{k,i} \gamma_{k,i} p^{\text{S0}}_{k,i} \quad \forall k,i \tag{2.24}$$

$$\sum_i x_{k,i} = 1 \quad \forall k \tag{2.25}$$

$$\sum_i y^{\text{eq}}_{k,i} = 1 \quad \forall k \tag{2.26}$$

$$\sum_i \begin{pmatrix} V_{k-1,i} - V_{k,i} & \text{for } k > 1 \\ -V_{k,i} & \text{for } k = 1 \end{pmatrix} \Delta H^{\text{V}}_i$$
$$- \sum_j \Delta H^{\text{R}}_j P_{k,j} + Q^{\text{Heat}}_k - Q^{\text{Cool}}_k = 0 \quad \forall k \tag{2.27}$$

For simplicity, the heat of reaction of reaction j, ΔH^{R}_j, and the heat of vaporization ΔH^{V}_i of component i are assumed to be constant. The elements are equipped with external heating Q_k^{Heat} and cooling Q_k^{Cool} to represent a reboiler or a condenser. The liquid and vapor component streams,

$L_{k,i}$ and $V_{k,i}$, as well as the liquid withdrawal streams $D_{k,i}$ are calculated from the molar fractions in the liquid and in the vapor phase, $x_{k,i}$ and $y_{k,i}$, and the total material streams L_k, V_k and D_k:

$$L_{k,i} = x_{k,i} L_k \quad \forall k,i \tag{2.28}$$

$$V_{k,i} = y_{k,i} V_k \quad \forall k,i \tag{2.29}$$

$$D_{k,i} = x_{k,i} D_k \quad \forall k,i \tag{2.30}$$

Equations of the substance system

The feed streams $F_{k,i}$, the production rates $P_{k,i}$, the vapor pressure $p^{S0}_{k,i}$, the fugacity coefficients $\varphi_{k,i}$, and the activity coefficients $\gamma_{k,i}$ depend on the substance system and are described in the examples section for methyl tertiary butyl ether and methyl acetate (section 2.4.3).

Equations of the superstructure

The existence of an optional element, that is a column tray, k in the optimal solution depends on the value of a binary variable δ_k. If the binary variable is 1, the element exists. If the binary variable is 0, all functions of the element are switched off, that is, there is neither separation nor reaction, and neither feed streams nor liquid withdrawals exist. The separation can be switched off by the separation efficiency E_k of the vapor phase (eq. (2.31)). It is assumed that the phase equilibrium is reached in the reboiler (eq. (2.32)).

$$E_k = \frac{y_{k,i} - y_{k-1,i}}{y^{eq}_{k,i} - y_{k-1,i}} \quad \forall k > 1, i \tag{2.31}$$

$$y_{k=1,i} = y^{eq}_{k=1,i} \quad \forall k = 1, i \tag{2.32}$$

If δ_k equals 1, the separation efficiency E_k is set to $P^{\text{Efficiency}}$, if δ_k equals 0, E_k is set to 0 (eq. (2.33)).

The value of the parameter $P^{\text{Efficiency}}$ depends on the substance system. It is assumed that the separation efficiency is the same on all trays for all components. The values of the parameters in the following equations are given in section 2.6.

$$E_k = P^{\text{Efficiency}} \delta_k \quad \forall k \tag{2.33}$$

If E_k is set to 0, the composition of the vapor stream leaving the respective element is equal to the composition of the vapor stream entering the element ($y_{k,i} = y_{k-1,i}$) (eq. (2.31)). As furthermore neither reaction takes place ($P_{k,j} = 0$) nor external streams enter or leave the elements ($F_{k,i} = 0$, $D_{k,i} = 0$), the composition of the liquid remains also unaffected, thus leading to a non-existing element (eq. (2.23)). Switching off reaction as well as liquid withdrawal is achieved by using so-called big-M constraints (eq. (2.34) – (2.35)). Chemical reaction takes place on a tray k if the effective holdup H^{eff}_k, which is modeled as a continuous variable, takes on a positive value. The definition of the effective holdup depends on the reaction kinetics, since the chemical reaction can be based on the mass or on the number of active sites on the heterogeneous catalyst, for example.

$$H^{\text{eff}}_k \leq 10\,000 \delta_k \quad \forall k \tag{2.34}$$

$$D_k \leq P^{\text{max Dist}} \delta_k \quad \forall k \tag{2.35}$$

Similarly, feed streams are switched off. The number of feed streams with different compositions depends on the substance system; the corresponding equations are given in section 2.4.3.

Since reboiler and condenser always exist, the corresponding binary variables are fixed: $\delta_{k=1} = 1$, $\delta_{k=N} = 1$. Furthermore, no vapor stream leaves the total condenser ($V_{k=N} = 0$, $V_{k=N,i} = 0 \; \forall i$) and no liquid stream leaves the reboiler ($L_{k=1} = 0$, $L_{k=1,i} = 0 \; \forall i$). As in (Ciric and Gu 1994) and (Jackson and Grossmann 2001) equivalent solutions with equal total number of elements are excluded from the feasible space by logic constraints: A tray can only exist if the tray below it exists. The only exception is the condenser ($k = N$), which is always existent.

$$\delta_{k+1} \leq \delta_k \quad \forall k < N-1 \tag{2.36}$$

Constraints on the geometry and on the operating conditions

The maximal reactive holdup H_k^{eff} on a tray is constrained by its maximum height, which is assumed to be 0.2 m, and the column diameter Dia (eq.

(2.37)). The value of the parameter P^{Holdup} depends on the catalyst density and its volume fraction.

$$H_k^{\text{eff}} \leq P^{\text{Holdup}} Dia^2 \quad \forall k \tag{2.37}$$

The diameter is constrained by the operating range of the column. On each column tray bounds on the minimum and the maximum vapor velocities are given (eq. (2.38) and (2.39)). The values of the parameters $P^{\text{min Dia}}$ and $P^{\text{max Dia}}$ depend on the substance system and the column pressure.

$$Dia^2 \geq P^{\text{min Dia}} V_k \quad \forall k \tag{2.38}$$

$$Dia^2 \leq P^{\text{max Dia}} V_k \quad \forall k < N \tag{2.39}$$

Cost model

The total annual costs C_a^{Total} of the process are minimized. They are calculated from the annual operating cost for cooling and heating (with parameters $P^{\text{Cooling Water}}$ and P^{Steam} for the cost of cooling water and steam), the annualized investment costs for column shell C^{Shell} and internals $C^{\text{Internals}}$, catalyst (with parameter P^{Catalyst}), condenser and reboiler (with parameters $P^{\text{Condenser}}$ and P^{Reboiler}), and the annual raw material costs $C_a^{\text{Raw Materials}}$ (which is a parameter, if the total amount of feed is given), reduced by the annual product revenues R_a^{Products} (eq. (2.40)). Most of the parameters were taken from (Cookbook 2003). The product revenues R_a^{Products}, which depend on the substance system, are given in section 2.4.3.

$$
\begin{aligned}
C_a^{\text{Total}} = {} & P^{\text{Cooling Water}} \sum_k Q_k^{\text{Cool}} + P^{\text{Steam}} \sum_k Q_k^{\text{Heat}} \\
& + 0.3 \left(C^{\text{Shell}} + C^{\text{Internals}} + P^{\text{Catalyst}} \sum_k H_k^{\text{eff}} \right. \\
& \left. + P^{\text{Condenser}} \left(\frac{Q_{k=N}^{\text{Cool}}}{\Delta T^{\text{Condenser}}} \right)^{0.6} + P^{\text{Reboiler}} \left(\frac{Q_{k=1}^{\text{Heat}}}{\Delta T^{\text{Reboiler}}} \right)^{0.6} \right) \\
& + C_a^{\text{Raw Materaials}} - R_a^{\text{Products}}
\end{aligned}
\tag{2.40}
$$

In the condenser cost calculation the logarithmic mean temperature difference $\Delta T^{\text{Condenser}}$ is used, which is calculated from the temperature in the condenser $T_{k=N}$ and the temperature of the cooling water entering and leaving the condenser ($T_{\text{in}}^{\text{Water}}, T_{\text{out}}^{\text{Water}}$). In the reboiler cost calculation the difference of the temperatures of steam T^{Steam} and the reboiler $T_{k=1}$ is used:

$$\Delta T^{\text{Condenser}} = \frac{T_{\text{out}}^{\text{Water}} - T_{\text{in}}^{\text{Water}}}{\ln\left(\dfrac{T_{k=N} - T_{\text{in}}^{\text{Water}}}{T_{k=N} - T_{\text{out}}^{\text{Water}}} \right)} \tag{2.41}$$

$$\Delta T^{\text{Reboiler}} = T^{\text{Steam}} - T_{k=1} \tag{2.42}$$

The cost correlations and parameter values for the calculation of the investment costs of column shell and internals are taken from (Cookbook 2003):

$$C^{\text{Shell}} = P^{\text{Shell}} H^{0.81} Dia^{1.03} \tag{2.43}$$

$$\begin{aligned} C^{\text{Internals}} &= 3.5 \sum_k \delta_k \\ &\cdot \left(0.522 + 0.059 Dia + 0.258 Dia^2 - 0.021 Dia^3 \right) \end{aligned} \tag{2.44}$$

The height H of the column is given by the number of trays, assuming a fixed tray height of 0.6 m, an additional height at the top and bottom of the column of 3 m, plus a variable height for the reactive holdup up to 0.2 m:

$$H = 0.6\left(\sum_k \delta_k + 3 \right) + \frac{0.2 \sum_k H_k^{\text{eff}}}{P^{\text{Holdup}} Dia^2} \tag{2.45}$$

2.4.2 Solution method

The modeling environment GAMS is used as an interface to commercial optimization algorithms. A classical solution approach for MINLP is branch-and-bound in conjunction with nonlinear continuous optimization algorithms. The solvers which were used in this work are the branch-and-bound solver SBB (SBB 2003) and the generalized reduced gradient solver CONOPT3 (CONOPT 2003).

Starting from the root node of the branch-and-bound tree in which all binary variables are integer-relaxed, that is, in which they are defined on the interval [0, 1] rather than on the discrete set, one binary variable is fixed on each layer of the search tree. During the solution procedure the results of the relaxed problems at the nodes generate lower bounds on the optimal solution in the subsequent sub-tree. If all binary variables are set to discrete values in a leaf node, an upper bound on the optimal solution is found. If the relaxed solution at a node is larger than the best upper bound, the respective sub-tree is fathomed. The algorithm is terminated if the best solution found equals the best bound on the objective (zero optimality gap). This procedure is capable of finding the best, that is global, solution if the relaxed problems are solved to global optimality. Since the nonlinear part of the problem is non-convex, it is unlikely that a global solution is found by a local solver like CONOPT3. Thus, this approach will in general only find locally optimal solutions.

It should be mentioned that the solver parameters of CONOPT3 were set to non-standard values to increase the efficiency of the solution procedure. Numerical studies with trivial initial values (that is all initial values were set to one or the nearest lower or upper bounds) showed that for the first example the following solver options are a good choice (Sand et al. 2004a):

– rtnwma = 10^{-6}: decrease of the feasibility tolerance,
– lfscal = 1: increase of the frequency of automatic scaling of variables and constraints,
– lslack = true: introduction of slack variables to infeasible constraints in the first iteration step,
– lmmxsf = 1: alternative method for determining the step length,
– lfstal = 10^{5}: increase of the number of iterations with zero step length or with worse objective.

For the second example the following options improved the solution:
– rtnwma = 10^{-6} and rtnwmi = 10^{-11}: decrease of feasibility tolerances,
– lfscal = 1: increase of the frequency of automatic scaling of variables and constraints,
– lstcrs = true: optional method for improving the initial point.

Due to the non-convexity of the problem, the initialization is a crucial part in the solution of design problems. In addition to trivial initial values two initialization strategies were studied. In the first approach process-specific knowledge is used to derive a simpler optimization model. According to the qualitative design suggestion some of the variables are fixed, leading to a nonlinear continuous problem without discrete decisions. The solution of the reduced NLP serves as initial point for the original MINLP. Although two different problems have to be solved in this case, the computational time does not increase (Sand et al. 2004b).

The second approach is the evolutionary generation of initial values by scatter search. This algorithm generates a population of candidate initial points for the continuous part of the problem. Based on two filter criteria, the merit filter and the distance filter, promising points are chosen as initial values for the optimization. The NLP is then solved not only from one but from a series of initial points. Numerical studies have shown that the use of scatter search increases the probability of a good solution, however along with a pronounced increase in computational time (Sand et al. 2005b). Figure 2.26 illustrates the different initialization strategies.

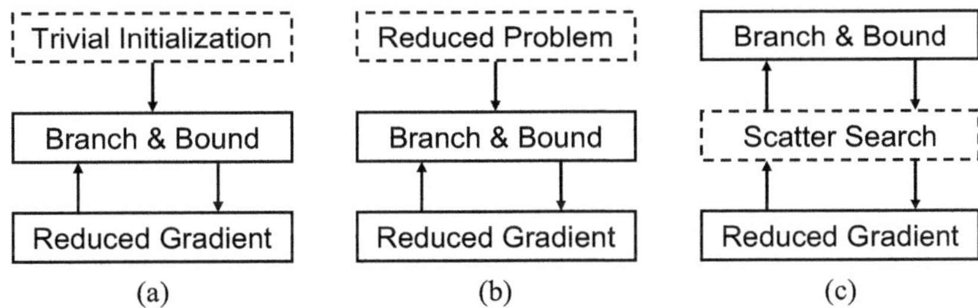

Fig. 2.26: Different initialization strategies (a: trivial initialization, b: knowledge-based initialization, c: initialization by scatter search)

Efficient model formulation

The model was implemented in the modeling environment GAMS (**G**eneral **A**lgebraic **M**odeling **S**ystem) (Brooke et al. 2003). An efficient implementation of the model supports a numerically stable optimization. Often different optimization solvers require different model formulations. Since the solver CONOPT3 was used, the following issues concerning good model formulations were considered (CONOPT 2003):
− complex non-linear terms were simplified by intermediate variables,
− degeneracy was avoided by implementing only necessary bounds on variables,
− very small as well as very large gradients and values of functions were avoided, for example by lower and upper bounds on variables which appear in logarithmic or exponential expressions.

In addition, bounds on variables were introduced based on insight in the process to prevent the solver from searching in regions were no good solution could be expected. These bounds are often not rigorous in a physical sense, but known to be non-binding in the optimal solution. An example is a lower bound of 1 kW on the cooling (heating) load in the condenser (reboiler).

2.4.3 Examples

Example 1: Production of methyl tertiary butyl ether (MTBE)

The conceptual design of a reactive distillation column for the production of methyl tertiary butyl ether, which was tackled by heuristic methods in section 2.3.11, was optimized by mathematical programming techniques. In the following, the characteristic equations of the system are given.

Equations of the substance system

For a specific substance system, the feed streams, the production rates and the phase equilibrium must be specified, leading to additional constraints.

A lower bound on the product purity is introduced: $x^{lo}_{k=1,i=\text{MTBE}} = 0.99$. It is assumed that no reaction takes place in the reboiler and in the condenser: $H^{eff}_{k=1} = H^{eff}_{k=N} = 0$. Furthermore, the superstructure is reduced: Liquid withdrawal is only allowed in the condenser and in the reboiler and is set to zero on the column trays: $D_k = 0 \ \forall \ 1 < k < N, \ D_{k,i} = 0 \ \forall \ i, 1 < k < N$. Similarly, heating is only allowed in the reboiler and cooling is only allowed in the condenser: $Q^{Heat}_k = 0 \ \forall \ k > 1, \ Q^{Cool}_k = 0 \ \forall \ k < N$.

The annual revenues are composed of revenues of the main product MTBE at the bottom and revenues of methanol and isobutene/n-butane at the top of the column.

$$R^{\text{Products}}_a = 825.08 D_{k=1} + 138.42 D_{k=N,i=\text{MeOH}} \\ + 410.8\left(D_{k=N,i=\text{IB}} + D_{k=N,i=\text{n-butane}}\right) \tag{2.46}$$

Two different types of feed streams are defined: $F1_k$ is a methanol stream ($x^F_{\text{MeOH}} = 1$) and $F2_k$ is a mixture of isobutene and n-butane ($x^F_{\text{IB}} = 0.652, \ x^F_{\text{n-butane}} = 0.348$, taken from (Stichlmair and Frey 2001)), and no MTBE is fed to the column. The total amounts of both feed streams are given in eq. (2.51) and (2.52).

$$F_{k,i=\text{MeOH}} = F1_k \, x^F_{\text{MeOH}} \quad \forall k \tag{2.47}$$

$$F_{k,i=\mathrm{IB}} = F2_k\, x_{\mathrm{IB}}^{\mathrm{F}} \quad \forall k \tag{2.48}$$

$$F_{k,i=\mathrm{n-butane}} = F2_k\, x_{\mathrm{n-butane}}^{\mathrm{F}} \quad \forall k \tag{2.49}$$

$$F_{k,i=\mathrm{MTBE}} = 0 \quad \forall k \tag{2.50}$$

$$\sum_k F1_k = 6.375 \tag{2.51}$$

$$\sum_k F2_k = 8.625 \tag{2.52}$$

As mentioned in section 2.4.1, the possible existence of a feed on a tray k is switched on or off by the following big-M constraints:

$$F1_k \le 6.375\delta_k \quad \forall k \tag{2.53}$$

$$F2_k \le 8.625\delta_k \quad \forall k \tag{2.54}$$

The chemical equilibrium constant K^{a}_k (eq. (2.55)), the kinetic parameters $k_{k,j}^{\mathrm{reac}}$ of the forward reaction R1 and the backward reaction R2 (eq. (2.56) and (2.57)) and the activity based calculation of the production rates $P_{k,j}$ (eq. (2.58) and (2.59)) are taken from (Rehfinger and Hoffmann 1990a, 1990b). The value of the parameter T_{ref} is 295.15 K.

$$\begin{aligned}
K^{\mathrm{a}}_k = 284\exp\Bigg(&-1.49277\cdot 10^3\left(\frac{1}{T_k} - \frac{1}{T_{\mathrm{ref}}}\right) \\
&- 7.74002\cdot 10^1 \ln\!\left(\frac{T_k}{T_{\mathrm{ref}}}\right) + 5.07563\cdot 10^{-1}\left(T_k - T_{\mathrm{ref}}\right) \\
&- 9.12739\cdot 10^{-4}\left(T_k^2 - T_{\mathrm{ref}}^2\right) + 1.10649\cdot 10^{-6}\left(T_k^3 - T_{\mathrm{ref}}^3\right) \\
&- 6.27996\cdot 10^{-10}\left(T_k^4 - T_{\mathrm{ref}}^4\right)\Bigg) \quad \forall k
\end{aligned} \tag{2.55}$$

$$k_{k,j=\mathrm{R1}}^{\mathrm{reac}} = 0.0155\exp\left(33.3731 - \frac{92400}{R_{\mathrm{m}}T_k}\right) \quad \forall k \tag{2.56}$$

$$k^{\text{reac}}_{k,j=\text{R2}} = \frac{k^{\text{reac}}_{k,j=\text{R1}}}{K^{\text{a}}_{k}} \quad \forall k \tag{2.57}$$

$$P_{k,j=\text{R1}} = H^{\text{eff}}_{k} k^{\text{reac}}_{k,j=\text{R1}} \frac{a_{k,i=\text{IB}}}{a_{k,i=\text{MeOH}}} \quad \forall k \tag{2.58}$$

$$P_{k,j=\text{R2}} = H^{\text{eff}}_{k} k^{\text{reac}}_{k,j=\text{R2}} \frac{a_{k,i=\text{MTBE}}}{a^{2}_{k,i=\text{MeOH}}} \quad \forall k \tag{2.59}$$

Due to the non-ideality of the liquid phase, the activities $a_{k,i}$ are calculated from the molar fractions of the liquid phase $x_{k,i}$ and the activity coefficients $\gamma_{k,i}$, which are calculated by Wilson's approach (eq. (2.60)-(2.62)). The vapor pressures $p^{\text{S0}}_{k,i}$ are given by the Antoine equation (eq. 2.63)). The thermodynamic property data (parameters v^{L}_{i}, $a^{\text{Wilson}}_{i,m}$ and $A_{i,1}$, $A_{i,2}$, $A_{i,3}$) are taken from (Beßling 1998). The vapor phase is assumed to be ideal, leading to a fugacity coefficient $\varphi_{k,i}$ of 1.

$$a_{k,i} = \gamma_{,k,i} x_{k,i} \quad \forall k,i \tag{2.60}$$

$$\Lambda_{k,i,m} = \frac{v^{\text{L}}_{m}}{v^{\text{L}}_{i}} \exp\left(-\frac{a^{\text{Wilson}}_{i,m}}{1.9872 T_{k}}\right) \quad \forall k,i,m \tag{2.61}$$

$$\ln(\gamma_{k,i}) = 1 - \ln\left(\sum_{m} x_{k,m} \Lambda_{k,i,m}\right) - \sum_{n} \frac{x_{k,n} \Lambda_{k,n,i}}{\sum_{m} x_{k,m} \Lambda_{k,i,m}} \quad \forall k,i \tag{2.62}$$

$$p^{\text{S0}}_{k,i} = \exp\left(A_{i,1} - \frac{A_{i,2}}{A_{i,3} + T_{k} - 273.15}\right) \quad \forall k,i \tag{2.63}$$

Initialization based on process knowledge

Based on problem specific knowledge for the MTBE example described in section 2.3.11, the following variables were fixed in the initial problem:

- Assuming a typical column height of 30 m, the number of existent column trays is set to 38 by fixing the binary variables: $\delta_k = 1$ $\forall$ $k \leq 37 \vee k = N$, $\delta_k = 0$ $\forall$ $k \geq 38 \wedge k < N$.
- No reaction takes place in the lower part of the column. In the upper integrated part, methanol is fed above the mixture of isobutene and n-butane. Thus, the column is divided into three parts: the lower part without reaction: $H_k^{\text{eff}} = 0$ $\forall$ $k \leq 12$, $P_{k,j} = 0$ $\forall$ $k \leq 12$, the middle integrated part with possible isobutene/n-butane feeds, that is, the isobutene/n-butane feed is fixed to zero in the lower and upper part: $F2_k = 0$ $\forall$ $k \leq 12 \vee k \geq 26$ and the upper integrated part with possible methanol feeds, that is, methanol feed is fixed to zero in the middle part and the lower part: $F1_k = 0$ $\forall$ $k \leq 25 \vee k \geq 38$.

For simplicity, the conversion of isobutene was maximized instead of a minimization of the total annualized costs. Since the separation of n-butane and isobutene, which can be achieved by a complete conversion of isobutene, is one of the objectives of this process, a maximization of the conversion is plausible.

Optimization results

Figure 2.27 depicts the objective function values obtained for 51 problem instances with maximal tray numbers between 10 and 60. The solutions obtained by trivial initialization are on average inferior to solutions which were initialized by the scatter search. The solutions obtained by knowledge-based initialization are quite similar to those initialized by the scatter search.

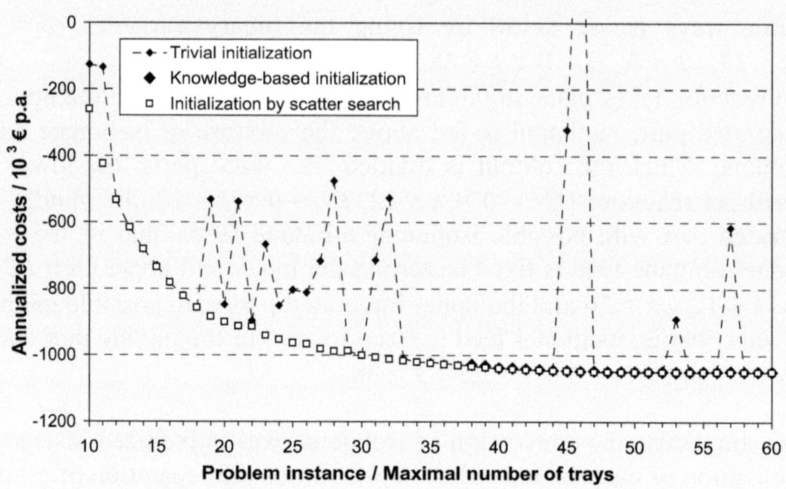

Fig. 2.27: MINLP solutions of the MTBE column

The improvement of the values of the objective function in case of an initialization by scatter search is achieved at the expense of an increase of the computational time by about two orders of magnitude, as can be seen from Figure 2.28. In case of knowledge-based initialization the computational time is even less compared to the case of trivial initialization.

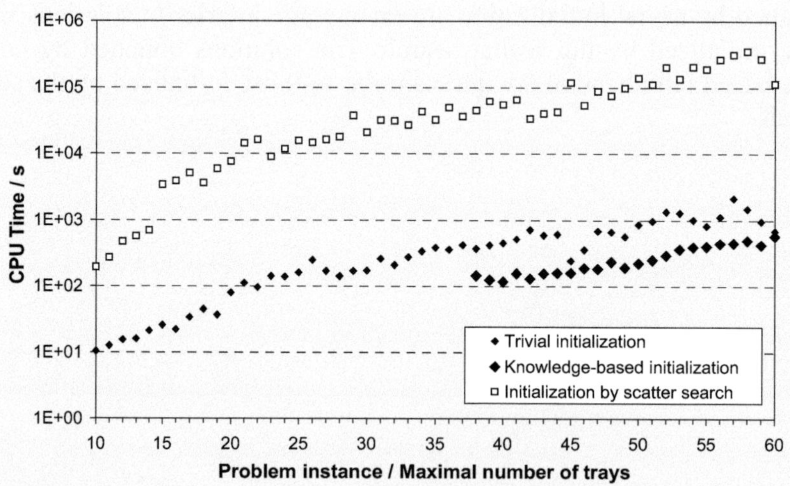

Fig. 2.28: Computational effort for MINLP and NLP solutions of the MTBE column problem

In Table 2.6 the values of the objective function and some characteristics of the optimal solutions are given for a problem instance with $N = 60$ optional trays. The optimal solutions obtained by trivial initialization and initialization by scatter search are the same. For the knowledge-based initialization the objective function value is very close, the column structure is quite similar, only the number of feed streams differs (Figure 2.29). The scatter search requires significantly more computation time.

Table 2.6: Optimal solution for the MTBE column

		Trivial Initialization	Knowledge-based initialization	Initialization by scatter search
No. of variables	[-]	6793	6793	6793
No. of constraints	[-]	7031	7031	7031
Solution time	[min]	11	9	1893
Maximal no. of trays	[-]	60	60	60
Optimal no. of trays	[-]	50	49	50
Diameter	[m]	0.39	0.39	0.39
Reflux ratio	[-]	4.57	4.51	4.57
Heat duty	[kW]	231.24	231.60	231.24
Cool duty	[kW]	447.28	446.77	447.28
Product flow rate	$[\text{mol s}^{-1}]$	5.57	5.54	5.57
Investment costs	$[10^3\,\text{€ p.a.}]$	144.01	140.74	144.01
Operating costs	$[10^3\,\text{€ p.a.}]$	64.13	64.21	64.13
Total annualized costs	$[10^3\,\text{€ p.a.}]$	-1050.53	-1049.55	-1050.53

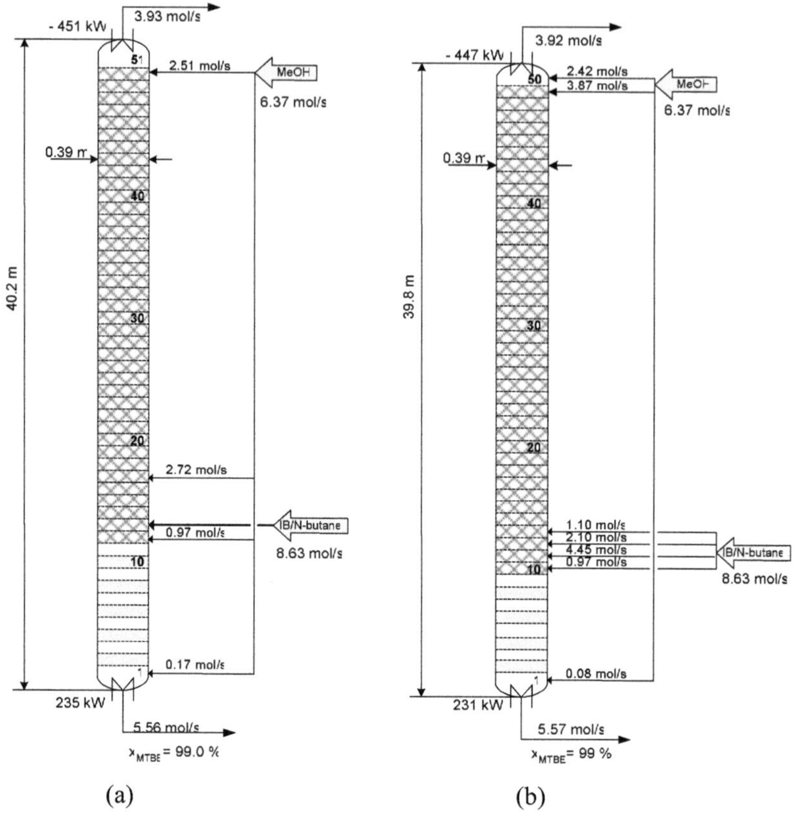

Fig. 2.29: Optimal solution for the MTBE column
(a: knowledge-based initialization, b: initialization by scatter search)

Extension of the superstructure by an external reactor

An analysis of the optimal solution showed that the geometric inequality constraints for the effective holdup and the diameter are active in the optimal solution (Sand et al. 2005a). Therefore, the superstructure of the column was extended by an optional external reactor and each column tray is (potentially) connected to the reactor. This has two advantages: the reactor provides additional holdup for the reaction and it offers the possibility to remove the heat of reaction by external cooling.

The existence of the CSTR-type reactor is characterized by its continuous reactive holdup. The optimal solutions for an instance with 60 optional trays are given in Table 2.7.

Table 2.7: Optimal solution of MTBE column with external reactor

		Trivial initialization	Knowledge-based initialization	Initialization by scatter search
No. of variables	[-]	7482	7482	7482
No. of constraints	[-]	7832	7832	7832
Solution time	[min]	16	15	4658
Maximal no. of trays	[-]	60	60	60
Optimal no. of trays	[-]	39	28	37
Diameter	[m]	0.34	0.38	0.33
Reflux ratio	[-]	1.95	2.51	1.63
Heat duty	[kW]	187.62	218.84	187.24
Cool duty	[kW]	356.63	379.96	213.73
Cool duty reactor	[kW]	0	0	272
Product flow rate	[mol s^{-1}]	5.59	5.57	5.58
Investment costs	[10^3 € p.a.]	118.80	105.11	108.46
Operating costs	[10^3 € p.a.]	51.95	60.09	53.63
Total annualized costs	[10^3 € p.a.]	-1098.15	-1089.43	-1102.27

In this case, the solutions based on knowledge-based initialization and initialization by scatter search differ both in terms of the value of the objective function as well as in terms of the structure as can be seen in Figure 2.30. The three main differences are the distribution of the reaction zone, the location of the isobutene/n-butane feed, and the cooling duty in the reactor. Whereas in the solution for a knowledge-based initialization the reactor is connected to several non-reactive trays, it is only connected to the trays at the boundary between the separating and the integrated column zone for the result obtained by scatter search. In that solution the total feed of isobutene/n-butane is fed to the reactor, which is cooled by 272 kW. The structure of the solution obtained by trivial initialization is similar to the structure obtained by knowledge-based initialization.

By including an external reactor, the optimal number of column trays is significantly decreased. Despite the extension by an additional apparatus, both the investment cost as well as the operating cost could be reduced.

As can be seen from Figure 2.31,b the external reactor is used to cross the distillation boundary. In contrast, in case of the integrated column without reactor (Figure 2.31,a), the chemical equilibrium is approached within the reaction zone. In both processes, the lower separating zone in the column is used to purify the main product MTBE.

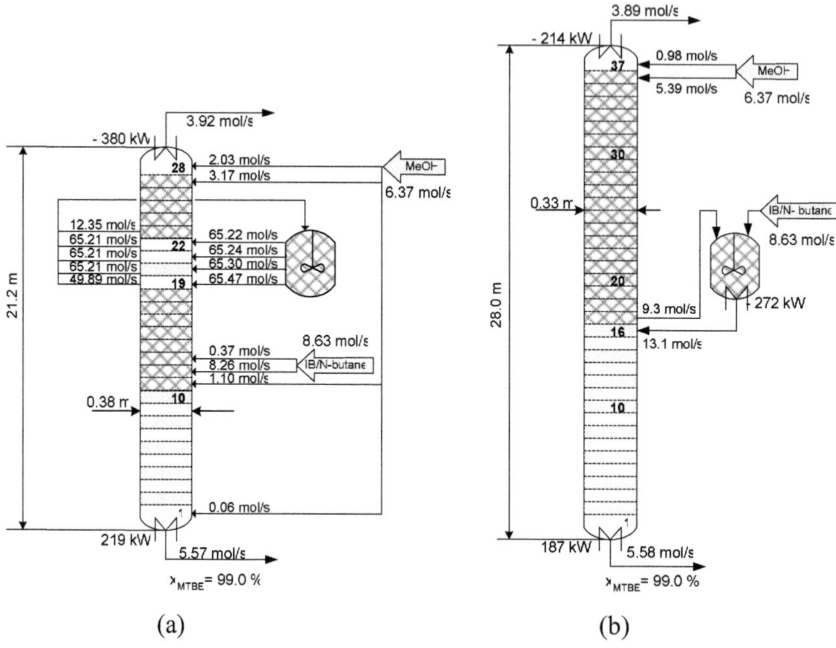

(a)   (b)

Fig. 2.30: Optimal solution for the MTBE column with external reactor (a: knowledge-based initialization, b: initialization by scatter search)

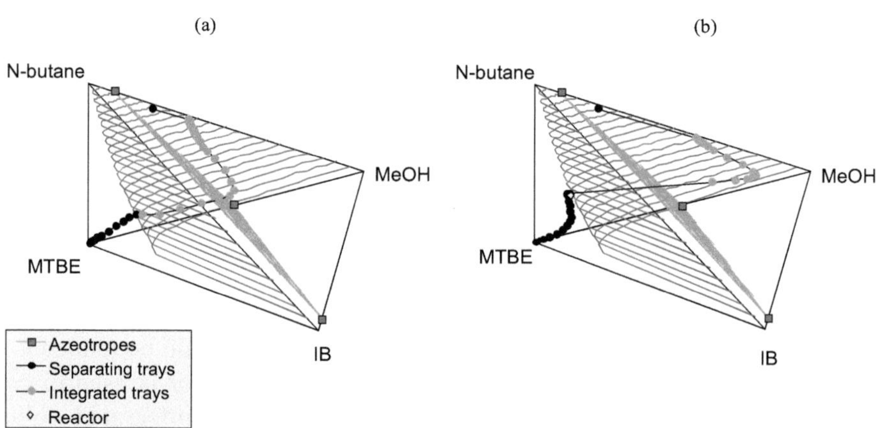

Fig. 2.31: Liquid molar fractions (a: column without external reactor, b: column with external reactor)

Example 2: Production of methyl acetate

The optimization procedure was applied to a second system: the production of methyl acetate from methanol and acetic acid in a reactive distillation column. The characteristic equations, knowledge-based initialization and the optimization results are given below.

Equations of the substance system

The purity of the top product stream is constrained by a lower bound: $x^{\mathrm{lo}}_{k=N,i=\mathrm{MeAc}} = 0.99$. The superstructure is reduced as in example 1, so that liquid withdrawal as well as external heat exchange is only allowed in the condenser and in the reboiler.

The product revenues are given by the top and bottom flow rates, whereas the water effluent at the bottom makes a negative contribution to the revenues.

$$R_{\mathrm{a}}^{\mathrm{Products}} = 1810.5 D_{k=N} - 2.6 D_{k=1} \tag{2.64}$$

Two different types of feed streams are defined: $F1_k$ is a methanol stream ($x^{\mathrm{F}}_{\mathrm{MeOH}} = 1$) and $F2_k$ consists of acetic acid ($x^{\mathrm{F}}_{\mathrm{HAc}} = 1$), neither methyl acetate nor water is fed to the column:

$$F_{k,i=\mathrm{MeOH}} = F1_k\, x^{\mathrm{F}}_{\mathrm{MeOH}} \quad \forall k \tag{2.65}$$

$$F_{k,i=\mathrm{HAc}} = F2_k\, x^{\mathrm{F}}_{\mathrm{HAc}} \quad \forall k \tag{2.66}$$

$$F_{k,i=\mathrm{MeAc}} = 0 \quad \forall k \tag{2.67}$$

$$F_{k,i=\mathrm{Water}} = 0 \quad \forall k \tag{2.68}$$

The total amounts of the feed streams are fixed to 10 mol/s:

$$\sum_k F1_k = 10 \tag{2.69}$$

$$\sum_k F2_k = 10 \tag{2.70}$$

As in the first example, the possible existence of a feed on a tray k is switched on or off by big-M constraints:

$$F1_k \leq 10\delta_k \quad \forall k \tag{2.71}$$

$$F2_k \leq 10\delta_k \quad \forall k \tag{2.72}$$

The pseudo-homogeneous kinetic approach for the kinetic parameters and the production rates is taken from (Pöpken et al. 2000).

$$k^{\text{reac}}_{k,j=\text{R1}} = 2.961 \cdot 10^7 \exp\left(-\frac{49\,190}{R_{\text{m}}T_k}\right) \quad \forall k \tag{2.73}$$

$$k^{\text{reac}}_{k,j=\text{R2}} = 1.348 \cdot 10^9 \exp\left(-\frac{69\,230}{R_{\text{m}}T_k}\right) \quad \forall k \tag{2.74}$$

$$P_{k,j=\text{R1}} = H^{\text{eff}}_k k^{\text{reac}}_{k,j=\text{R1}} \, a_{k,i=\text{HAc}} a_{k,i=\text{MeOH}} \quad \forall k \tag{2.75}$$

$$P_{k,j=\text{R2}} = H^{\text{eff}}_k k^{\text{reac}}_{k,j=\text{R2}} \, a_{k,i=\text{MeAc}} a_{k,i=\text{Water}} \quad \forall k \tag{2.76}$$

As in the first example, the activity is calculated from activity coefficients which are given by Wilson's approach, and the vapor pressure is given by the Antoine equation. The association of acetic acid requires to consider the non-ideality of the vapor phase. The fugacity coefficient $\varphi_{k,i}$ is calculated from the chemical theory (Gmehling and Kolbe 1992). The dimerization constant $K_k^{\text{D}*}$ depends on the temperature T_k and the pressure p (eq. (2.77) – (2.78)). The values of the parameters ($f^0 = 1$ bar) are taken from (Gmehling and Kolbe 1992).

$$\ln K_k^{\text{D}} = f^0\left(-17.374 + \frac{7290}{T_k}\right) \quad \forall k \tag{2.77}$$

$$K_k^{\text{D}*} = \frac{K_k^{\text{D}} \cdot p}{f^0} \quad \forall k \tag{2.78}$$

The fugacity coefficient φ^{AC}_k of the associating component is calculated from the true molar fraction of the associating component z_k^{AC} and the vapor molar fraction of the associating component $y_k^{\text{AC}} = y_{k,i=\text{HAc}}$:

$$z_k^{\mathrm{AC}} = \frac{\sqrt{1 + 4 K_k^{\mathrm{D^*}} y_k^{\mathrm{AC}} \left(2 - y_k^{\mathrm{AC}}\right)} - 1}{2 K_k^{\mathrm{D^*}} \left(2 - y_k^{\mathrm{AC}}\right)} \quad \forall k \tag{2.79}$$

$$\varphi_k^{\mathrm{AC}} = \frac{z_k^{\mathrm{AC}}}{y_k^{\mathrm{AC}}} \quad \forall k \tag{2.80}$$

Similarly, the fugacity coefficient φ_k^{NAC} of the non-associating components depends on the true molar fraction of the non-associating components z_k^{NAC} and the vapor molar fraction of the non-associating components $y_k^{\mathrm{NAC}} = 1 - y_{k,i=\mathrm{HAc}}$ (eq. (2.81) – (2.82)).

$$z_k^{\mathrm{NAC}} = y_k^{\mathrm{NAC}} \frac{1 + 4 K_k^{\mathrm{D^*}} \left(2 - y_k^{\mathrm{AC}}\right) - \sqrt{1 + 4 K_k^{\mathrm{D^*}} y_k^{\mathrm{AC}} \left(2 - y_k^{\mathrm{AC}}\right)}}{2 K_k^{\mathrm{D^*}} \left(2 - y_k^{\mathrm{AC}}\right)^2} \tag{2.81}$$

$$\forall k$$

$$\varphi_k^{\mathrm{NAC}} = \frac{z_k^{\mathrm{NAC}}}{y_k^{\mathrm{NAC}}} \quad \forall k \tag{2.82}$$

Initialization based on process knowledge

The process synthesis strategy was applied to the example of the production of methyl acetate similar to the first example leading to the following design suggestion which was implemented by bounds on variables:

- The number of column trays is chosen to be 38, leading to the following bounds on the binary variables: $\delta_k = 1 \;\; \forall \, k \leq 37 \lor k = N$, $\delta_k = 0 \;\; \forall \, k \geq 38 \land k < N$.
- The upper part of the column is purely separating, and the lower part is integrated: $H_k^{\mathrm{eff}} = 0 \;\; \forall \, 20 \leq k \leq N$, $P_{k,j} = 0 \;\; \forall \, 20 \leq k \leq N$.
- Methanol is fed to the integrated lower part of the column, and acetic acid is fed to the purely separating upper part: $F1_k = 0 \;\; \forall \, 20 \leq k \leq N$, $F2_k = 0 \;\; \forall \, 1 \leq k \leq 19$.

Optimization results

35 problem instances with fixed numbers of trays between 26 and 60 were solved with initialization by scatter search (Figure 2.32). For all instances feasible solutions were obtained. The graph of the value of the objective function over the number of trays is very flat. A minimum is reached for 50 trays.

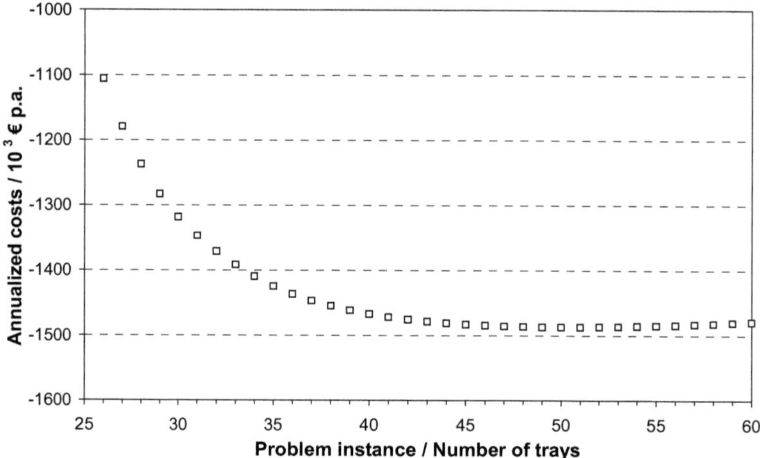

Fig. 2.32: Objective values of NLP problems (solution by OQNLP-CONOPT)

The results of two MINLP optimizations for an instance with 54 optional trays with knowledge-based initialization and initialization by scatter search are given in Table 2.8. In case of trivial initialization no feasible solution was found. The objective values differ only slightly by 0.3 % and the structures are very similar as can be seen from Figure 2.33.

Table 2.8: Optimal solution of methyl acetate column

		Knowledge-based initialization	Initialization by scatter search
No. of variables	[-]	6387	6387
No. of constraints	[-]	6601	6601
Solution time	[min]	2.2	260
Maximal no. of trays	[-]	54	54
Optimal no. of trays	[-]	42	50
Diameter	[m]	0.99	0.95
Reflux ratio	[-]	3.9	3.5
Heat duty	[kW]	1472.11	1351.60
Cool duty	[kW]	1502.15	1381.64

Product flow rate	[mol s^{-1}]	10.06	10.06
Investment costs	[10^3 € p.a.]	241.46	256.91
Operating costs	[10^3 € p.a.]	325.46	298.85
Total annualized costs	[10^3 € p.a.]	-1475.97	-1487.19

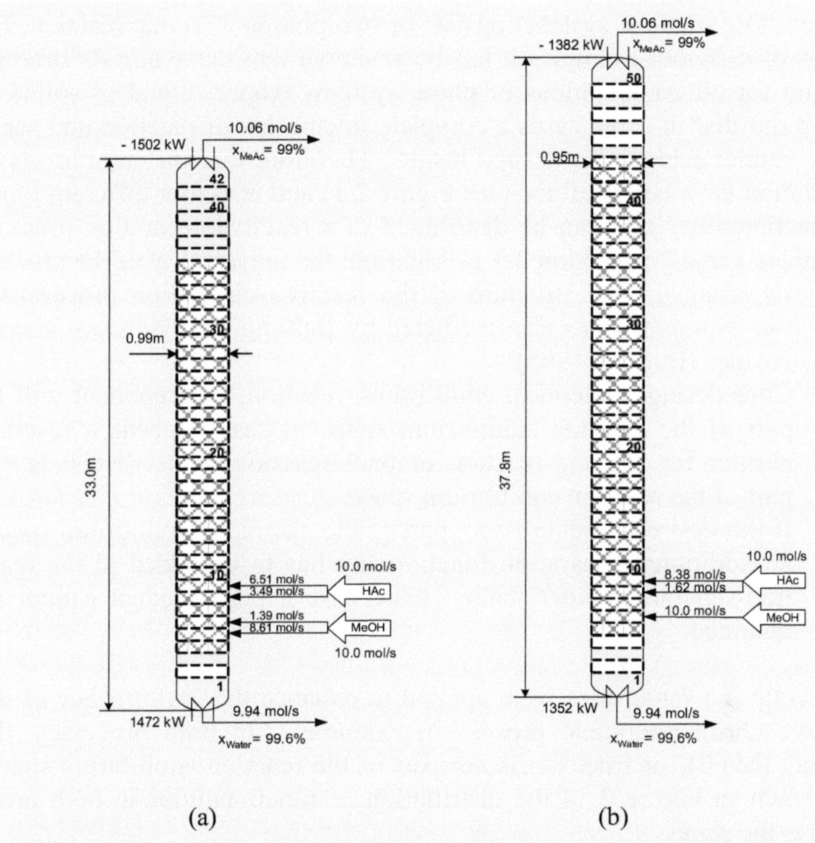

Fig. 2.33: Optimal solution for the methyl acetate column
(a: knowledge-based initialization, b: initialization by scatter search)

In contrast to the knowledge-based design suggestion, acetic acid is fed to the integrated part of the column. Also there are a few purely separating trays at the bottom of the column which were not included in the knowledge-based design.

2.5 Conclusions

In this chapter, a generic process synthesis strategy for reactive separation processes has been introduced. In this strategy, integrated reaction and separation processes are considered as multiple-phase systems with N_R reactions. The smallest system consists of two phases with one reaction. By means of different examples it has been shown that the synthesis strategy is valid for all combinations of phase systems (vapor, liquid or solid). It turned out that in some cases a complete integration of reaction and separation cannot achieve the desired results. Therefore four different levels of integration have been defined (see Figure 2.1) and also four different types of functionalities that can be distributed in a reactive separation process. Examples 1 and 2 in section 2.3.11 highlight the universality of the process synthesis strategy. The structure of the reactive distillation process described in example 1 was also predicted by Beßling, applying two simple heuristic rules (Beßling 1998):

1. Considering a chemical equilibrium reaction, a component will be part of the reaction equilibrium space in case it needs a reaction partner for forward-reaction or back-reaction. Otherwise it is not part of the reaction equilibrium space.
2. If the desired product is not part of the chemical equilibrium space, an additional separation functionality has to be added to the reaction-separation functionality. Otherwise a pure product cannot be obtained.

Exactly the same rules were applied to enhance the performance of the reactive chromatographic process in example 2. In both processes, the product (MTBE or fructose) is not part of the reaction equilibrium space. As shown in Figure 2.34 the distribution of functionalities in both processes is the same.

This is valid, although the two examples differ in three significant aspects:

− The reaction type is different ($A + B \leftrightarrows C$ and $A \leftrightarrows B$),
− the phase system is different (vapor-liquid and liquid-solid) and,
− the phase-creating components are different (Example 1: both phases consist of reaction components; Example 2: both phases (water and adsorbent) are added to the reaction system).

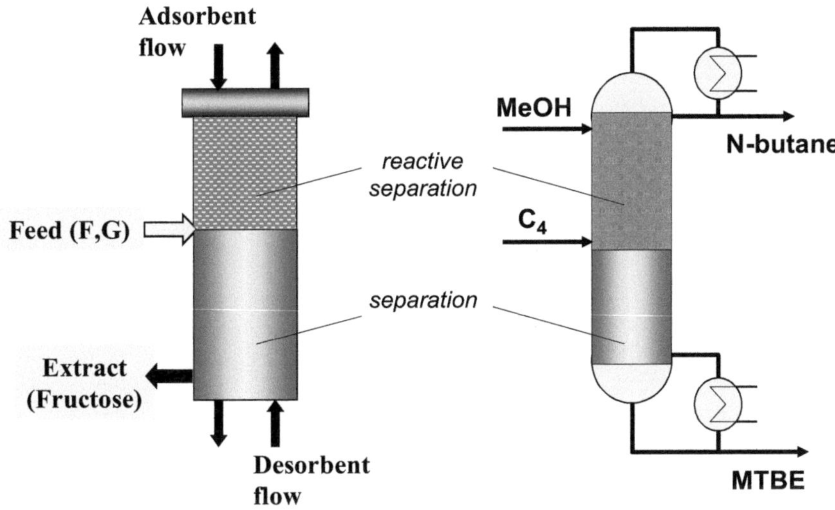

Fig. 2.34: Comparison of functionality distribution in reactive chromatography and reactive distillation

By mathematical programming techniques, the suggested process (resulting from the synthesis strategy) can be quantified and cost-optimal solutions can be found. Numerical studies showed that the use of process specific knowledge to generate good initial values can enhance the robustness of the solution compared to the use of trivial initial values. However, only local optima may be computed. Initialization by scatter search can improve the robustness of the solution with respect to the initialization at the price of a considerable increase in solution time.

2.6 Notation

Symbols

Symbol	Description	Unit
a	Activity	misc.
C	Costs;	10^3 €;
	Annualized costs	10^3 € p.a.
D	Liquid withdrawal	mol s^{-1}
Dia	Diameter	m
E	Vapor phase efficiency	-
F	Feed flow rate	mol s^{-1}

$F1$	Feed 1	mol s^{-1}
$F2$	Feed 2	mol s^{-1}
H	Column height	m
H^{eff}	Reactive holdup	misc.
k^{reac}	Kinetic parameter	mol s^{-1}l^{-1}
K^{a}	Activity based chemical equilibrium constant	misc.
L	Liquid flow rate	mol s^{-1}
N	Maximal number of column trays	-
N_C	Number of components	-
N_F	Number of degrees of freedom	-
N_P	Number of phases	-
N_R	Number of independent variables	-
p	Pressure	bar
P	Production rate	mol s^{-1}
p^{S0}	Vapor pressure	bar
Q^{Cool}	Cooling duty	kW
Q^{Heat}	Heating duty	kW
R	Product revenue	10^3 € p.a.
T	Temperature	K; °C
T_B	Boiling temperature	K; °C
V	Vapor flow rate	mol s^{-1}
VP	Vapor pressure	bar
X_i	Transformed molar fraction of component i	-
x	Molar fraction (liquid)	-
y	Molar fraction (vapor)	-

Greek Symbols

Symbol	Description	Unit
γ	Activity coefficient	-
δ	Existence of a tray	-
φ	Fugacity coefficient	-
ν	Stoichiometric coefficient	-
$\Delta h^0_{R\_298}$	Standard reaction enthalpy at 298 K	kJ mol^{-1}
ΔT	Temperature difference	K
Λ	Wilson parameter	-

Subscripts

Subscript	Description
a	Activity based
a	Annual costs; annual revenues
i, m, n	Component
j	Reaction

k	Column tray; condenser; reboiler
ref	Reference condition

Superscripts

Superscript	Description
AC	Associating component
eq	Phase equilibrium
lo	Lower bound
NAC	Non-associating component
up	Upper bound

Abbreviations

Abbreviation	Description
BuAc	Butyl acetate
BuOH	Butanol
But	Butadiene
CSTR	Continuously stirred tank reactor
DA	Distillation area
D/F	Distillate to feed ratio
DMD	Dortmunder Datenbank
EG	Ethylene glycol
EO	Ethylene oxide
EtAc	Ethylene acetate
EtOH	Ethanol
H_2O	Water
IB	Isobutene
LLE	Liquid-liquid equilibrium
MeAc	Methyl acetate
MeOH	Methanol
MT	Mono-telomere
MTBE	Methyl tertiary butyl ether
RD	Reactive distillation
RDA	Reactive distillation area
RR	Reflux ratio
TPP	Triphenylphosphine
TPPTS	Triphenylphosphinetrisulfonate

Parameters

Parameter	Description	Value	Unit
K_{eq}	Equilibrium constant for isomerization	1.079	-
T_{ref}	Reference temperature	298.15	K

Parameter	MTBE Value	Unit	MeAc Value	Unit
$C_a^{\text{Raw Material}}$	4719.72	10^3 € p.a.	16140.0	10^3 € p.a.
P^{Catalyst}	0.01	10^3 € p.a.	0.01	10^3 € p.a.
$P^{\text{Condenser}}$	5.77	10^3 € p.a.	5.77	10^3 € p.a.
$P^{\text{Cooling Water}}$	0.0138	10^3 € p.a. kW^{-1}	0.0138	10^3 € p.a. kW^{-1}
$P^{\text{Efficiency}}$	0.7	-	1.0	-
P^{Holdup}	103.6726	kg m^{-2}	302.4	kg m^{-2}
$P^{\text{min Dia}}$	0.0077	m^2 s mol^{-1}	0.0188	m^2 s mol^{-1}
$P^{\text{max Dia}}$	0.0192	m^2 s mol^{-1}	0.0471	m^2 s mol^{-1}
$P^{\text{min Dist}}$	15.0	mol s^{-1}	20.0	mol s^{-1}
P^{Reboiler}	5.56	10^3 € p.a.	5.56	10^3 € p.a.
P^{Shell}	45.092	-	33.39	-
P^{Steam}	0.251	10^3 € p.a. kW^{-1}	0.207	10^3 € p.a. kW^{-1}
T^{Steam}	457.0	K	416.0	K
$T^{\text{Water}}_{\text{in}}$	293.15	K	293.15	K
$T^{\text{Water}}_{\text{out}}$	303.15	K	303.15	K

2.7 Literature

Abrams DS, Prausnitz JM (1975). Statistical thermodynamics of liquid mixtures: a new expression for the excess Gibbs energy of partly or completely miscible systems. AIChE Journal 21, 116-128.

Agreda VH, et al. (1990). High-purity methyl acetate via reactive distillation. Chemical Engineering Progress, 40-46.

Basmadjian D, Coroyannakis P (1987). Equilibrium-theory revisited - Isothermal fixed-bed sorption of binary-systems. 1. Solutes obeying the binary langmuir isotherm. Chemical Engineering Science 42, 1723-1735.

Basmadjian D, et al. (1987a). Equilibrium-theory revisted - Isothermal fixed-bed sorption of binary-systems. 2. Non-langmuir solutes with type-I parent isotherms - Azeotropic systems. Chemical Engineering Science 42, 1737-1752.

Basmadjian D, et al. (1987b). Equilibrium-theory revisited - Isothermal fixed-bed sorption of binary-systems. 3. Solutes with type-I, type-II and type-IV parent isotherms - Phase-separation phenomena. Chemical Engineering Science 42, 1753-1764.

Baur R, Krishna R (2004). Distillation column with reactive pump arounds: An alternative to reactive distillation. Chemical Engineering and Processing 43, 435-445.

Bearns M, et al. (1999). Chemische Reaktionstechnik, 3rd edn. New York: Georg Thieme Verlag.

Beckmann A, et al. (2002). Industrial experience in the scale-up of reactive distillation with examples from C_4-chemistry. Chemical Engineering Science 57, 1525-1530.

Bedenik NI, et al. (2004). An integrated strategy for the hierarchical multilevel MINLP synthesis of overall process flowsheets using the combined synthesis/analysis approach. Computers & Chemical Engineering 28, 693-706.

Behr A, Urschey M (2003). Palladium-catalyzed telomerization of butadiene with ethylene glycol in liquid single phase and biphasic systems: control of selectivity and catalyst recycling. Journal of Molecular Catalysis a-Chemical 197, 101-113.

Beßling B, et al. (1997). Experiments of the homogeneous transesterifcation of ethyl acetate with methanol. In Technical Report of the Brite EuRam Project "Reactive Distillation" (Project No BE95-1335).

Beßling B (1998). Zur Reaktivdestillation in der Prozeßsynthese. Dr.-Ing. Dissertation, Fachbereich Chemietechnik, Universität Dortmund.

Bessling B, et al. (1997). Design of processes with reactive distillation line diagrams. Industrial and Engineering Chemistry Research 36, 3032-3042.

Bessling B, et al. (1998). Investigations on the synthesis of methyl acetate in a heterogeneous reactive distillation process. Chemical Engineering Technology 21, 393-400.

Blass E (1997). Entwicklung verfahrenstechnischer Prozesse, 2. Auflage Berlin, Heidelberg: Springer Verlag.

Borren T, Schmidt-Traub H (2004). Vergleich chromatographischer Reaktorkonzepte. Chemie Ingenieur Technik 76, 805-814.

Brooke A, et al. (2003). GAMS - A User's Guide. Edited by GAMS Development Corp. W.

Bühner C (2001). Ein Beitrag zur Auswahl von Reaktoren für mehrphasige Reaktionssysteme. Dr.-Ing. Dissertation, Fachbereich Chemietechnik, Universität Dortmund.

Camacho-Rubio F, et al. (1996). A comparative study of the activity of free and immobilized enzymes and its application to glucose isomerase. Chemical Engineering Science 51, 4159-4165.

Cardoso MF, et al. (2000). Optimization of reactive distillation processes with simulated annealing. Chemical Engineering Science 55, 5059-5078.

Ciric AR, Gu DY (1994). Synthesis of Nonequilibrium Reactive Distillation Processes by Minlp Optimization. AIChE Journal 40, 1479-1487

Citro F, Lee JW (2004). Widening the applicability of reactive distillation technology by using concurrent design. Industrial and Engineering Chemistry Research 43, 375-383.

Clark FD, Lorenzoni AB (1978). Applied cost engineering. New York: Dekker.

CONOPT (2003). ARKI Consulting and Development. http://www.gams.com.

Converti A, Del Borghi M (1998). Kinetics of glucose isomerization to fructose by immobilized glucose isomerase in the presence of substrate protection. Bioprocess Engineering 18, 27-33.

Cookbook (2003). Process Design Center B. V. Breda, The Netherlands.

Cronewitz T, et al. (2000). Fructose. In Ullmann's Enyclopedia of Industrial Chemistry. Weinheim: Wiley-VCH.

Daichendt MM, Grossmann IE (1998). Integration of hierarchical decomposition and mathematical programming for the synthesis of process flowsheets. Computers & Chemical Engineering 22, 147-175.

Doherty MF, Perkins JD (1978a). On the dynamics of distillation processes - I. The simple distillation of multicomponent non-reacting, homogeneous liquid mixtures. Chemical Engineering Science 33, 281-301.

Doherty MF, Perkins JD (1978b). On the dynamics of distillation processes - II. The simple distillation of model solutions. Chemical Engineering Science 33, 569-578.

Doherty MF, Buzad G (1992). Reactive distillation by design. Transactions of the Institution of Chemical Engineers 70, 448-458.

Douglas JM (1988). Conceptual design of chemical processes. New York: McGraw-Hill.

Dröge T (1996). Auswahl technisch einsetzbarer Reaktoren. Dr.-Ing. Dissertation, Fachbereich Chemietechnik, Universität Dortmund.

Dye SR (1995). Fractional crystallization: Design alternatives and tradeoffs. AIChE Journal 41, 2427-2438.

Fredenslund A, et al. (1975). Group-contribution estimation of activity-coefficients in nonideal liquid-mixtures. AIChE Journal 21, 1086-1099.

Frey T, Stichlmair J (2000). MINLP optimization of reactive distillation columns. In ESCAPE (European Symposium on Computer Aided Process Engineering), pp. 115-120. Edited by Pierucci S.

Frey T (2001). Synthese und Optimierung von Reaktivrektifikationsprozessen. Dr.-Ing. Dissertation, Technische Universität München.

Fricke J (2005). Entwicklung einer Auslegungsmethode für chromatographische SMB-Reaktoren. Dr.-Ing. Dissertation, Fachbereich Bio- und Chemieingenieurwesen, Universität Dortmund.

Fried B (1991). Regelbasierte Auswahl grundlegender Reaktortypen mittels wissensbasierter Programmierung. Dr.-Ing. Dissertation, Fachbereich Chemietechnik, Universität Dortmund.

Gmehling J, et al. (1979). Selection of solvents for the extractive rectification by precalculated equilibrium data. Berichte der Bunsen-Gesellschaft-Physical Chemistry Chemical Physics 83, 1133-1136.

Gmehling J, Kolbe B (1992). Thermodynamik: VHC Verlagsgesellschaft.

Grassmann P (1971). Physical principles of chemical engineering, 1st english edn. Oxford, New York: Pergamon Press.

Grossmann IE, et al. (2005). Optimal synthesis of complex distillation columns using rigorous models. Computers & Chemical Engineering 29, 1203-1215.

Grüner SK, A. (2004). Equilibrium theory and nonlinear waves for reactive distillation columns and chromatographic reactors. Chemical Engineering Science 59, 901-918.

Harmsen GJ (2004). Industrial best practices of conceptual process design. Chemical Engineering and Processing 43, 677-681.

Hashimoto K, et al. (1983). A new process combining adsorption and enzyme reaction for producing higher-fructose syrup. Biotechnology and Bioengineering 25, 2371-2393.

Hostrup M, et al. (2001). Integration of thermodynamic insights and MINLP optimization for the synthesis, design and analysis of process flowsheets. Computers & Chemical Engineering 25, 73-83.

Huang YS, et al. (2004). Residue curve maps of reactive membrane separation. Chemical Engineering Science 59, 2863-2879.

Hugo P (1965). Die Berechnung des chemischen Umsatzes von Mehrkomponenten-Gasgemischen an porösen Katalysatoren - I. Das Strömungsrohr ohne Diffusionseinfluss. Chemical Engineering Science 20, 187-194.

Ismail SR, et al. (1999). Synthesis of reactive and combined reactor separation systems utilizing a mass heat exchange transfer module. Chemical Engineering Science 54, 2721-2729.

Jackson JR, Grossmann IE (2001). A disjunctive programming approach for the optimal design of reactive distillation columns. Computers & Chemical Engineering 25, 1661-1673.

Jakobsson K, et al. (2002). Modelling of a side reactor configuration combining reaction and distillation. Chemical Engineering Science 57, 1521-1524.

Jupke A, et al. (2002). Optimal design of batch and simulation moving bed chromatographic separation processes. Journal of Chromatography A 944, 93-117.

Kienle A, et al. (2005). Zur Integration von Reaktion und Stofftrennung. Chemie Ingenieur Technik 77, 1417-1429.

Levenspiel O (1996). The chemical reactor omnibook. Corvallis, Or.: Distributed by OSU Book Stores.

Levenspiel O (1999). Chemical reaction engineering, 3rd edn. New York: Wiley.

Myerson AS (2002). Handbook of industrial crystallization, 2nd edn. Boston: Butterworth-Heinemann.

Onken U, Behr A (1996). Chemische Prozeßkunde, 1. Auflage Stuttgart: Georg Thieme Verlag.

Perry RH, et al. (1997). Perry's chemical engineers' handbook, 7th edn. New York: McGraw-Hill.

Peters U, et al. (2003). Methyl Tert-Butyl Ether. In Ullmann's Enyclopedia of Industrial Chemistry. Weinheim: Wiley-VCH.

Pöpken T, et al. (2000). Reaction kinetics and chemical equilibrium of homogeneously and heterogeneously catalyzed acetic acid esterification with methanol and methyl acetate hydrolysis. Industrial & Engineering Chemistry Research 39, 2601-2611.

Pöpken T (2001). Reaktive Rektifikation unter besonderer Berücksichtigung der Reaktionskinetik am Beispiel von Veresterungsreaktionen. Dr. rer. nat. Dissertation, Fachbereich Chemie, Carl von Ossietzky Universität, Oldenburg.

Qi ZS, et al. (2002). Reactive separation of isobutene from C4 crack fractions by catalytic distillation processes. Separation and Purification Technology 26, 147-163.

Rautenbach R (1997). Membranverfahren. Berlin: Springer.

Rehfinger A, Hoffmann U (1990a). Kinetics of methyl tertiary butyl ether liquid phase synthesis catalyzed by ion exchange resin - I. Intrinsic rate expression in liquid phase activities. Chemical Engeneering Science 45, 1605-1617.

Rehfinger A, Hoffmann U (1990b). Kinetics of methyl tertiary butyl ether liquid phase synthesis catalyzed by ion exchange resin - II. Macropore diffusion of methanol as rate-controlling step. Chemical Engeneering Science 45, 1619-1626.

Renon H, Prausnitz JM (1968). Local compositions in thermodynamic excess functions for liquid mixtures. AIChE Journal 14, 135-144.

Rohm, Haas (2005). Product Information Ion Exchanger Resins.

Samant KD, Ng KM (1998a). Synthesis of extractive reaction processes. AIChE Journal 44, 1363-1381.

Samant KD, Ng KM (1998b). Design of multistage extractive reaction processes. AIChE Journal 44, 2689-2702.

Sand G, et al. (2004a). Structuring of reactive distillation columns for non-ideal mixtures using MINLP-techniques. In ESCAPE, European Symposium on Computer Aided Process Engineering, 493-498. Edited by Barbosa-Povoa A, Matos H. Lisbon, Portugal: Elsevier.

Sand G, et al. (2004b). Robust and efficient MINLP optimization of reactive distillation columns. In FOCAPD, Conference on Foundations of Computer-Aided Process Design, 319-322. Edited by Floudas CA, Agrawal R. Princeton/New Jersey.

Sand G, et al. (2005a). MINLP-Optimization in the integration of reaction and separation processes. In Sustainable (Bio)Chemical Process Technology, 317-324. Edited by Jansens P, et al. Delft: BHR Group.

Sand G, et al. (2005b). Global optimization in the conceptual design of reactive distillation columns. In WCCE, World Congress of Chemical Engineering, pp. C3-004. Glasgow: http://chemengcongress.somcom.co.uk.

SBB (2003). ARKI Consulting and Development. http://www.gams.com.

Schembecker G (1998a). Heuristisch-Numerische Prozeßsynthese unter Berücksichtigung der Energieintegration. Habilitation, Fachbereich Chemietechnik, Universität Dortmund.

Schembecker G (1998b). State-of-the-art and future development of computer-aided process synthesis. Weinheim: DECHEMA Monographien Wiley-VCH Verlag GmbH.

Schembecker G, Tlatlik S (2003). Process synthesis for reactive separations. Chemical Engineering and Processing 42, 179-189.

Schenck FW (2003). Glucose and glucose-containing syrup. In Ullmann's Enyclopedia of Industrial Chemistry. Weinheim: Wiley-VCH.

Schmidt-Traub H (2005). Preparative chromatography of fine chemicals and pharmaceutical agents. Weinheim: Wiley-VCH.

Schoenmakers H, Bessling B (2003). Reactive and catalytic distillation from an industrial perspective. Chemical Engineering and Processing 42, 145-155.

Schubert S, et al. (2001). Enhancement of carbon dioxide absorption into aqueous methyldiethanolamine using immobilised activators. Chemical Engineering Science 56, 6211–6216.

Seader JD, Henley EJ (1998). Separation process principles. New York: Wiley.

Seider WD, et al. (2004). Product and process design principles: Synthesis, analysis, and evaluation, 2nd edn. New York: Wiley.

Smith AS (1950). Solutropes. Industrial and Engineering Chemistry 42, 1206-1209.

Song W, et al. (1997). Discovery of a reactive azeotrope. Letters to Nature 388, 561-563.

Sørensen JM, Arlt W (1979). Liquid-liquid equilibrium data collection. Frankfurt/Main; Great Neck, N.Y.: DECHEMA; distributed exclusively by Scholium International.

Steinigeweg S (2003). Zur Entwicklung von Reaktivrektifikationsprozessen am Beispiel gleichgewichtslimitierender Reaktionen. Dr.-Ing. Dissertation, Fakultät für Mathematik und Naturwissenschaften, Carl von Ossietzky Universität, Oldenburg

Stichlmair J (1988). Zerlegung von Dreistoffgemischen durch Rektifikation. Chemie Ingenieur Technik 60, 747-754.

Stichlmair J, Frey T (2001). Mixed-integer nonlinear programming optimization of reactive distillation processes. Industrial and Engineering Chemistry Research 40, 5978-5982.

Tlatlik S (2004). Beitrag zur Prozesssynthese integierter Reaktions- und Trennoperationen. Dr.-Ing. Dissertation, Fachbereich Chemietechnik, Universität Dortmund.

Tlatlik S, Schembecker G (2005). To integrate or not to integrate? A systematic method to identify benefits of integrated reaction and separation processes. In Sustainable (Bio) Chemical Process Technology in Cooperation with 6th International Conference on Process Identification, 131-147. Delft, NL.

Toumi A, Engell S (2004). Optimization-based control of a reactive simulated moving bed process for glucose isomerization. Chemical Engineering Science 59, 3777-3792.

Ung S, Doherty MF (1995a). Synthesis of reactive distillation systems with multiple equilibrium chemical reactions. Industrial and Engineering Chemistry Research 34, 2555-2565.

Ung S, Doherty MF (1995b). Calculation of residue curve maps for mixtures with multiple equilibrium chemical reactions. Industrial and Engineering Chemistry Research 34, 3195-3202.

Ung S, Doherty MF (1995c). Necessary and sufficient conditions for reactive azeotropes in multireaction mixtures. AIChE Journal 41, 2383-2392.

Urschey M (2004). Telomerisation von Butadien mit mehrfunktionellen Nukleophilen. Dr. rer. nat. Dissertation, Fachbereich Bio- und Chemieingenieurwesen, Universität Dortmund.

Vriens GN, Medcalf EC (1953). Correlation of ternary liquid-liquid equilibria (An explanation of solutropy). Industrial and Engineering Chemistry 45, 1098-1104.

Vuilleumier S (1996). World outlook for high fructose syrups to the year 2000. International Sugar Journal 98, 467.

Wahnschafft OM, et al. (1992). The product composition regions of single-feed azeotropic distillation-columns. Industrial & Engineering Chemistry Research 31, 2345-2362.

Westhaus U (1995). Beitrag zur Auswahl chemischer Reaktoren mittels heuristisch-numerischer Verfahren. Dr.-Ing. Dissertation, Fachbereich Chemietechnik, Universität Dortmund.

Widagdo S, Seider WD (1996). Azeotropic distillation. AIChE Journal 42, 96-130.

Wilson GM (1964). Vapor-liquid equilibrium XI. A new expression for the excess free energy of mixing. Journal of the American Chemical Society 86, 127-130.

3 Catalytic distillation

Joachim Richter, Andrzej Górak, Eugeny Y Kenig

3.1 Introduction

The synthesis of chemical products from selected feed stocks is based on a variety of chemical transformations. The reaction extent is often limited by the chemical equilibrium between the reactants and products, thus reducing the conversion and selectivity towards the main product. In order to separate the components leaving the reactor, usually several unit operations are used, resulting in high investment costs, considerable energy consumption and a large number of recycle streams. The unit operations are commonly sequentially arranged and each unit operation is carried out in a separate vessel or apparatus. Hence, the energy and equipment costs for these major process steps add up. The integration of two or more unit operations into one single apparatus leads to the processes often referred to as "integrated reaction and separation processes" (IRSIRS). Such an integration offers the advantage of reducing the equipment and energy costs; in addition, it may lead to an increased efficiency of the whole process with respect to the consumption of reactants since conversion and selectivity of the reaction system in use can be improved.

During the last decades, IRSIRS have become more and more popular (Doherty a. Malone 2001). Due to the demand for more efficient production processes, the use of IRSIRS has been extended to more and more suitable reaction systems. Since different functionalities are implemented into one reactive separation process, IRSIRS are often referred to as multifunctional reactors (Agar 1999). Compared with other processes, catalytic distillation (CD) represents one of the most important applications of the concept of IRSIRSs in today's chemical industry.

3.2 Basics of catalytic distillation

Catalytic distillation is a combination of heterogeneously catalyzed chemical reactions and separation via distillation. This stands in contrast to a reactive distillation (RD) process in which usually homogeneous catalysis is applied. In a catalytic distillation column, reaction and separation occur simultaneously within the part of the column containing a heterogeneous catalyst; in addition, columns of this type often have an enrichment section above and a stripping sections below the catalytic section (see Figure 3.1). This integration of reaction and separation into one apparatus provides promising process alternatives to traditional sequential operations. Among potential advantages of CD are (Towler a. Frey 2001):

- Shifting of chemical equilibrium,
- Reduction of plan costs,
- Heat integration benefits,
- Avoidance of azeotropes,
- Improved selectivity,
- Definition of multiple reaction zones,
- Avoidance of hot spots inside of catalyst beds compared to trickle bed reactors,
- No catalyst removal from product streams due to the immobilization of catalyst inside the apparatus.

Drawbacks for the application of CD processes are:

- Volatility constraints,
- Residence time requirements depending on the reaction rate that can be realized in the apparatus,
- Process conditions for reaction and separation (e.g. temperature, pressure) have to match.

Summarizing these advantages, the combination of a reactor with a distillation column offers great potential for overall savings.

The synthesis of methyl acetate from methanol and acetic acid has been first suggested by Agreda et al. (Agreda a. Partin 1984; Agreda a.

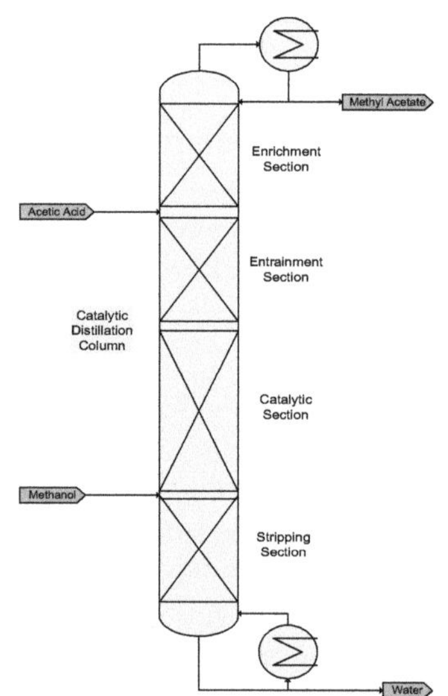

Fig. 3.1: CD process for MeAc synthesis (Noeres 2003)

Lilly 1990; Agreda, Partin et al. 1990) as a homogeneously catalyzed reactive distillation process. Nowadays it is considered to be a classic example of the useful application of catalytic distillation. The production of other esters similar to methyl acetate, like ethyl acetate (Kolena, Lederer et al. 1999; Kenig, Bäder et al. 2001; Klöker, Kenig et al. 2004), butyl acetate (Davies a. Jeffreys 1973; Steinigeweg a. Gmehling 2002; Wang, Wong et al. 2003) or propyl propionate (Buchaly, Lauterbach et al. 2005), provided interesting applications of CD processes in the past years.

CD processes can successfully be used to handle other chemical reactions. Especially for fuel ethers like methyl tertiary butyl ether (MTBE), ethyl tertiary butyl ether (ETBE) and tertiary amyl methyl ether (TAME), which are widely used as modern gasoline components, CD has become a major alternative to conventional production processes (Thiel 1997; Stichlmair a. Frey 1998; Mohl, Kienle et al. 1999; Qi, Sundmacher et al. 2002; Klöker, Kenig et al. 2003; Kolodziej, Jaroszynski et al. 2004). Reaction systems that have been discussed as possible CD applications also include ester hydrolysis (e.g. methyl acetate hydrolysis (Fuchigami 1990; Loning, Horst et al. 2000; Pöpken, Steinigeweg et al. 2001)), transesterifications (e.g. n-butyl acetate production) (Steinigeweg a. Gmehling 2003), isomerizations (e.g. paraffin and isoamylenes, (Lebas, Jullian et al. 1997; Linnekoski a. Rihko-Struckmann 1999)) and alkylations (e.g. Cumol synthesis (Shoemaker a. Jones 1987; Giessler, Danilov et al. 1999)).

Among the constraints limiting the successful use of CD is its high complexity. This includes multiple phases (three to four, including the catalyst), multicomponent mixtures, and complex thermodynamics with strongly non-ideal behavior, including phenomena like conventional azeotropes, reactive azeotropes and miscibility gaps. Along with other parameters, the viscosity and density can change drastically with the composition, influencing the hydrodynamics inside the column. These changes can significantly affect the separation efficiency of column internals and have been intensely discussed in literature (Stitt 2001; Leet a. Kulprathipanja 2002; Stitt 2003).

The design of catalytic distillation is currently based on expensive and time-consuming series of laboratory and pilot plant experiments. There is no design software that is offered commercially on the open market and is capable of describing all phenomena taking place inside a CD column, including the parallel occurrence of chemical reactions (reaction kinetics, catalyst behavior, hold-up of gaseous and liquid phases, residence time distribution, etc.) and distillation (vapor-liquid equilibria, mass transfer issues, hydrodynamic behavior of plates and packings). Even if all of these effects could be taken into account, the resulting model would become too complex to be solved by today's numerical methods in an appropriate time (Taylor a. Krishna 2000; Noeres, Kenig et al. 2003; Taylor, Krishna et al. 2003).

In addition, the catalytic functionality in CD has not yet been extended beyond conventional, well known catalysts and internals. There is a need for more active, improved catalysts as well as for more effective column internals that can only be achieved through deeper understanding of the micro-scale behavior inside a catalytic bag or pocket (Klöker, Kenig et al. 2004).

3.2.1 Catalyst

The choice of catalyst significantly influences chemical reactions and thus the performance of IRSIRS. Different homogeneous and heterogeneous chemical catalysts as well as enzymatic catalysts are currently used in organic synthesis. All of them have advantages and disadvantages, which determine their field of useful application. For instance, homogeneous catalysis is applied in carbonylation and hydroformylation processes (Cornils a. Herrmann 1996) whereas biocatalysts have successfully been used in several reactions under mild reaction conditions with high chemo-, regio- and enantioselectivity (Drauz a. Waldmann 1995; Schuchardt, Sercheli et al. 1998; Bornscheuer 2002).

Homogeneous catalysts usually have faster overall reaction rates, while for heterogeneous catalysts it may be slowed down by the mass transfer. Heterogeneous catalysts have the advantage of an easy recovery in slurry CD processes (Barreira, de Toledo et al. 2003; Judzis 2004), can be immobilized inside the process units, and are readily amenable to continuous processing. They have already been applied in a wide range of processes, therefore detailed experience and understanding for this class of catalysts is available (Sheldon a. van Bekkum 2001).

Moreover, the application of catalysts like zeolites, hydrotalcites and other related types offer an additional advantage of shape-selective catalysis. Considering a model reaction, in which the molecules of two components A and B react to form a linear product molecule of type C or a branched product molecule D, the formation of D can be reduced and the selectivity can be significantly shifted towards C, if the specific catalyst only allows the formation of the linear product molecule. This shape selectivity can be achieved through several modifications of the catalyst: cage structure and size in zeolites or the use of special ligands (Cornils a. Herrmann 1996).

Catalytic reactions in organic chemistry for which CD processes have been applied can be divided into five reaction types:

− solid-acid catalysis,
− solid-base catalysis,
− catalytic hydrogenation,
− catalytic dehydrogenation and
− catalytic formation of C-C bonds.

A wide range of acidic solid catalysts is available on the market: acidic ion exchanger resins, zeolites, silica-occluded heteropoly acids, nafion-silica composites and others. In principle, all of them can be applied to manifold acid-catalyzed processes, e.g. esterifications, transesterifications, etherifications, ester hydrolysis, alkylations, etc. The number of heterogeneous solid-base catalysts is by far lower than the numbers of solid-acid catalysts. Nevertheless, anionic hydrotalcite clays and mesoporous silica have successfully been used in aldol and Knoevenagel condensations (Figueras, Tichit et al. 1998; Choudary, Kantam et al. 2000; Weitkamp, Hunger et al. 2001).

Catalytic hydrogenation is one of the oldest applications of heterogeneous catalysts in which nowadays a variety of functional groups can be hydrogenated. Supported noble metal catalysts have been applied for this type of reactions (Piironen, Haario et al. 2001). Selective oxidations, e.g. alcohol oxidations and olefin epoxidations, can be carried out using a variety of heterogeneous catalysts, e.g. modified zeolites like the well known TS-1, hydrotalcites impregnated with redox metals and silica impregnated with redox metals (Sheldon a. van Bekkum 2001; Heijnen, de Bruijn et al. 2003). A totally different approach is the use of the so-called "ship-in-a-bottle" complexes, in which organometallic compounds are encapsulated within zeolite structures.Using this "catalysts in a cage" concept, active organometallic compound can be immobilized within a process through the heterogeneous zeolites cage. However, leaching of the catalytic active species and diffusion inside the zeolites pores can become a problem (Dartt a. Davis 1994). The generation of C-C bonds includes the condensation of carbonyl compounds mentioned above as well as olefin metathesis reactions and Heck couplings. The latter involve transition metal catalysis, which allows the application of heterogeneous catalysts. Although these reaction classes offer a broad spectrum of useful applications, in particular for IRSIRS, the relatively fast deactivation of the catalysts currently commercially available complicates the application of IRS for this type of reactions (Sheldon a. van Bekkum 2001).

The integration of a chemical reaction and the separation via distillation into one single unit operation leads to certain requirements for a heterogeneous catalyst (Kenig, Górak et al. 2004):

- High catalytic activity

  The activity of a heterogeneous catalyst affects the reaction rate and, therefore, determines the required liquid hold-up inside a CD unit to achieve a given conversion rate. Hold-up must be large enough to cover all active sites of the catalyst. However, from an economical point of view there exists an upper limit to the liquid volume fraction that can be reserved inside a column (Stute 1995; Yin, Afacan et al. 2002), and under certain circumstances, a side reactor or pre-reactor in addition to a catalytic distillation is preferable (Baur and Krishna 2004; Ojeda Nava, Baur et al. 2004). Catalysts that have already been used in integrated reaction and separation processes comprise ion exchanger resins and zeolites.

- No or very slow deactivation

  Independent of the chosen method of immobilization, the replacement of a heterogeneous catalyst in a CD column is rather complex, time-consuming and, consequently, costly. Therefore, it is necessary that the catalyst has a possibly small deactivation rate. The acceptable loss of activity strongly depends on the process and product specifications, the initial reaction rate of the fresh catalyst, the economical background of the overall process and additional factors, e.g. the type of internals used.

- Operating temperature within the range of the boiling point temperatures of the reactants and products

  Since the boiling point temperatures of the pure components and those of their resulting azeotropes determine the temperature range inside a catalytic distillation column, the chosen catalyst should be working properly under these conditions (Schembecker a. Tlatlik 2003).

  For example, some catalysts like zeolites have an optimal operating temperature between 180-200°C, and others like ion exchanger resins degrade above temperatures between 60°C and 125°C, depending on the catalytic active group used. Thus, the range of suitable catalysts for a certain reaction system is relatively narrow. In some cases, pressurizing the CD column can help to adjust the temperature range, but this also affects the investment and operating costs of the unit.

- Mechanical stability

  Since the catalyst inside a CD column is usually immobilized, using various containers made for example from wire gauze sheets, the mechanical stress on the catalyst particles is by far lower than in other reactors, e.g. in a fluidized bed reactor. Nevertheless, a certain mechanical stability is necessary to keep the catalyst inside these containments.

Despite some drawbacks, heterogeneous catalysis shows some advantages over the homogeneous one (Kenig, Górak et al. 2004):
- easy catalyst recovery and recycling,
- reduced amount of waste streams,
- less corrosion in pipings and vessels,
- influencing of selectivity (also shape selectivity),
- safer handling of catalysts compared to organic or inorganic acids and bases used for homogeneous catalysis.

3.2.2 Internals

For IRS and, in particular, for the efficient design of CD processes, properly selected internals play an important role. They must allow for the intense mixing of counter-current liquid and vapor streams and, hence, high separation efficiency. Their design must also has a significant impact on the chemical reaction because of the liquid hold-up or the residence time distribution inside the column packing.

At a first glance, column internals for catalytic and non-catalytic separation processes may look similar. However, catalytic internals have to combine catalytic and separation function. Today three basic concepts are used to manufacture catalytic internals:
- Immobilization of commercial heterogeneous catalysts inside small bags or bales that are placed inside or next to a conventional internal's structure (trays, downcomers, structured catalytic packing),
- Immobilization of catalyst on or catalytic activation of the surfaces of conventional internals,
- Production of catalytic active internals from the catalyst material itself.

When utilizing solid catalyst particles for counter-current flow operations, the following additional conditions should be fulfilled:
- narrow particle size distribution,
- uniform liquid flow in the catalyst bed without stagnant zones and liquid bypassing,
- broad vapor and liquid loading ranges without flooding,
- limited catalyst abrasion,
- simple catalyst exchange.

Structures immobilizing catalyst

A common way for immobilizing catalysts is to form objects of a certain shape from wire gauze mesh and to fill these with the catalyst. The geometry of these objects depends on their required hydrodynamic behavior as well as the possibly long residence time of the liquid inside them. Examples of this type of structures are different random packings made of cylindrical baskets (Johnson a. Dallas 1994), wire gauze boxes (van Hasselt, Calis et al. 1999) or wire mesh bales (Smith 1984). Wire gauze envelopes filled with catalyst pellets can also be placed on trays, either across the tray (Jones 1985) or in the downcomer section (Yeoman, Pinaire et al. 1994). Catalyst beads can be immobilized like in a fixed-bed reactor between two non-reactive distillation trays (Nocca, Leonard et al. 1991; Adrian, Bessling et al. 1996) as well as in an external side stream reactor (Jacobsson, Pyhälathi et al. 2001; Schoenmakers a. Bessling 2003).

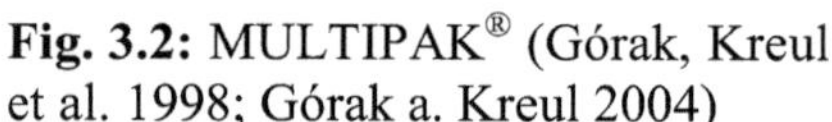

Fig. 3.2: MULTIPAK® (Górak, Kreul et al. 1998; Górak a. Kreul 2004)

Fig. 3.3: KATAPAK® SP 11 (Götze a. Bailer 2000; Götze, Bailer et al. 2000)

Additionally to this type of packings, a significant effort has been made in the last decade towards the development of structured sandwich-type packings. Here, catalyst pellets are immobilized between two sheets of corrugated wire gauze forming a sandwich. A parallel arrangement of these layers results into open and closed channels for the vapor and liquid flow respectively, which show hydrodynamic characteristics similar to those of traditional structured packings, e.g. reduced pressure drop and optimum flow conditions within a wide operating range (KATAMAX® (Gelbein a. Buchholz 2000), KATAPAK-S® (Shelden a. Stringaro 1995)). Furthermore, it is possible to combine catalyst sandwiches with conventional corrugated wire gauze layers. This hybrid sandwich structure meets the demands of flexible catalyst amount and separation efficiency

(MULTIPAK® (Górak, Kreul et al. 1998; Górak a. Kreul 2004), KATAPAK-SP® (Götze a. Bailer 2000; Götze, Bailer et al. 2000), see Figure 3.2 and Figure 3.3 (reproduced from Kenig, Gorák and Bart 2004).

Catalytic active structures

The second concept for catalytic column internals is the use of catalytically active structures instead of those filled with a catalyst. Catalytically active structures represent either carrier supported catalysts or solid catalysts prepared in a certain geometric form. Carrier supports can be coated with any kind of catalyst (e.g. GPP-rings (Kunz a. Hoffmann 1995; Kunz 1998), KATAPAK-M® (Sulzer 2000)). It is noteworthy that for these structures abrasion becomes a critical operating parameter that limits their operating time. It is also possible to develop solid catalytic structures without any carrier. This concept has been demonstrated by the ICVT rings made from a catalytic active polymer material. Abrasion does not compulsorily deactivate the internals, since during the process of abrasion also fresh catalytic material is disclosed. Up to now, all catalytic internals mentioned are primarily used for CD processes. However, there is also some potential for their application in other IRS like reactive extraction, catalytic membrane reactors or reactive stripping (Schildhauer, Kapteijn et al. 2005).

3.3 Modeling

Integrated reaction and separation processes are influenced not only by complex multicomponent thermodynamic behavior and the simultaneous chemical reaction, but also the hydrodynamic behavior of the internals. To describe these phenomena adequately, sophisticated mathematical models have been developed that take into account fluid dynamics, mass transfer phenomena and complex chemical reaction schemes (Kenig, Klöker et al. 2000; Taylor a. Krishna 2000; Noeres, Kenig et al. 2003; Taylor, Krishna et al. 2003).

In case of auto-catalytic and homogeneously catalyzed reaction systems, where the reaction takes place inside the liquid phase, a two-phase system with a single interface is present. For these reaction systems, the reaction inside the liquid bulk and, depending on the velocity of the reaction, inside the liquid film has to be taken into account.

Due to the heterogeneous catalyst, a third solid phase and an additional interface are present in CD units. In literature, there are several models of different complexity available to account for the phenomena that occur at the liquid-solid interface (Higler, R. Krishna et al. 2000; Nijhuis, Kreutzer et al. 2001; Zheng, Flora T.T. Ng et al. 2001; Zheng, Rempel et al. 2003). The most complex models take into account the intrinsic kinetics and cover the mass transfer inside the catalyst pores (Mohl, Kienle et al. 2001). Nevertheless, it is often assumed that all internal and external resistances can be lumped together into one relatively simple kinetic description called "pseudo-homogeneous" kinetic approach. Its basic assumption is that each single active site on the catalyst particles is imposed on the surrounding liquid bulk and can be described solely by the bulk composition and temperature (Steinigeweg 2003). In addition, very fast chemical reactions are often described by their chemical equilibrium constants; for these cases it is assumed that the reaction does not depend on the reaction kinetics.

Modeling of hydrodynamics in vapor/liquid contactors includes an appropriate description of axial dispersion, liquid hold-up and pressure drop. The correlations giving such a description have been published in numerous papers and are collected in several reviews and textbooks (e.g. (Wesselingh a. Krishna 1990; Billet 1995; Higler, R. Krishna et al. 2000; Bird, Stewart et al. 2001; Mackowiak 2003)). Nevertheless, there is still a need for a better modeling of hydrodynamics in catalytic column internals; this is being reflected by research activities in progress (Yuxiang a. Xien 1992; Noeres, Hoffmann et al. 2002; Klöker, Kenig et al. 2003; Larachi, Petre et al. 2003; Petre, Larachi et al. 2003; Hoffmann, Noeres et al. 2004; Klöker, Kenig et al. 2004; Egorov, Menter et al. 2005; Klöker, Kenig et al. 2005).

The description of thermodynamics and chemical properties of the IRS is very process-specific, so that its detailed discussion would require a separate chapter. Therefore, we will offer only a brief discussion of these topics in the context of the following case studies (see Chap. 3.5). Further related details can be found in (Reid, Prausnitz et al. 1987; Froment a. Bischoff 1990; Billet 1995; Kenig, Kloeker et al. 2001; Mackowiak 2003).

In order to model large industrial reactive separation units, a proper subdivision of a column apparatus into smaller elements is usually necessary. These elements (the so-called stages) are identified with real trays or segments of a packed column. They can be described by using different theoretical concepts, with a wide range of physicochemical assumptions and accuracy.

3.3.1 Equilibrium stage model

The so-called equilibrium model has been the model of choice during the last century. The first publication on this aspect was made by Sorel in 1893 (Sorel 1893); numerous additional papers followed. The basic assumptions made in this model are (Wesselingh 1990; Seader a. Henley 2005):
- The vapor and liquid streams leaving the stage are in thermodynamic equilibrium
- Both the vapor and liquid phase are homogeneously mixed
- The vapor stream leaving the stage carries no liquid with it (no entrainment); therefore there is no axial backmixing.

The equilibrium model was developed for the calculation of tray columns. To transfer the concept of the equilibrium stage to packed columns, the idea of the Height Equivalent to a Theoretical Stage (HETS) was proposed. A HETS value represents a certain bed length of a packing equivalent to one theoretical stage. To account for the chemical reactions taking place within IRS, usually reaction rate expressions are integrated into the mass and energy balances. The reaction rate itself can be calculated via chemical equilibrium equations or rate expressions. In this respect, much depends on the relation between the mass transfer and reaction rates in IRS. The definition of the Hatta number representing the reaction rate in reference to that of the mass transfer helps to discriminate between very fast, fast, average and slow chemical reactions (Danckwerts 1970; Doraiswamy a. Sharma M. M 1984; Baerns, Hofmann et al. 1992).

If a fast reaction system is present, the IRS can be described assuming chemical reaction equilibrium. In this case, a proper modeling approach is based on the non-reactive equilibrium stage model extended by the chemical equilibrium equations.

Such descriptions can be appropriate for instantaneous and very fast reactions. However, if the reaction is slow, it becomes time-related and dominates the whole process. In this case, the reaction kinetics have to be integrated into the mass and energy balances. This concept has been used in a number of studies for CD (e.g., (Baur, Taylor et al. 2004; Schmitt, Hasse et al. 2004)).

In reality, thermodynamic equilibrium is never reached within one theoretical stage. Tray efficiencies and other relevant correction factors have been introduced to cope with this fact. However, for multicomponent mixtures this concept sometimes shows non-consistent results, since the diffusion interactions in these mixtures lead to unusual phenomena like osmotic or reversible diffusion and mass transfer barriers. In these cases, efficiency factors can vary from plus to minus infinity within the same stage and the application of these models becomes difficult (Toor 1964; Taylor a. Krishna 1993; Gorak 1995).

The equilibrium stage model, extended by the consideration of chemical reactions, seems to be suitable for esterification reactions in CD processes (see (DeGarmo, Parulekar et al. 1992; Taylor a. Krishna 2000; Taylor, Krishna et al. 2003)). However, it cannot be recommended for reactions with a higher rate.

3.3.2 Rate-based approach

In order to take the actual rates of multicomponent mass and heat transfer into account directly, the so-called rate-based approach was developed. This approach is physically more consistent than the equilibrium stage model.

For the description of vapor/liquid mass transfer, different models have been developed:
– Two-film model,
– Penetration model,
– Surface-renewal model,
– Film-penetration model.

The exact description of the heterogeneous catalysis is only possible if relevant parameters, the pore diffusion coefficient or the mass transfer resistance between liquid phase and catalyst particle, are available. Since the measuring of these parameters is in most cases not accurate enough or simply impossible, hydrodynamic non-idealities in catalytic packings have been studied to determine the necessary data indirectly. (Ellenberger a. Krishna 1999; Moritz a. Hasse 1999; Hoffmann a. Górak 2000). The easiest way to incorporate the chemical reaction is the use of a quasi-homogeneous approach (see e.g. (Schneider, Noeres et al. 2001)).

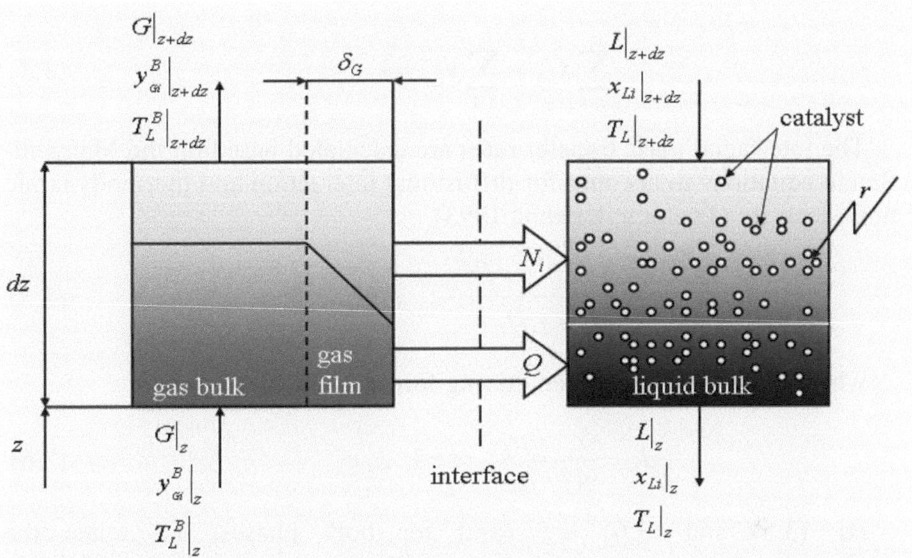

Fig. 3.4: Film model for a differential packing element with heterogeneous catalyst (Noeres, Kenig et al. 2003)

The rate-based concept using the two-film model is illustrated in Figure 3.4 for a column segment (a stage). The component balances for the gas and liquid phase on a stage j are written as

$$\frac{dn_{G,i}}{dt} = G_{j-i}y_{i,j-1} - G_j y_{i,j} - n_{i,j}a_j A_C dz; \quad i = 1,\ldots,n_C \tag{3.1}$$

and

$$\frac{dn_{L,i}}{dt} = L_{j+1}x_{i,j+1} - L_j x_{i,j}$$
$$+ \left(n_{i,j}a_j + r_{i,j}\psi_{cat}\rho_{cat} \right) A_C \Delta z; \quad i = 1,\ldots,n_c \tag{3.2}$$

whereas the interfacial molar fluxes $n_{i,j}$ are related to the diffusional fluxes by

$$n_{i,j} = J_{i,j} + x_{i,j}n_{i,j}; \quad i = 1,\ldots,n_c - 1 \tag{3.3}$$

The following summation conditions are valid for both the liquid and the vapor phase:

$$\sum_{i=1}^{n_c} x_{i,j} = \sum_{i=1}^{n_c} y_{i,j} = 1 \tag{3.4}$$

The interfacial mass transfer rates are calculated based on the Maxwell-Stefan equations to account for diffusional interaction and thermodynamic non-idealities (Taylor a. Krishna 1993).

$$d_i = \sum_{j=1}^{n_c} \frac{x_i N_{Lj} - x_j N_{Li}}{c_{Lt} Ð_{ij}}; \quad i = 1,\ldots,n_c \tag{3.5}$$

where d_i is the generalized driving force:

$$d_i = \frac{x_i}{\Re T} \frac{\partial \mu_i}{\partial z}; \quad i = 1,\ldots,n_c \tag{3.6}$$

eq. (3.5) and (3.6) are valid for both phases. To relate the multicomponent mass transfer rates to binary mass transfer coefficients, the method of Krishna and Standart (Krishna a. Standart 1979) is used. The diffusional fluxes can be calculated from

$$(J) = -c_{t,G}^{av} \cdot \left[k_G^{av}\right] \cdot \left(y_{i,j}^I - y_{i,j}\right) = -c_{t,L}^{av} \cdot \left[k_L^{av}\right] \cdot \left[\Gamma_L^{av}\right]\left(x_{i,j} - x_{i,j}^I\right) \tag{3.7}$$

for which the matrix of mass transfer coefficients is defined as:

$$\left[k^{av}\right] = \left[R^{av}\right]^{-1} \tag{3.8}$$

with

$$R_{ii} = \frac{y_i^{av}}{\kappa_{in}} + \sum_{\substack{k=i \\ k \neq j}}^{n_C} \frac{y_k^{av}}{\kappa_{ik}}; \quad i = 1,\ldots,n_C - 1 \tag{3.9}$$

$$R_{ij} = -y_i^{av}\left(\frac{1}{\kappa_{ij}} - \frac{1}{\kappa_{in_c}}\right); \quad i = 1,\ldots,n_c - 1 \tag{3.10}$$

The binary mass transfer coefficients κ_{ij} can be extracted from suitable binary mass transfer correlations, using the appropriate Maxwell-Stefan diffusion coefficients $Ð_{ij}$. A number of mass transfer correlations for both catalytic and non-catalytic column internals are available in the literature (Bravo a. Fair 1982; Kolodziej, Jaroszynski et al. 2003).

According to the linearized theory of (Stewart a. Prober 1964; Toor 1964; Toor 1964), the matrices $\left[R^{av}\right]$ and $\left[\Gamma^{av}\right]$ are evaluated using an average mole-fraction defined as

$$y_{ij}^{av} = \frac{y_{ij}^{I} + y_{ij}}{2}; \quad i = 1, \ldots, n_c \tag{3.11}$$

All required physical properties such as diffusivities are calculated on the same basis. At the interface, phase equilibrium is assumed:

$$y_{ij} = K_{ij}^{eq} x_{ij}^{I}; \quad i = 1, \ldots, n_c \tag{3.12}$$

where vapor-liquid equilibrium constants K_{ij}^{eq} are determined using the selected thermodynamic models, such as UNIQUAC or NRTL, and the extended Antoine equation for the vapor pressure. The energy balances for both phases are formulated as follows:

$$\frac{d\left(V\overline{H}_V\right)}{dz} = G_{j-1}H_{G,j-1} - G_j H_{G,j} - \left(q_j^v + q_j^I \frac{a_j d}{4}\right)\pi d\Delta z \tag{3.13}$$

$$\frac{dE_{L,z}}{dz} = L_{j+1}H_{L,j+1} - L_j H_{L,j} + q_j^I a_j A_c \Delta z \tag{3.14}$$

The heat flux across the vapor-liquid interface comprises of a convective and a conductive part:

$$\begin{aligned}
q_j^I &= h_{G,j} a_j \left(T_{G,j} - T_j^I\right) + \sum_{i=1}^{n_c} n_{ij} H_{G,ij} \\
&= h_{L,j} a_j \left(T_j^I - T_{L,j}\right) + \sum_{i=1}^{n_c} n_{ij} H_{G,ij}
\end{aligned} \tag{3.15}$$

whereas $h_{L,j}$ and $h_{G,j}$ are heat transfer coefficients determined via the Chilton-Colburn analogy (Kreul, Górak et al. 1998; Hoffmann, Noeres et al. 2004).

The influence of the process hydrodynamics is directly taken into account via correlations for mass transfer coefficients, specific contact area, liquid hold-up and pressure drop. The necessary hydrodynamic and mass-transfer correlations for the catalytic packings can be obtained from literature. The discretisation of the reactive separation column in axial direction in combination with the detailed rate-based approach is a suitable way to model such a unit, but it results in a large and highly non-linear algebraic system of equations. Further details are given in (Noeres, Kenig et al. 2003; Kenig, Górak et al. 2004).

3.4 Model parameters

Model parameters play an important role in modeling and simulation of IRS. Therefore, the influence of different parameters is discussed in this chapter. For further reading, see (Noeres, Kenig et al. 2003; Kenig, Górak et al. 2004).

3.4.1 Vapor-liquid equilibrium

In distillation processes, the correct modeling of the vapor-liquid equilibrium (VLE) is necessary to gain reliable simulation results. Thermodynamic models of different complexity are given in (Reid, Prausnitz et al. 1987; DECHEMA 2005).

Non-ideal thermodynamic behavior is a widespread phenomenon in CD processes. For the liquid phase, the real behavior can be described employing activity coefficients calculated via different methods. The models currently in use include predictive methods like UNIFAC and models based on the regression of experimental data like Wilson, NRTL or UNIQUAC.

Non-idealities in the vapor phase are very common among mixtures of alcohols or components with carboxylic acid or similar functional groups. The non-idealities result from the dimerization of two or more molecules carrying the carboxylic acid functional group (Nothnagel, Abrams et al. 1973). (Hayden a. O'Connell 1975) have implemented a virial equation of state to include the dimerization. The latter method provides reliable results over a wide range of pressure and temperature.

It is recommended that the data calculated by a selected thermodynamic model has to be checked against experimental data, which can be obtained e.g. from the DETHERM data base (DECHEMA 2005) or other sources (Gmehling, Onken et al. 1977-1984; Gmehling, Menke et al. 1994).

3.4.2 Reaction kinetics

The correct implementation of equations describing the reaction kinetics is indispensable for a catalytic distillation process. For heterogeneously catalyzed reactions it is often assumed that both internal and external mass transfer resistances can be lumped together with the micro-kinetic behavior of the catalyst into one relatively simple kinetic description called "pseudo-homogeneous" kinetic approach. The basic assumption of the pseudo-homogeneous approach is that every single active site on the catalyst particles is in contact with the surrounding liquid bulk and can be described by the bulk composition and temperature.

Assuming the following reaction system

$$v_A A + v_B B \leftrightarrow v_C C + v_D D \qquad (3.16)$$

one can write using pseudo-homogeneous reaction kinetics:

$$r = \frac{1}{v_i}\frac{\partial n_i}{\partial t} = m_{CAT}k_1\left(a_A a_B - \frac{1}{K}a_C a_D\right) \qquad (3.17)$$

In this equation, a_i is the activity coefficient of component i and K is the adsorption-based equilibrium constant of the reaction given above.

In this model no sorption effects that may influence the kinetic behavior in multicomponent mixtures are not taken into account explicitly. This can result in an incorrect description of reaction kinetics, especially for polymeric catalysts like ion exchanger resins, which are widely used in CD processes. On the other hand, this model is advantageous since only few measurements have to be carried out to fit the relevant parameters.

In accordance to the theory of Flory and Huggins (Flory 1941; Huggins 1941; Flory 1942; Huggins 1942; Mazzotti 1997), the activity of a single component in a polymer depends on its volumetric fraction inside the matrix. Since smaller molecules diffuse into polymers more easily than bigger molecules, a larger amount of small-size molecules inside the polymer can be found. If one assumes equilibrium between the polymer and the liquid phase, the activities of both have to be equal. Simple kinetic models based on molar fractions do not take into account these phenomena and can predict a wrong reaction rate. Additional factors like polar reactants or products, e.g. water, and more complex molecules also lead to a non-accurate description of reaction rates when simple models are used. Even the detailed Flory-Huggins model is unable to cope with all of these factors. A model that allows to take into account at least the most important effects is the modified Langmuir-Hinshelwood-Hougen-Watson model (LHHW), that has been proposed by Song et al. (Song, Venimadhavan et al. 1998). The LHHW equation for a second order reaction with four components (see eq. (3.16)) can be written as

$$r = \frac{1}{v_i}\frac{dn_i}{dt} = m_{KAT}\frac{k_1\left(a_A' a_B' - \dfrac{1}{K}a_C' a_D'\right)}{\left(1 + a_A' + a_B' + a_C' + a_D'\right)^2} \tag{3.18}$$

with

$$a_A' = \frac{K_i a_i}{M_i} \tag{3.19}$$

where K_i is the adsorption constant and M_i is the molar mass of the component i.

Assuming that all catalytic active groups are occupied, eq. (3.18) can be written without the unit in the denominator (Song, Venimadhavan et al. 1998):

$$r = \frac{1}{v_i}\frac{\partial n_i}{\partial t} = m_{CAT}\frac{k_1\left(a_A' a_B' - \dfrac{1}{K}a_C' a_D'\right)}{\left(a_A' + a_B' + a_C' + a_D'\right)^2} \tag{3.20}$$

3.4.3 Hydrodynamics and mass transfer

Modeling of hydrodynamics in vapor/liquid contactors must include an appropriate flow-pattern description of the inside the apparatus. Especially axial dispersion, liquid hold-up and pressure drop are of great interest. Non-ideal flow resulting e.g. in stagnant zones can affect reaction rates, change the effective driving force and thus influence the conversion and selectivity of the overall process. For catalytic internals macromixing phenomena play an important role. They are caused by molecular and turbulent diffusion (eddy diffusion), velocity fluctuations, non-uniform velocity profiles, by-passing and channeling in packings (Shah, Stiegel et al. 1978), as well as by entrainment and weeping in tray units (Bell 1972).

To describe CD processes and their non-ideal flow patterns adequately, all necessary parameters influencing the liquid and vapor streams inside the column should be considered, but the derived model should be as simple as possible. Two model types, namely the cell model and the differential model, were developed to cope with these requirements.

Cell model

Cell models are based on a stage-wise discretisation of a certain volume within a reactor, a catalytic packing or a catalytic tray. These single volumes or cells are connected with each other via heat and mass transfer equations (Baerns, Hofmann et al. 1992).

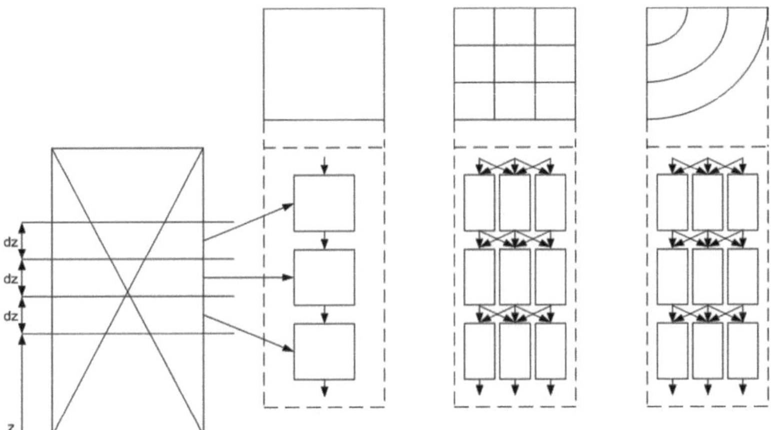

Fig. 3.5: Cell Models; left: cascade (1-dim); middle: square structure (2-dim); right: radial network (2-dim) (Baerns, Hofmann et al. 1992)

To adjust the cell model to a certain reactor geometry, the number of cells n_C and the way they interact with each other need to be specified. Therefore, the easiest cell model consists of a cascade of n_C cells but they can be combined into square and radial structures (Figure 3.5).

Alejski (Alejski 1991) studied the influence of backmixing on the performance of RD tray columns. In this work, an extended equilibrium model is used in which the flow path of the liquid phase is split into several cells (see Figure 3.5, left). Higler, Krishna and Taylor extended the approach of Alejski to include backmixing of the vapor and liquid phase. The combination of a rate-based approach model with a fluid dynamic model to study the influence of flow patterns on the performance of a catalytic distillation process was introduced in this paper (Higler, Krishna et al. 1999).

3.4.4 Differential models

In 1953, Danckwerts proposed the axial dispersion model (ADM) to account for flow patterns that deviate from standard plug flow (Danckwerts 1953). The model is based on the general continuity equation:

$$\frac{\partial C_i}{\partial t} + \nabla\left(C_i u\right) + \nabla J_i = R_i \tag{3.21}$$

Assuming a non-reactive and one-dimensional system as well as molar fluxes defined on the ground of Fick's law, eq. (3.21) can be transformed to

$$\frac{\partial C_i}{\partial t} = D_{ax}\frac{\partial^2 C_i}{\partial z^2} - u\frac{\partial C_i}{\partial z} \tag{3.22}$$

The axial dispersion coefficient D_{ax} can be calculated with the help of the Bodenstein number:

$$Bo = \frac{u_L l_c}{D_{ax}} \tag{3.23}$$

The Bodenstein number is determined from correlations obtained for specific reactor or catalytic packing types. The boundary conditions for the axial dispersion model have been introduced by Danckwerts (Danckwerts 1953) and have been intensively discussed in literature (Kreft a. Zuber 1978; Michelsen 1994).

Another model for the description of non-ideal flow patterns, the so-called piston-flow model with axial dispersion and mass exchange (PDE), was suggested by van Swaaij et al. (van Swaaij, Charpentier et al. 1969). The PDE model extends the ADM model by stagnant zones, which are connected to the dynamic axial dispersed zones by diffusion type streams. The balances for the dynamic zones can be written as:

$$\frac{\partial C_D}{\partial t} = \frac{1}{Bo}\frac{\partial^2 C_D}{\partial z^2} - u_D\frac{\partial C_D}{\partial z} - \frac{k_{eff}\, a^{D/S}}{\varepsilon h_L \phi_D}\left(C_D - C_S\right) \tag{3.24}$$

whereas the stagnant zones can be described via the following equation:

$$\frac{\partial C_S}{\partial t} = \frac{k_{eff}\, a^{D/S}}{\varepsilon h_L\left(1 - \phi_D\right)}\left(C_S - C_D\right) \tag{3.25}$$

The PDE model has been used for several studies (Bennett a. Grimm 1991; Kooijman 1995; Bart a. Landschützer 1996) and extended to include intra-particle diffusion (Illiuta, Larachi et al. 1999; Nigam, Illiuta et al. 2002).

A more detailed insight into the flow patterns in catalytic packings can be achieved with the help of computational fluid dynamics (CFD) (Kenig, Kloeker et al. 2001; van Baten a. Krishna 2002; Klöker, Kenig et al. 2003; Klöker, Kenig et al. 2005). (Egorov, Menter et al. 2002) performed a detailed numerical analysis of flow patterns within the catalytic beds of Katapak-S. In this work, the catalytic particles in an pocket of a catalytic packing were modeled as single spheres in a periodic subvolume to study the retention time distribution within one pocket. The results were achieved by using the commercial CFD code CFX5 by ANSYS. However, this study included only single-phase flows with water and a tracer, assuming that during the operation of such a packing vapor does not enter a pocket. This assumption needs to be confirmed by experimental studies.

3.5 Case studies

In the following chapter, different case studies for the use of CD are presented. The reaction systems reach from relatively simple esterification reactions to esterifications with phase splitting and transesterification reactions with consecutive and side reactions.

3.5.1 Methyl acetate synthesis

Methyl acetate is a weakly polar solvent that is used as an intermediate to manufacture pharmaceuticals, artificial leather and synthetic flavoring and added as a solvent to consumer products like glue or nail polish. The synthesis of methyl acetate from methanol and acetic acid has been one of the first commercially successful applications of a CD process on an industrial scale (Agreda a. Partin 1984; Agreda a. Lilly 1990; Agreda, Partin et al. 1990).

Chemical system

The synthesis of methyl acetate from methanol and acetic acid is a slightly exothermic equilibrium limited liquid-phase reaction:

$$CH_3OH + CH_3COOH \leftrightarrow CH_3COOCH_3 + H_2O$$

$$\Delta H_R^0 = -3.01\frac{kJ}{mol} \tag{3.26}$$

The chemical system is charaterised by the low equilibrium constant and the strongly non-ideal thermodynamic behavior which causes the forming of the two binary azeotropes methyl acetate/methanol and methyl acetate/water. Since the process is carried out under atmospheric pressure, the reaction temperatures are low and no side reactions in the liquid phase (e.g. etherification of methanol) occur.

Process set-Up and operation

The catalytic packing MULTIPAK® applied in this case study consists of corrugated wire gauze sheets and catalyst bags from the same material

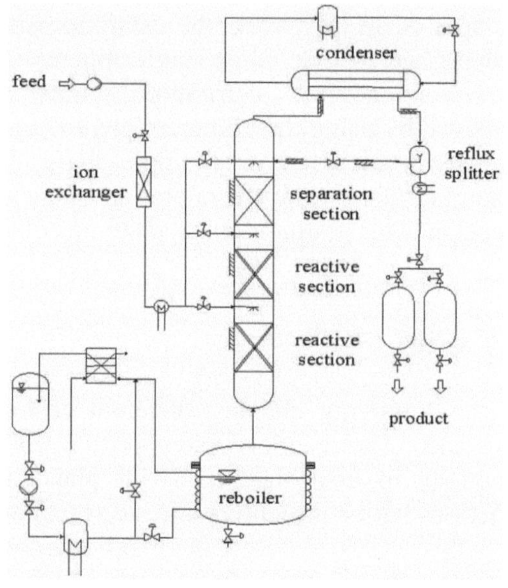

Fig. 3.6: Flow sheet for the catalytic distillation column wit 100mm I.D. (reproduced from Noeres 2003)

assembled in an alternating sequence (see Figure 3.2). Sufficient mass transfer between gas and liquid phase is guaranteed by segmentation of the catalyst bags and numerous contact spots with the wire gauze sheets. For the experimental study, the packing has been filled with an ion exchanger resin known as an effective catalyst for esterification processes.

A batch distillation column with a diameter of 100 mm, a reactive packing height of 2 m (MULTIPAK I®) in the bottom section and one meter of non-catalytic packing (ROMBOPAK 6M®) in the enrichment section has been used. The flow sheet of the column is shown in Figure 3.6.

Each experiment was performed according to the following procedure. At the beginning the distillation was still charged with the low-boiling reactant methanol and heated up under total reflux, until steady-state conditions were achieved. From this moment on, the high-boiling reactant acetic acid was fed at a constant rate above the reaction zone into the column. After 30 min. the reflux ratio was changed from infinity to two, thus leading to a continuous withdrawal of product at the top of the column. During the column operation, the liquid phase concentration profiles along the column and the temperature profiles were measured. For the determination of the liquid phase composition, two methods were applied simultaneously: on the one hand, samples were taken along the whole column and analyzed by gas chromatography to determine column profiles at discrete time intervals; on the other hand, a near-infrared spectrometer (NIR) was used to determine the concentration in the outlet stream at the top as well as in the reboiler online, without taking any samples. The NIR measurements were imported into the process control system and used as target values for the implemented quality control.

Results and discussion

The liquid phase composition as a function of time are displayed in Figure 3.7 for the reboiler and Figure 3.8 for the condenser as functions of time. After the start-up of the plant, the concentration of methanol decreases continuously, whereas the mole fraction of methyl acetate in distillate reaches about 90 %. A comparison of the calculated and measured liquid-phase compositions at the column top and in the column reboiler demonstrates their satisfactory agreement (Figure 3.7 and Figure 3.8). Figure 3.9 provides a comparison of the simulation results obtained on the basis of different modeling approaches, after an operation time of 10000 s.

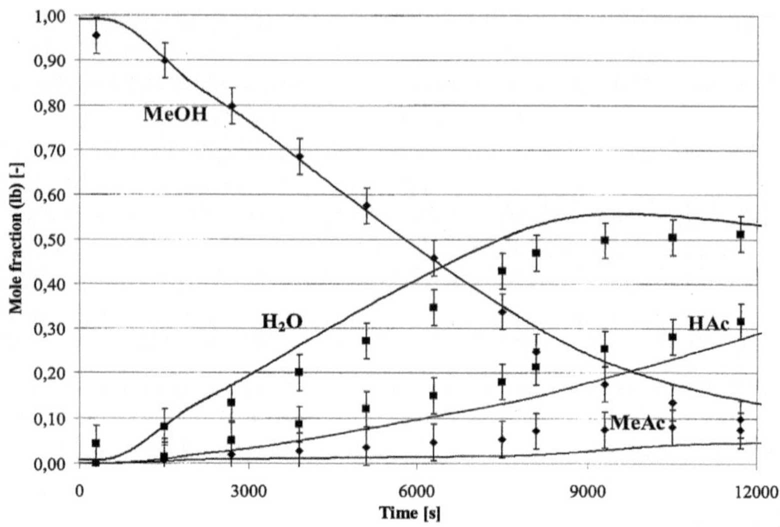

Fig. 3.7: Time related liquid-phase composition profiles for the methyl acetate synthesis (reboiler)

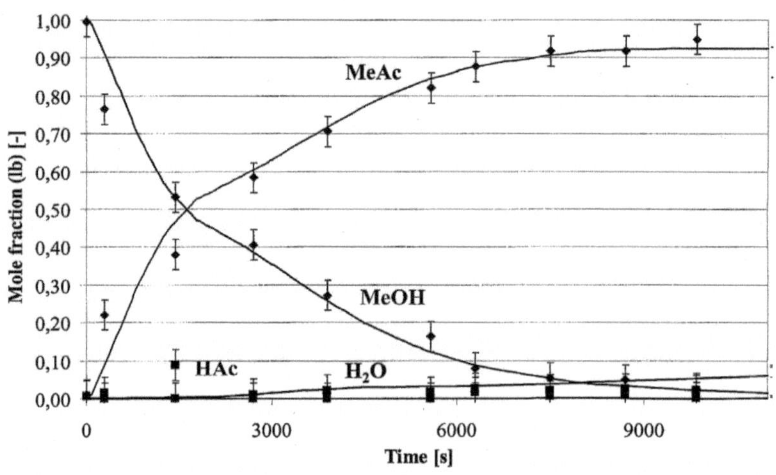

Fig. 3.8: Time related liquid phase composition profiles for the methyl acetate synthesis (condenser)

The reference model employs the rate-based approach and the Maxwell-Stefan diffusion equations. Another rate-based model assumes effective diffusion coefficients instead of the Maxwell-Stefan equations. The third model used is based on the equilibrium stage concept. Both the reference model and effective diffusion model show similar results. This can be explained by the low reaction rate which dominates the whole process kinetics (Noeres 2003).

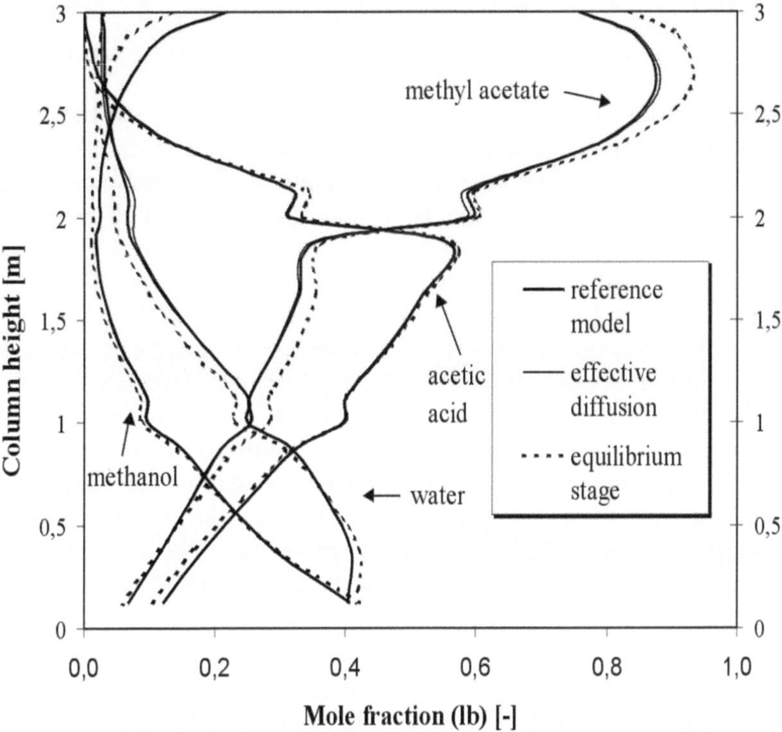

Fig. 3.9: Influence of simulation model on composition profiles for the methyl acetate synthesis (reproduced from Noeres, 2003)

3.5.2 Ethyl acetate synthesis

Ethyl acetate is a commodity chemical substance produced and used in facilities worldwide. The production in Western Europe was estimated about 246,000 tons/year in 1997 (SRI 1999). Ethyl acetate is primarily used as a solvent. Other applications of ethyl acetate as well as conventional production processes are highlighted in (Ullmann 1985; SRI 1999).

Chemical system

The formation of ethyl acetate is equilibrium limited and hence conversion can be increased via reactive separation. The main chemical reaction is given as:

$$CH_3COOH + CH_3CH_2OH \leftrightarrow CH_3COOCH_2CH_3 + H_2O \qquad (3.27)$$

In recently developed processes for the ethyl acetate synthesis, CD has already been applied. Kolena et al. (Kolena, Lederer et al. 1999) and Wu (Wu a. Lin 1999) suggested the combination of a pre-reactor and a CD column to carry out the reaction. The only difference between these processes is the location of the feed to the CD column. The process of Kolena et al. (Kolena, Lederer et al. 1999) is nowadays commercialized by Sulzer Chemtech Ltd. (Moritz 2002).

The boiling points of the pure components at atmospheric pressure are as follows: ethyl acetate (ETAC) 77.2°C; ethanol (ETOH) 78.3°C; water (H2O) 100.0°C; acetic acid (HAC) 118.0°C. There are three binary azeotropes and one ternary azeotrope summarized in Table 3.1, with respective boiling points at atmospheric pressure. The normal boiling points for the pure components as well as the compositions of the azeotropes are obtained from ASPEN Properties Plus® using UNIQUAC and show satisfactory agreement with the data available elsewhere (Kenig, Bäder et al. 2001).

Table 3.1: Calculated azeotropic data for the ethyl acetate system (UNIQUAC method, 1 bar) (Kenig, Bäder et al. 2001)

Azeotrope	ETAC [wt%]	ETOH [wt%]	H2O [wt%]	T [°C]
ETOH-H2O	----	95.5	4.5	78.1
ETAC-ETOH	70.2	29.8	----	71.8
ETAC-H2O	91.3	----	8.7	70.9
ETAC-ETOH-H2O	82.0	9.8	8.2	70.4

Reactive distillation lines were analyzed also by Kenig et al. (Klöker, Kenig et al. 2005). Due to the presence of minimum-boiling azeotropes it is not possible to obtain pure ethyl acetate at the top of the column. A liquid-liquid phase separation at the top of the column allows a further enrichment of ethyl acetate. Phase separation is only possible at low ethanol concentrations, since a higher ethanol content prevent phase splitting. Thus, a sufficient ethanol conversion is important for the process. This can be achieved by an excess of acetic acid in the column feed (Klöker, Kenig et al. 2002).

Process set-up and operation

Catalytic distillation experiments were conducted in a 50 mm diameter laboratory scale column (Klöker, Kenig et al. 2005). The basic column configuration is illustrated in Figure 3.10. The laboratory scale column with 50 mm inner diameter consists of three packed sections, of 1m height each. Structured wire gauze packings are applied for the non-catalytic sections of the column. The rectifying section at the top of the column is equipped with Sulzer DX, the stripping section at the bottom of the column with Sulzer BX®. For the catalytic section, different catalytic internals, KATAPAK-S® or MULTIPAK®, each in two different modifications, were tested. The decanter is operated at 20° C.

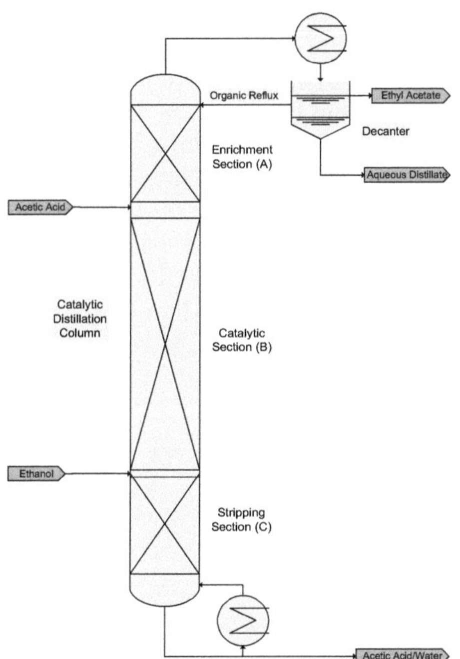

Fig. 3.10: Column configuration for the synthesis of EtAc (50mm I.D.) (Klöker, Kenig et al. 2004)

Acetic acid is fed above and ethanol below the catalytic section of the column; the enrichment section above the catalytic section is used to reach the azeotrope as a top product, while the stripping section below the catalytic section is used to purify the bottom stream. Since acetic acid is fed in excess into the process, the bottom product mainly consists of acetic acid, whereas the condensate consists of ethyl acetate, water and non-converted ethanol due to the presence of minimum boiling azeotropes. This mixture is fed to a liquid-liquid separator where a part of the organic phase is withdrawn and the remaining part is fed back to the column as reflux. This phase consists mainly of the product ethyl acetate. The aqueous phase is also withdrawn from the process. Further details on the experimental work as well as an analysis of the influence of the applied catalytic packings are given in (Klöker, Kenig et al. 2004).

Results and discussion

A series of simulations show a very good agreement (Figure 3.11) between simulated and experimental data (Klöker, Kenig et al. 2004). In the given example, the operating conditions correspond to an experiment where the total feed rate is 832 g/h at a molar feed ratio acetic acid to ethanol of 1.03, a reflux ratio of 3.16 and 629 g/h organic distillate. For the description of the vapor phase, the Hayden-O'Connell equation of state was applied, which accounts for the non-idealities due to the dimerization of acetic acid. The UNIQUAC method was adopted to calculate the activity coefficients. The reaction kinetics for the applied heterogeneous catalyst PUROLITE® CT179 were determined and adjusted to a pseudo-homogeneous approach by Hangx et al. (Hangx, Kwant et al. 2001).

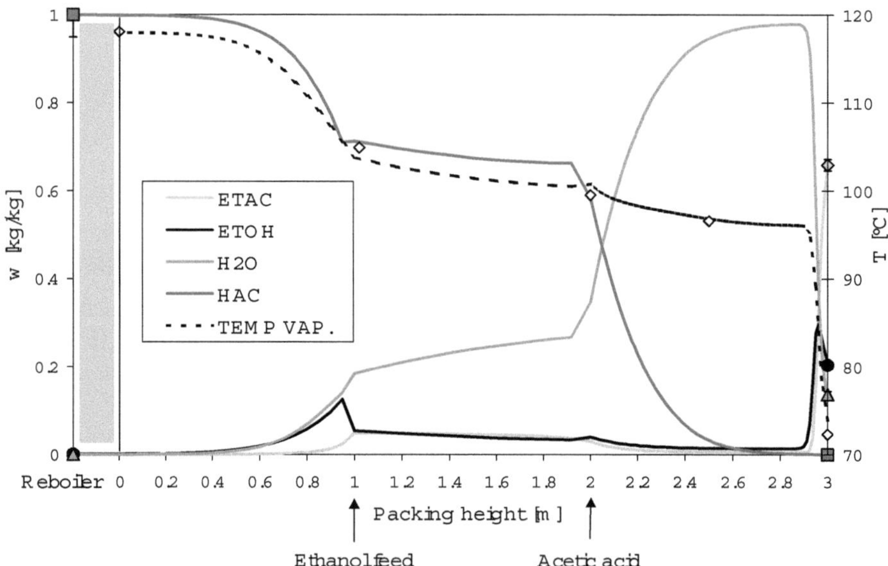

Fig. 3.11: Simulated (lines) and experimental (solid symbols) liquid-phase column profiles and vapor phase temperature profile (simulation: dotted line, experiments: empty symbols) (ethyl acetate system, no liquid-liquid phase separation, KATAPAK®-S, experiments by DSM Research)

The very good agreement between simulations and experiments is achieved not only for bottom and distillate concentrations, but also for the temperature profile. The ethanol conversion observed in the experiments is only around 64%, whereas the prediction is slightly lower and no liquid-liquid phase separation is achieved in the experiment, nor is it predicted. Conversion is limited by the amount of catalyst implemented, since the height of the catalytic section is only one meter.

3.5.3 Ethyl acetate transesterification

The reaction systems discussed before can be considered as single reaction step systems. Thus only the influence of CD processes on the conversion of such systems can be studied. To determine the impact of CD on the process selectivity, an additional reaction step has to be present.

The transesterification of ethyl acetate with a mixed stream of methanol and n-butanol offers the opportunity to explore how the selectivity can be shifted between the two main products methyl acetate an n-butyl acetate by use of a CD process. For the synthesis of ester, the remaining reactions are undesired and have to be suppressed. Instead of using a highly selective catalyst, it should be accomplished by distributing the two functionalities "reaction" and "separation" inside the CD column (see Chapter 2 for definitions).

Chemical system

The chemical reaction system consists of six reactants, namely ethyl acetate (EtAc), methanol (MeOH), n-butanol (BuOH) and the three possible products ethanol (EtOH), methyl acetate (MeAc) and n-butyl acetate (BuAc) as well as five binary homogeneous azeotropes (see Tab. 3.2).

Table 3.2: Normal boiling points of azeotropes and pure components and (UNIFAC-DMD, 1 bar)

T_B [K]	Components
326.88	MeAc-MeOH
330.20	MeAc
330.20	MeAc-EtOH
335.66	EtAc-MeOH
337.68	MeOH
345.80	EtAc-EtOH
350.35	EtAc
351.46	EtOH
390.39	BuAc-BuOH
390.83	BuOH
399.16	BuAc

$$\qquad\qquad\qquad\qquad\qquad\qquad 3.28$$

$$3.29$$

$$3.30$$

The reaction kinetics as well as the equilibrium constants for all reaction steps were obtained from (Pöpken 2000; Steinigeweg 2003), (BriteEuram 1997). In a first step, a survey on published thermodynamic data revealed that only UNIFAC-DMD as a predictive method is capable to describe the vapor-liquid equilibrium properly, including all five homogeneous binary azeotropes (see Tab. 3.2). In a second step, binary parameters for the NRTL equation were determined with the help of 154 experimental data sets from the DECHEMA data series (Gmehling, Onken et al. 1977-1984; Gmehling, Menke et al. 1994) to establish a more reliable description of the VLE behavior of the reaction system.

Process set-up

After the pre-studies described above a process synthesis strategy was applied based on heuristic rules developed in (Schembecker a. Tlatlik 2003). Starting with a detailed thermodynamic analysis of the reaction system and using the software tool PROSYN® (Schembecker a. Tlatlik 2003; Tlatlik 2004), topographies of the distillation areas as well as the reactive distillation areas were determined for the six-component system and for several three- and four-component sub-

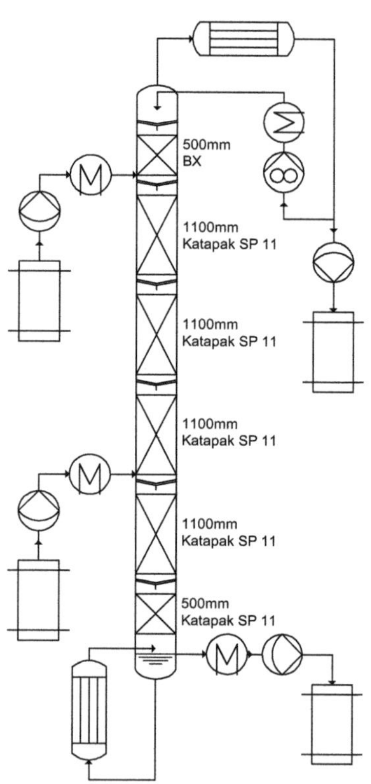

Fig. 3.12: Configuration for the catalytic distillation column with 50 mm I.D.

systems. Two distillation areas and two reactive distillation areas were found in a five-dimensional component space. These areas intersect only in some regions. For example, pure EtOH is a saddle node in each of the two distillation areas, but only in one of the reactive distillation areas. Another interesting point is that only three azeotropes remain under reactive conditions out of five azeotropes without reaction. No reactive azeotropes were found under conditions studied. Utilizing this information as a starting point for the heuristic-based process synthesis, an advantageous column design could be developed. In this design, the main goal to influence selectivity towards either one (MeAc) or the other product (BuAc) is fulfilled by only changing the reflux ratio and the distillate-to-feed ratio. The adjustment of the column design is not necessary. This simplifies experimental studies, since no reconstruction of the pilot plant is required. However, some thermodynamic limitations could not be overcome. A pure product stream is only possible in case of the production of butyl acetate, but not for that of methyl acetate. In case that MeAc is the main product, only the azeotrope between MeAc and MeOH can be obtained as a top product, because this azeotrope is the only unstable node in all distillation and reactive distillation areas (see chapter 2).

Based on this theoretical analysis, the experimental setup with the proposed configuration was built up and operated for eighteen experimental runs under different operating conditions. This configuration is presented in Figure 3.12.

Results and discussion

A series of experiments have been performed with a stoichiometric feed ratio $\chi = 1$ between EtAc and the total alcohol stream as well as between the EtAc stream and each additional reactant. The reflux ratio was varied in the range between 1 and 14, while the distillate-to-feed ratio was changed from 0.24 to 0.93. In all experiments, the feed rate was kept constant at a value of 2.0 kg/h.

From the results presented in Figure 3.13 and Figure 3.14 it is clearly visible that the guidelines for the process synthesis have been fulfilled. During the experiments, the operating point for n-butyl acetate (prediction from process synthesis: $D/F \geq 0.67$, reflux ratio $v \leq 2$) showed conversion rates X_{BuOH} of 70% with respect to BuOH, while the selectivity $S_{BuOH,BuAc}$ was 70% due to the production of di-n-butyl ether (DBE) in the lower part of the catalytic section.

The column profile in Figure 3.13 shows that the molar fraction of BuOH is decreasing towards the bottom, while BuAc is enriched. Besides, an additional side product DBE is produced in the lower part of the catalytic section and, as the highest boiling component, it is enriched towards the bottom of the column. All other components leave the column with the distillate stream.

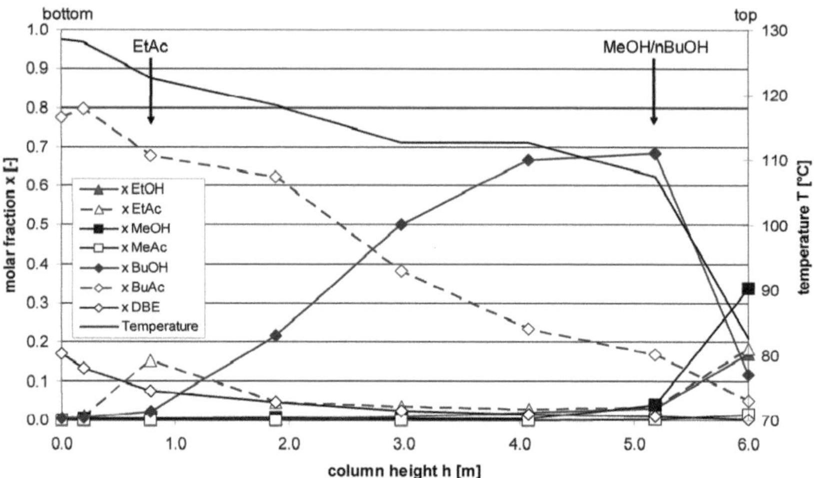

Fig. 3.13: Column profiles for the BuAc operating point (D/F ratio=0.74; reflux ratio =2; molar feed ratio EtAc:MeOH:nBuOH=1:1:1)

For the experiments with the operating conditions, under which MeAc should be produced (prediction from the process synthesis: $D/F \leq 0.33$, $v \geq 8$), the average conversion rate X_{MeOH} was nearly 50 % while the selectivity $S_{MeOH,MeAc}$ was nearly 100 %. Figure 3.14 shows that the MeAc-MeOH azeotrope is reached as the top product, while the n-butanol leaves the column with the bottom product stream.

Figure 3.13 and Figure 3.14 illustrate a clear change in selectivity within the CD column, which is achieved only by switching the operating conditions distillate-to-feed ratio (via changing the reboiler heat duty Q_H) and the reflux ratio.

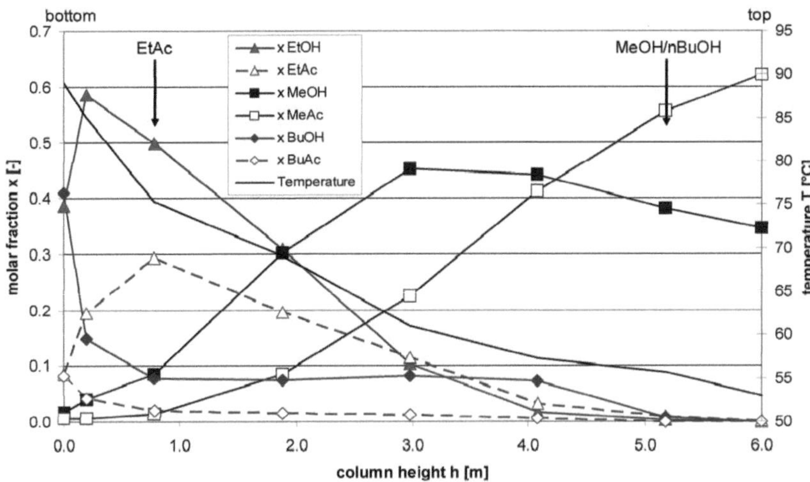

Fig. 3.14: Column profile for the MeAc operating point (D/F ratio=0.24; reflux ratio=14; molar feed ratio EtAc:MeOH:nBuOH=1:1:1)

3.5.4 Dimethyl carbonate transesterification

Similar to the transesterification of ethyl acetate, the reaction of di-methyl carbonate (DMC) with ethanol (EtOH) to produce ethyl methyl carbonate (EMC) and diethyl carbonate (DEC) offers the possibility to influence the selectivity of this reaction system via CD. Both products can be used as solvents in Lithium-Ion accumulators or as intermediates for the production of other products like polycarbonates and pesticides (Shaikh a. Sivaram 1996; Tundo 2001; Tundo a. Selva 2002).

Chemical system

The chemical reaction system consists of the two reaction steps shown by eq. (3.31) and (3.32). In a first step, the intermediate EMC is produced from the reactant DMC and EtOH, releasing methanol. The equilibrium constant of this reaction step changes with temperature within the range of the boiling point temperatures of all chemical components present (see Tab. 3.3) between 1.9 and 2.1 (Luo a. Xiao 2001).

In a second reaction step, the intermediate EMC again reacts with ethanol, producing DEC and, again, releasing methanol. The equilibrium constant for this reaction step varies between 0.44 and 0.46 as temperature changes within the range of the boiling point temperatures of the reactants (Luo a. Xiao 2001).

The second reaction step can be regarded as a parallel reaction to the first reaction with respect to EtOH or, alternatively, as a consecutive reaction with respect to the intermediate EMC (eq. (3.32)).

$$\tag{3.31}$$

$$\tag{3.32}$$

The transesterification of DMC to DEC represents a complex reaction system including five components and three binary azeotropes shown in Tab. 3.3. The characteristics of some binary mixtures of the substances involved in the reaction system have been investigated by Franchesconi et al. (Francesconi a. Comelli 1997), Comelli et al. (Comelli, Francesconi et al. 1996; Comelli a. Francesconi 1997; Comelli, Ottani et al. 1997; Comelli, Francesconi et al. 2001), Luo et al. (Luo, Xiao et al. 2000; Luo, Zhou et al. 2001; Luo, Zhou et al. 2002) and Rodriguez et al. (Rodriguez, Canosa et al. 2002; Rodriguez, Canosa et al. 2003; Rodrigues, Canosa et al. 2004), resulting in a good description of the required thermodynamic properties.

Due to the chemical equilibrium limitations, the reaction inside a batch stirred tank reactor would lead to a conversion rate of DMC $X_{DMC}=0.45$ and a selectivity of DMC towards DEC

Tab. 3.3: Boiling point temperatures of pure components and binary azeotropes (UNIFAC, 1 bar)

Temp (°C)	Components
63.8	DMC-MeOH
64.5	MeOH
74.9	EtOH-EMC
77.8	DMC-EtOH
78.5	EtOH
90.2	DMC
109.2	EMC
126	DEC

$S_{DMC,DEC}=0.27$. Especially the low selectivity is caused by the unfavorable equilibrium constant of the second reaction step. CD offers a possibility to increase this value through the removal of methanol from both reaction steps.

Process set-up

Starting with the given thermophysical and kinetic data, a heuristic-numeric process synthesis has been carried out. This procedure led to the column configurations displayed in Figure 3.15 for the production of DEC and the process design given in Figure 3.16 for the synthesis of EMC. Both processes have a catalytic distillation column as their main process unit, but in case of the EMC process additional unit operations are needed for optimal performance.

A CD process for the synthesis of DEC has been studied using a rate-based model implemented into the simulation environment gPROMS. Dynamic modeling techniques have been used to describe the time-depended behavior of the process. As a first step, a model database was established, containing submodels for catalytic and conventional stages, redistributors, condenser and reflux drum. The complexity of the implemented process models ranges from simple equilibrium stage considerations up to detailed models on the grounds of the rate-based approach (Noeres, Kenig et al. 2003). Furthermore, models of different complexity for the calculation of physical properties (activity coefficients, diffusion coefficients, etc.) are applied. Using these models, extensive sensitivity studies on different column

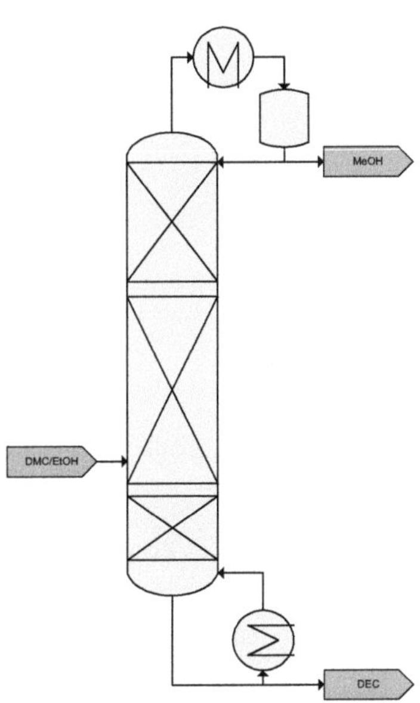

Fig. 3.15: Column configuration for the synthesis of DEC

configurations were performed; e.g. the impact of the reflux ratio and the heat duty on conversion and selectivity of this reaction system were investigated.

Results and discussion

As a result of the extended simulation studies, it was found that the conversion of DMC increases with increasing heat duty at the reboiler, reaches a nearly constant value and then decreases. The selectivity for the main product DEC also rises with increasing heat duty but, in contrast to the conversion rate, a minimum can be observed (Figure 3.17, Figure 3.18).

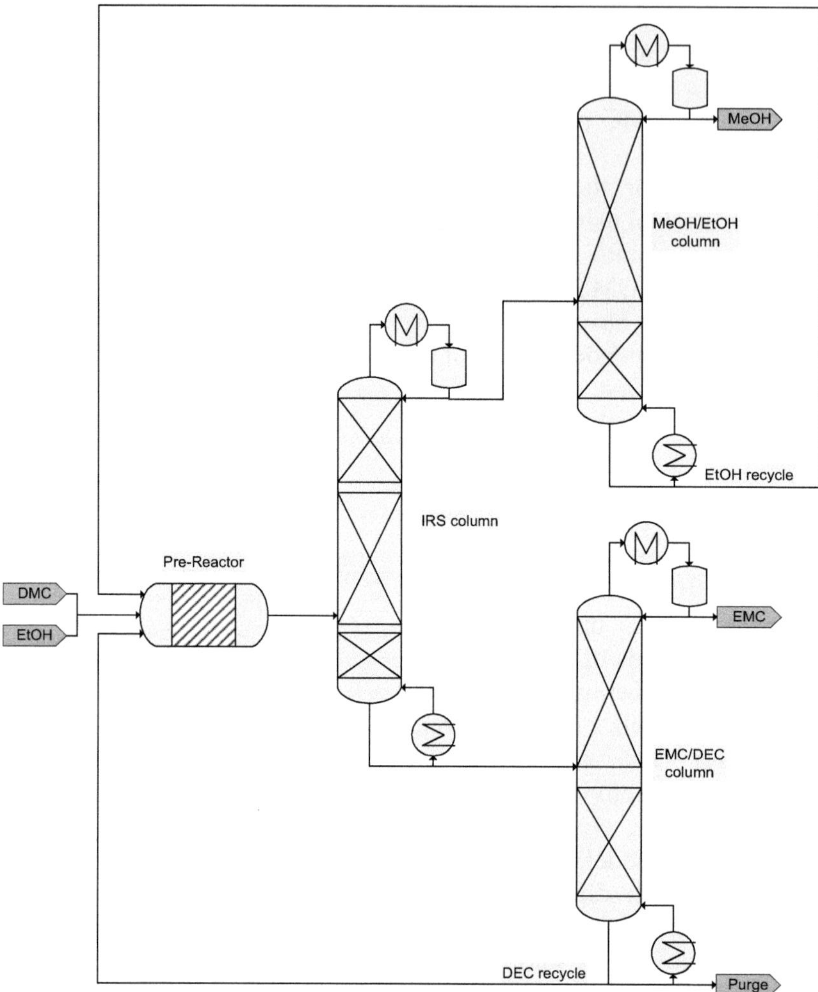

Fig. 3.16: Column configuration for the synthesis of EMC

This process behavior can be understood if one takes a closer look at the composition profiles along the column at the respective operating points. While operating the column under a given and constant reflux ratio, increasing the heat duty leads to a shift in those concentrations which first favors both reaction steps, since DMC and EtOH are enriched inside the catalytic section. Accordingly, a high conversion rate and relatively low selectivity can be observed in Figure 3.17.

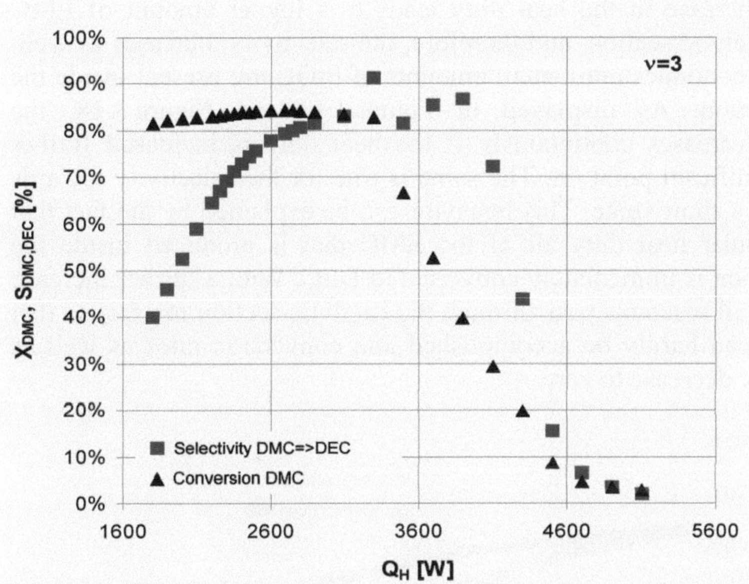

Fig. 3.17: Conversion and selectivity for the synthesis of DEC (ν=3) (Richter a. Górak 2004)

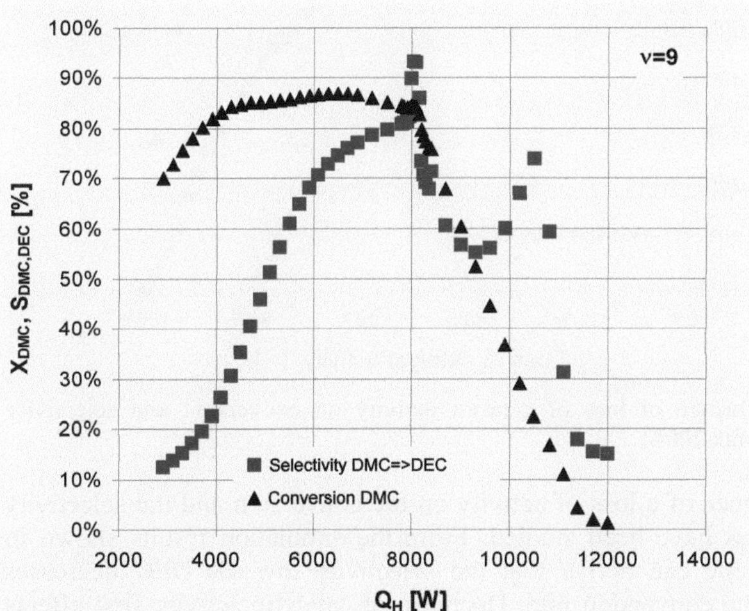

Fig. 3.18: Conversion and selectivity for the synthesis of DEC (ν=9) (Richter a. Górak 2004)

A slight increase in the heat duty leads to a higher amount of EMC inside the catalytic section, and therefore, the selectivity increases as well. This process continues until small amounts of EtOH are present inside the catalytic section. As displayed in Figure 3.17 and Figure 3.18, the conversion decreases continuously if the heat duty is increased further from this significant point on. The same is true for the selectivity towards DEC, except a short spike. This behavior can be explained by the fact that at this particular heat duty all of the EMC that is produced inside the catalytic section is immediately converted to DEC. With a further increase in heat duty, all reactants pass through the catalytic section too fast so that the reaction can hardly be accomplished and conversion rates as well as the selectivity decrease to zero.

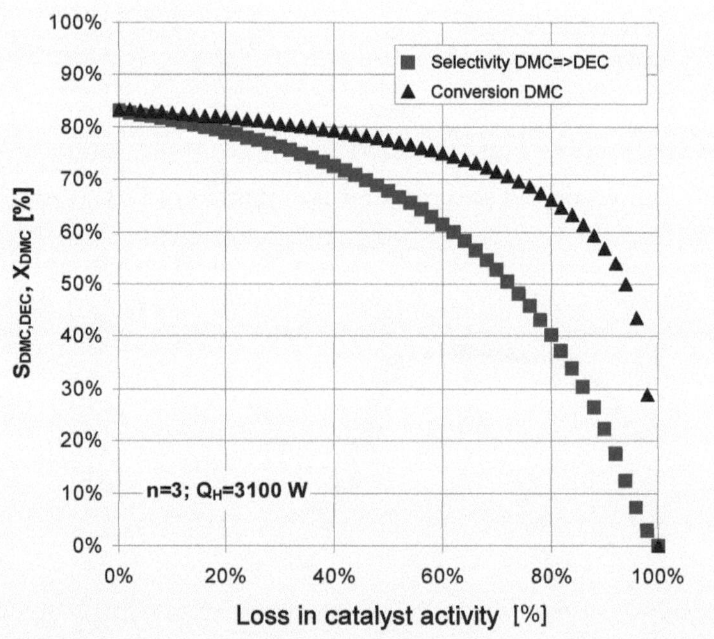

Fig. 3.19: Influence of loss of catalyst activity on conversion and selectivity (Richter a. Górak 2004)

The influence of a loss of activity on the conversion and the selectivity of the process have been studied. From the simulation results shown in Figure 3.19 one can derive that the selectivity towards DEC decreases faster than the conversion rate. Decrease of catalytic activity first affects the second reaction step and thus influences the overall selectivity towards DEC. As displayed in Figure 3.19, a slight decrease in activity significantly influences the selectivity of DMC towards DEC, whereas the conversion rate remains nearly constant. This is true for a loss in activity up to nearly 20%, at which the conversion decreases by 2% only, whereas the selectivity by 8%.

The results demonstrate that the two-step transesterification represents a suitable reaction system for a CD process. The catalyst screening shows that there are several heterogeneous catalysts available to produce EMC and DEC with sufficient yields for this application. The selectivity studies provide an opportunity to identify optimal operational conditions.

3.6 Conclusions

Catalytic distillation is one of the best known examples of an integrated reaction and separation process. Integrating a chemical reaction and the simultaneous separation via distillation of the products formed, this process concept offers an excellent approach towards the idea of process intensification in the chemical industry. Although the advantages of CD processes have been presented throughout the last decades in literature, industrial acceptance is still limited. In most cases, CD processes have to compete economically with sequentially arranged unit operations, which are highly optimized and very cost effective. Although IRS may offer economic benefits compared to these conventional processes, the chemical industry still tends to trade off the advantages of an IRS for a conventional process that is well known, more robust and easier to handle.

In this chapter, methods and examples have been presented which demonstrate that the research on CD processes during the last decades has led to a more detailed understanding of this particular IRS, and of its advantages and limitations. The rate-based models have been validated through experimental results for several reaction systems. This reliable simulation tool allows us to study the feasibility of CD for a new, not yet tested reaction system, and reduces the necessary number of costly pilot plant experiments and the time-to-market for new products. The examples described in this chapter prove that the former uncertainties connected with implementing CD into an industrial process have been reduced to a level that is comparable with that of a conventional unit operation.

The models can handle dynamic processes of synthesis of methyl acetate within a CD batch process and offer a basis for further investigations of other reaction systems. It was proven for ethyl acetate synthesis. This chemical system exhibits three binary azeotropes and a ternary heterogeneous azeotrope between ethyl acetate, ethanol and water.

Considering reaction systems with a single reaction, only the impact of CD processes on the conversion can be studied. For the analysis of selectivity issues, a reaction system with at least two reaction steps should be available. This can be either a reaction system which can be extended by an additional reaction step or a reaction system in which one reactant can interact with one of the products to form an additional product. Examples for both options are given in this chapter with the transesterification of EtAc and the transesterification of DMC, respectively. The results prove that for both systems CD can influnce the selectivity towards one main product.

The different case studies presented in this chapter show that the applicability of CD as a process integrating reaction and separation into one unit operation has evolved over the last years from single reaction systems, with few thermodynamic non-idealities, to more complex reaction systems with two or three reaction steps and highly non-ideal thermodynamic behavior. In parallel, the methods for designing CD process as well as the modeling and simulation tools available to study these systems, have become more elaborated and robust. Most of the simulation results presented have been validated with experimental data. For all of these reaction systems validated, the experimental and simulation data are in good agreement.

The future development of CD processes will be driven by the need for a more economical development and application of new CD processes, since catalytic distillation has to offer significant benefits compared to the competing conventional unit operations. To achieve this goal, future research needs to focus on two major research areas: micro-scale events inside a CD column and simplification of the application of CD processes beyond the current status.

Micro-scale research includes the examination of flow patterns inside the catalytic bags, including the chemical reactions, via validated CFD models. This future work, combined with the existing research results of flow patterns on non-catalytic packings, should lead to the design and virtual testing of new catalytic internals via modelling and simulation. This would allow the manufacturer to adapt the attributes (e.g. surface roughness, shape and size of catalyst bags) of a catalytic packing exactly to the thermodynamic properties and kinetic data available for the specific reaction system for which the packing will be used.

To simplify the application of CD in industrial processes can be regarded as the macro-scale side of future research. The methods for studying the feasibility of CD for a certain reaction system, and for performing the synthesis of a CD process from thermodynamic data and reaction kinetics available for this system, have been developed over the past years. But the application of these methods and the interpretation of the results received still require detailed knowledge of CD processes. The same, though not to to same extent, holds true for detailed modelling and simulation: the degree of robustness of RBA models needs to be increased without any loss of accuracy. Moreover, the user interfaces of modelling tools have to become standardized for easy interaction and to be made more user-friendly.

3.7 Notation

A_c	column cross section	m^2
a_I	specific gas-liquid interfacial area	m^2/m^3
a_i	activity of component i	
Bo	Bodenstein number	
c	molar concentration	mol/m^3
d	column diameter	m
$Đ$	Maxwell-Stefan diffusion coefficient	m^2/s
D_{ax}	axial dispersion coefficient	m^2/s
E	specific energy hold-up	J/(mol s)
G	gas molar flow rate	mol/s
ΔH_R^0	reaction enthalpy	J/mol
H	molar enthalpy	J/mol
J	molar flux	$mol/(m^2 \, s)$
K	second order reaction rate constant	$m^3/(mol \, s)$
k_1	rate constant of reaction 1	$1/(mol^2 \, s)$
k_{eff}	effective diffusion coefficient	m^2/s
K_i	adsorption constant	
K_{ij}^{eq}	Vapor-liquid equilibrium constant	
$[k^{av}]$	mass transfer coefficient matrix	m/s
l	specific length	m
L	liquid molar flow rate	mol/s

m	Mass	kg
M	molar mass	kg/mol
n	number of components of mixture	
n	molar hold-up	mol
N	molar flux	mol/(m^2 s)
q	heat flux	W/m^2
r	equivalent reaction rate	mol/(m^3 s)
R	total component reaction rate	mol/(m^3 s)
$[R^{av}]$	R matrix	s/m
$\Re$	gas constant	8.3144 J/(mol K)
t	time	sec
T	temperature	K
u	liquid velocity	m/s
V	molar vapor holdup	
x	liquid molar fraction	mol/mol
y	vapor molar fraction	mol/mol
z	film coordinate	m

Greek letters

χ	feed ratio	mol/mol
ε	void fraction	m^3/m^3
ϕ	volumetric hold-up	m^3/m^3
ρ	density	kg/m3
ψ	catalyst volume fraction	m^3/m^3
Γ	thermodynamic correction matrix	
κ	binary mass transfer coefficients	m/s
μ	chemical potential	J/mol
ν	stoichiometric coefficient, reflux ratio	

Subscripts

°C	
	component, column
G	gas phase
i,j	component/ reaction indices
L	liquid phase

Superscripts

B	bulk phase
I	phase interface

Abbreviations

ADM	axial dispersion model
CD	catalytic distillation
IRS	integrated reactive separation process
PDE	piston flow with axial dispersion and mass exchange
RD	reactive distillation
RBA	rate-based approach
RTD	residence time distribution

3.8 Literature

Adrian T, Bessling B, Hallmann H, Niekerken V, Spindler A, Ohligschlager A and Rumpf M (1996) Vorrichtung zur Durchführung von Destillationen und heterogen katalysierten Reaktionen (Patent DE19869598A1). Deutschland.

Agar D W (1999) "Multifunctional reactors: Old preconceptions and new dimensions." Chem. Eng. Sci. 54: 1299-1305.

Agreda V H and Lilly R D (1990) Preparation of Ultra High Purity Methyl Acetate. United States Patent 4,939,294. USA, Eastman Kodak Company, Rochester, N.Y.

Agreda V H and Partin L R (1984) Reactive Distillation Process For The Production Of Methyl Acetate. United States Patent 4,435,595. USA, Eastman Kodak Company, Rochester, N.Y.

Agreda V H, Partin L R and Heise W H (1990) "High-Purity Methyl Acetate via Reactive Distillation." Chem. Eng. Prog. 86: 40-46.

Alejski K (1991) "Computation of the Reacting Distillation Column Using a Liquid Mixing Model on the Plates." Comput. Chem. Eng. 15(5): 313-323.

Baerns M, Hofmann H and Renken A (1992) Chemische Reaktionstechnik. Stuttgart, Georg Thieme Verlag.

Barreira M N, de Toledo E C V, Filho R M and das Graças E M (2003) "Use of Different Numerical Solution Approaches for a Three-Phase Slurry Catalytic Reactor Model." Int. J. of Chem. Rec. Eng. 1: A53.

Bart H J and Landschützer H (1996) "Heterogene Reaktivdestillation mit axialer Rückvermischung." Chemie Ingenieur Technik: 944-946.

Baur R, Taylor R and Krishna R (2004) "Dynamic behaviour of reactive distillation columns described by a nonequilibrium stage model." Chemical Engineering Science 56: 2085-2102.

Bell R L (1972) "Residence Time and Fluid Mixing on Commercial Scale Sieve Trays." AIChE Journal 18: 498-505.

Bennett D L and Grimm H J (1991) "Eddy Diffusivity for Distillation Sieve Trays." AIChE Journal 37: 589-596.

Billet R (1995) Packed Towers. Weinheim, VCH.

Bird R B, Stewart W E and Lightfood E N (2001) Transport Phenomena. New York, John Wiley and Sons.

Bornscheuer U T (2002) "Microbial carboxyl esterases: classification, properties and application in biocatalysis." Fems Microbiology Reviews 26(1): 73-81.

Bravo J L and Fair J R (1982) "Generalized Correlation for Mass Transfer in Packed Distillation Columns." Ind. Eng. Chem. Process Des. Dev. 21: 162-170.

BriteEuram (1997) Technical Report: Transesterifcation of Methyl Acetate with Ethanol.

Buchaly C, Lauterbach S, Kreis P and Górak A (2005) Membrane Assisted Reactive Distillation and Batch Reaction. Aachener Membrankolloquium, Aachen.

Choudary B M, Kantam M L, Reddy C V, Aranganathan S, Santhi P L and Figueras F (2000) "Mg-Al-O-t-Bu hydrotalcite: a new and efficient heterogeneous catalyst for transesterification." Journal of Molecular Catalysis A - Chemical 159(2): 411-416.

Comelli F and Francesconi R (1997) "Isothermal vapor-liquid equilibria measurements, excess molar enthalpies, and excess molar volumes of dimethyl carbonate plus methanol, plus ethanol, and plus propan-1-ol at 313.15 K." Journal of Chemical and Engineering Data 42(4): 705-709.

Comelli F, Francesconi R and Castellari C (2001) "Excess molar enthalpies and excess molar volumes of binary mixtures containing dialkyl carbonates plus pine resins at (298.15 and 313.15) K." Journal of Chemical and Engineering Data 46(1): 63-68.

Comelli F, Francesconi R and Ottani S (1996) "Isothermal vapor-liquid equilibria of dimethyl carbonate plus diethyl carbonate in the range (313.15 to 353.15) K." Journal of Chemical and Engineering Data 41(3): 534-536.

Comelli F, Ottani S and Francesconi R (1997) "Excess molar enthalpies and excess molar volumes of dimethyl carbonate plus seven alkyl acetates at 298.15 K." Journal of Chemical and Engineering Data 42(6): 1208-1211.

Cornils B and Herrmann W A (1996) Applied Homogeneous Catalysis with Organometallic Compounds. Weinheim, VCH.

Danckwerts P V (1953) "Continuous Flow Systems - Distribution of Residence Times." Chem. Eng. Sci. 2: 1-13.

Danckwerts P V (1970) Gas-Liquid Reactions. New York, McGraw-Hill.

Dartt C B and Davis M E (1994) "Applications of Zeolites to Fine Chemicals Synthesis." Catalysis Today 19(1): 151-186.

Davies B and Jeffreys G V (1973) "The Continuous Trans-Esterification of Ethyl Alcohol and Butyl Acetate in a Sieve Plate Column - Part {III}: Trans-Esterification in a Six Plate Sieve Plate Column." Trans. Instn Chem. Engrs 51: 275-280.

DECHEMA (2005) DETHERM - Thermophysical properties of pure substances & mixtures. Frankfurt a. M., Germany, DECHEMA e.V. 2005.

DeGarmo J L, Parulekar V N and Pinjala V (1992) "Consider Reactive Distillation." Chem. Eng. Prog. 88(3): 42-50.

Doherty M F and Malone M F (2001) Conceptual Design of Distillation Systems. New York, McGraw Hill.

Doraiswamy L L and Sharma M. M (1984) Heterogeneous Reactions: Analysis, Examples and Reactor Design. New York.

Drauz K and Waldmann H (1995) Enzyme Catalysis in Organic Synthesis. Weinheim, VCH.

Egorov Y, Menter F, Kloeker M and Kenig E Y (2002) Detaillierte CFD Berechnung der Hydrodynamik in strukturierten Packungen. GVC-Fachausschuesse "Waerme- und Stoffaustausch" und "CFD - Computational Fluid Dynamics". Weimar.

Egorov Y, Menter F, Kloeker M and Kenig E Y (2005) "On the combination of CFD and rate-based modelling in the simulation of reactive separation processes." Chemical Engineering and Processing 44: 631–644.

Ellenberger J and Krishna R (1999) "Counter-current Operation of Structure Catalytically Packed Distillation Columns: Pressure Drop, Holdup and Mixing." Chem. Eng. Sci. 54: 1339-1345.

Figueras F, Tichit D, Naciri M B and Ruiz R (1998) "Selective Aldolisation of Acetone Into Diacetone Alcohol Using Hydrotalcites as Catalysts." Chemical Industries (Dekker) 75: 37-49.

Flory P J (1941) J. Chem. Phys. 9: 660.

Flory P J (1942) J. Chem. Phys. 10: 51.

Francesconi R and Comelli F (1997) "Excess molar enthalpies, densities, and excess molar volumes of diethyl carbonate in binary mixtures with seven n-alkanols at 298.15 K." Journal of Chemical and Engineering Data 42(1): 45-48.

Froment G F and Bischoff K B (1990) Chemical Reactor Analysis and Design. New York, John Wiley & Sons Inc.

Fuchigami Y (1990) "Hydrolysis of Methyl Acetate in Distillatiom Column Packed with Reactive Packing of Ion Exchange Resin." Journal of Chemical Engineering of Japan 23(3): 354-359.

Gelbein A P and Buchholz M (2000) Process and Structure For Effecting Catalytic Reactions in Distillation Structure (EP 0428265B2).

Giessler S, Danilov R Y, Pisarenko R Y, Serafimov L A, Hasebe S and Hashimoto I (1999) "Feasible Separation Modes for Various Reactive Distillation Systems." Ind. Eng. Chem. Res. 38: 4060-4067.

Gmehling J, Menke J, Krafczyk J and Fischer K (1994) Azeotropic Data (Part 1\& 2), VCH Verlagsgesellschaft mbH.

Gmehling J, Onken U, Arlt W, Grenzheuser P, Weidlich U and Kolbe B (1977-1984) Chemistry Data Series. Frankfurt a.M., DECHEMA.

Gorak A (1995) Simulation thermischer Trennverfahren fluider Vielkomponentengemische. Prozesssimulation. Schuler H, Wiley-VCH: 349-408.

Górak A and Kreul L (2004) Packung für Stoffaustausch-Kolonnen. EP.

Górak A, Kreul L U and Skowronski M (1998) Strukturierte Mehrzweckpackung. Deutsches Patent 19701045 A1.

Götze L and Bailer O (2000) "Katalysator-Sandwich - Reaktivdestillation mit einer neuen strukturierten Packung." Chemie Technik 29(2): 42-45.

Götze L, Bailer O, Moritz P and von Scala C (2000) "KATAPAK-SP: Baukastensystem für die Reaktivrektifikation." Chemie Ingenieur Technik 72(9): 1053-1054.

Hangx G, Kwant G, Maessen H, Markusse A P and Urseanu M I (2001) Reaction Kinetics of the Esterification of Ethanol and Acetic Acid Towards Ethyl Acetate, Intelligent Column Internals for Reactive Separations (INTINT), Technical Report to the European Commission: http://www.cpi.umist.ac.uk/intint/NonConf_Doc.asp.

Hayden J G and O'Connell J P (1975) "A Generalized Method for Predicting Second Virial Coefficients." Ind. Eng . Chem. Process Des. Dev. 14: 209-216.

Heijnen J H M, de Bruijn V G, van den Broeke L J P and Keurentjes J T F (2003) "Micellar catalysis for selective epoxidations of linear alkenes." Chemical Engineering and Processing 42(3): 223-230.

Higler A, Krishna R and Taylor R (1999) "Nonequilibrium Cell Model for Multicomponent (Reactive) Separation Processes." AiChE Journal 45(11): 2357-2370.

Higler A, R. Krishna and Taylor R (2000) "Nonequilibrium Modeling of Reactive Distillation: A Dusty Fluid Model for Heterogeneously Catalysed Processes." Ind.Eng. Chem. Res. 39: 1596-1607.

Hoffmann A and Górak A (2000) Methyl-Acetate via Catalytic Distillation: Characteristics of MULTIPAK and their Influence on the Process Performance. International Congress of Chemical and Process Engineering (CHISA). Prag.

Hoffmann A, Noeres C and Górak A (2004) "Scale-Up of Reactive Distillation Columns with Catalytic Packings." Chem. Eng. Process. 43: 383-395.

Huggins M L (1941) Chemical Phys. 9: 440.

Huggins M L (1942) J. Am. Chem. Soc. 64: 1712.

Illiuta I, Larachi F and Grandjean B P A (1999) "Residence Time, Mass Transfer and Back Mixing of the Liquid in Trickle Flow Reactors Containing Porous Particles." Chem. Eng. Sci. 54: 4099-4109.

Jacobsson K, Pyhälathi A, Pakkanen S, Keskinen K and Aittamaa J (2001) Modeling of a Configuration Combining Distillation and Reaction in a Side Reactor. International Symposium on Multifunctional Reactors (ISMR-2), Nürnberg, Germany.

Johnson K H and Dallas A B (1994) Catalytic Distillation Structure. US.

Jones E M J (1985) Contact Structure for Use In Catalytic Distillation. US, Chemical Research \& Licensing Company, Houston, TX.

Judzis A (2004) Advances in Process Intensification through Multifunctional Reactor Engineering. Energy U S D o. 2005.

Kenig E, Klöker M, Noeres C, Nijhuis T A, Beers A E W, Kapteijn F and Moulijn J A (2000) Detailed Rate-Based Model Including Extended Hydrodynamics Description (Deliverable 7), INTINT.

Kenig E Y, Bäder H, Górak A, Beßling B, Adrian T and Schoenmakers H (2001) "Investigation of Ethyl Acetate Reactive Distillation Process." Chem. Eng. Sci. 56: 6185-6193.

Kenig E Y, Górak A and Bart H-J (2004) Reactive Separations in Fluid Systems. Re-Engineering the Chemical Processing Plant: Process Intensification. Stankiewicz A a.Moulijn J A. New York, Marcel Dekker, Inc.: 309-377.

Kenig E Y, Kloeker M, Egorov Y, Menter F and Górak A (2001) Towards Improvement of Reactive Separation Performance Using Computational Fluid Dynamics. ISMR-2, 2nd International Symposium on Multifunctional Reactors. Nuernberg.

Kenig E Y, Kloeker M, Egorov Y, Menter F and Górak A (2001) "Towards Improvement of Reactive Separation Performance Using Computational Fluid Dynamics." Chemie-Ingenieur-Technik 73(6): 773.

Klöker M, Kenig E Y, Górak A, Egorov Y and Menter F (2003) Improved Design of Reactive Separation Internals via CFD and Process Simulation. ACHEMA 2003. Frankfurt / Main.

Klöker M, Kenig E Y, Górak A, Franczek K, Salacki W and Orlikowski W (2003) Experimental and Theoretical Studies of the TAME Synthesis by Reactive Distillation. European Symposium on Computer Aided Process Engineering - 13, Lappeenranta, Finland, Elsevier Science B.V.

Klöker M, Kenig E Y, Górak A, Markusse A P, Kwant G, Götze L and Moritz P (2002) Investigation of Different Column Configurations for the Ethyl Acetate Synthesis via Reactive Distillation. Distillation and Absorption, Baden Baden, Germany.

Klöker M, Kenig E Y, Górak A, Markusse A P, Kwant G and Moritz P (2004) "Investigation of Different Column Configurations for the Ethyl Acetate Synthesis via Reactive Distillation." Chem. Eng. Process. 43: 791-801.

Klöker M, Kenig E Y, Hoffmann A, Kreis P and Górak A (2005) "Rate-based modelling and simulation of reactive separations in gas/vapour–liquid systems." Chemical Engineering and Processing 44: 617–629.

Klöker M, Kenig E Y, Piechota R, Burghoff S and Egorov Y (2004) "CFD-gestützte Untersuchungen von Hydrodynamik und Stofftransport in Katalysatorschüttungen." Chem. Ing. Tech. 76: 236-242.

Klöker M, Kenig E Y, Piechota R, Burghoff S and Egorov Y (2005) "CFD-based Study on Hydrodynamics and Mass Transfer in Fixed Catalyst Beds." Chemical Engineering and Technology 28(1): 31-36.

Kolena J, Lederer J, Moravek P, Hanika J, Smejkal Q and Skala D (1999) Zpusob vyroby etylacetatu a zarizeni k provadeni tohoto zpusobu (Process for the production of ethyl acetate and apparatus for performing the process, Cz PV 3635-99). Czech Republic.

Kolodziej A, Jaroszynski M and Bylica I (2003) "Mass transfer and hydraulics for KATAPAK-S." Chemical Engineering and Processing 43(3): 457-464.

Kolodziej A, Jaroszynski M, Salacki W, Orlikowski W, Fraczek K, Klöker M, Kenig E Y and Górak A (2004) "Catalytic Distillation for the TAME Synthesis with Structured Catalytic Packings." Chem. Eng. Res. Des. 82: 175-184.

Kooijman H (1995) Dynamic Nonequilibrium Column Simulation. Chemical Engineering. Potsdam, N.Y., Clarkson University.

Kreft A and Zuber A (1978) "On the physical meaning of the dispersion equation and its solutions for different initial and boundary conditions." Chem. Eng. Sci. 33: 1471-1480.

Kreul L U, Górak A, Dittrich C and Barton P I (1998) "Dynamic Catalytic Distillation: Advanced Simulation and Experimental Validation." Computers and Chemical Engineering 22(Supplement 1): S371 - S378.

Krishna R and Standart G L (1979) "Mass and Energy Transfer in Multicomponent Systems." Chem. Eng. Com. 3: 201.

Kunz U (1998) Entwicklung neuartiger Polymer / Träger - Ionenaustauscher als Katalysatoren für chemische Reaktionen in Füllkörperkolonnen. Clausthal - Zellerfeld, Papierflieger Verlag.

Kunz U and Hoffmann U (1995) Perparation of Catalytic Polymer/Ceramic Ionexchange Packings for Reactive Distillation Columns. Preparation of Catalysts VI. Ponclet G, Elsevier Science B.V.: 299-308.

Larachi F, Petre C F, Iliuta I and Grandjean B (2003) "Tailoring the pressure drop of structured packings through CFD simulations." Chemical Engineering and Processing 42(7): 535-541.

Lebas E, Jullian S, Travers C, Capron P, Joly J-F and Thery M (1997) Process for the isomerisation of paraffins by reactive distillation (EP 000000787786A1), Inst. Francais du Petrol.

Leet W A and Kulprathipanja S (2002) Reactive Separation Processes. Reactive Separation Processes. Kulprathipanja S. London, Taylor & Francis: 1-16.

Linnekoski J A and Rihko-Struckmann L K (1999) "Simultaneous Isomerisation and Etherification of isoamylenes." Ind. Eng. Chem. Res. 38: 4563-4570.

Loning S, Horst C and Hoffmann U (2000) "Theoretical investigations on the quaternary system n-butanol, butyl acetate, acetic acid and water." Chemical Engineering & Technology 23(9): 789-794.

Luo H P and Xiao W D (2001) "A reactive distillation process for a cascade and azeotropic reaction system: Carbonylation of ethanol with dimethyl carbonate." Chemical Engineering Science 56(2): 403-410.

Luo H P, Xiao W D and Zhu K H (2000) "Isobaric vapor-liquid equilibria of alkyl carbonates with alcohols." Fluid Phase Equilibria 175(1-2): 91-105.

Luo H P, Zhou J H, Xiao W D and Zhu K H (2001) "Isobaric vapor-liquid equilibria of binary mixtures containing dimethyl carbonate under atmospheric pressure." Journal of Chemical and Engineering Data 46(4): 842-845.

Luo H P, Zhou J H, Xiao W D and Zhu K H (2002) " CORRECTION-Isobaric Vapor-Liquid Equilibria of Binary Mixtures Containing Dimethyl Carbonate under Atmospheric Pressure." J. Chem. Eng. Data 47: 113.

Mackowiak J (2003) Fluiddynamik von Füllkörpern und Packungen. Grundlagen der Kolonnenauslegung. Berlin Heidelberg, Springer.

Mazzotti M N, B.; Gelosa, D.; Kruglov, A.; Morbidelli, M. (1997) "Kinetics of liquid-phase esterification catalyzed by acidic resins." Ind. Eng. Chem. Res. 36: 3-10.

Michelsen M L (1994) "The axial dispersion model and orthogonal collocation." Chem. Eng. Sci. 49: 3675-3676.

Mohl K-D, Kienle A and Hoffman U (1999) "Steady-state multiplicities in reactive distillation columns for the production of fuel ethers MTBE and TAME: theoretical analysis and experimental verification." Chem. Eng. Sci. 54: 1029-1043.

Mohl K D, Kienle A, Sundmacher K and Gilles E D (2001) "A theoretical study of kinetic instabilities in catalytic distillation processes: influence of transport limitations inside the catalyst." Chemical Engineering Science 56(18): 5239-5254.

Moritz P (2002) Product Information 'Ethyl Acetate Production by Reactive Distillation'. Brochure of Sulzer Chemtech Ltd.

Moritz P and Hasse H (1999) "Fluid Dynamics in Reative Distillation Packing Katapak-S." Chem. Eng. Sci. 54: 1367-1374.

Nigam K D P, Illiuta I and Larachi F (2002) "Liquid Back-Mixing and Mass Transfer Effects in Trickle-Bed Reactors Filled with Porous Catalyst Particles." Chem. Eng. Process. 41: 365-371.

Nijhuis T A, Kreutzer M T, A.C.J. R, Kapteijn F and Moulijn J A (2001) "Monolithic catalysts as efficient three-phase reactors." Chemical Engineering Science 56: 823-829.

Nocca J L, Leonard J, Gaillard J F and Amigues P (1991) Apparatus for Reactive Distillation. US.

Noeres C (2003) Catalytic Distillation: Dynamic Modelling, Simulation and Experimental Validation. Department of Biochemical and Chemical Engineering. Dortmund, University of Dortmund.

Noeres C, Hoffmann A and Gorak A (2002) "Reactive distillation: Non-ideal flow behaviour of the liquid phase in structured catalytic packings." Chemical Engineering Science 57(9): 1545-1549.

Noeres C, Kenig E Y and Górak A (2003) "Modelling of Reactive Separation Processes: Reactive Absorption and Reactive Distillation." Chem. Eng. Process. 42: 157-178.

Nothnagel K H, Abrams D S and Prausnitz J M (1973) "Generalized Correlation for Fugacity Coefficients in Mixtures at Moderate Pressures." Ind. Eng. Chem. Process Des. Dev. 12(1).

Petre C F, Larachi F, Iliuta I and Grandjean P A (2003) "Pressure Drop Through Structured Packings: Breakdown into the Contributing Mechanisms by CFD modeling." Chem. Eng. Sci. 58: 163-177.

Piironen M, Haario H and Turunen I (2001) "Modelling of Katapak Reactor for Hydrogenation of Anthraquinones." Chem. Eng. Sci. 56: 859-864.

Pöpken T (2000) Reaktive Rektifikation unter besonderer Berücksichtigung der Reaktionskinetik am Beispiel von Veresterungsreaktionen. Oldenburg, University of Oldenburg.

Pöpken T, Steinigeweg S and Gmehling J (2001) "Synthesis and Hydrolysis of Methyl Acetate by Reactive Distillation Using Structured Catalytic Packings: Experiments and Simulation." Industrial Engineering Chemistry Research 40(6): 1566-1574.

Qi Z W, Sundmacher K, Stein E, Kienle A and Kolah A (2002) "Reactive separation of isobutene from C4 crack fractions by catalytic distillation processes." Separation and Purification Technology 26(2-3): 147-163.

Reid R C, Prausnitz J M and Poling B E (1987) The Properties of Gases and Liquids. New York, McGraw-Hill.

Richter J and Górak A (2004) Katalytische Rektifikation für ein System mit Folge- und Nebenreaktionen. DECHEMA/GVC-Jahrestagung. Karlsruhe.

Rodrigues A, Canosa J, Dominguez A and Tojo J (2004) "Viscosities of dimethyl carbonate with alcohols at several temperatures UNIFAC-VISCO interaction parameters (-OCOO-/alcohol=." Fluid Phase Equilibria 216: 167-174.

Rodriguez A, Canosa J, Dominguez A and Tojo J (2002) "Isobaric vapour-liquid equilibria of dimethyl carbonate with alkanes and cyclohexane at 101.3 kPa." Fluid Phase Equilibria 198(1): 95-109.

Rodriguez A, Canosa J, Dominguez A and Tojo J (2003) "Isobaric phase equilibria of diethyl carbonate with five alcohols at 101.3 kPa." Journal of Chemical and Engineering Data 48(1): 86-91.

Schembecker G and Tlatlik S (2003) "Process synthesis for reactive separations." Chemical Engineering and Processing 42(3): 179-189.

Schildhauer T J, Kapteijn F and Moulijn J A (2005) "Reactive Stripping in Pilot Scale Monolith Reactors - Application to Esterification." Chemical Engineering and Processing 44(6): 695-699.

Schmitt M, Hasse H, Althaus K, Schoenmakers H, Götze L and Moritz P (2004) "Synthesis of n-Hexyl Acetate by Reactive Distillation." Chem. Eng. Process. 43: 397-409.

Schneider R, Noeres C, Kreul L U and Gorak A (2001) "Dynamic modeling and simulation of reactive batch distillation." Computers & Chemical Engineering 25(1): 169-176.

Schoenmakers H G and Bessling B (2003) "Reactive and catalytic distillation from an industrial perspective." Chemical Engineering and Processing 42(3): 145-155.

Schuchardt U, Sercheli R and Vargas R M (1998) "Transesterification of vegetable oils: a review." Journal of the Brazilian Chemical Society 9(3): 199-210.

Seader J D and Henley E J (2005) Separation Process Principles, John Wiley & Sons.

Shah Y T, Stiegel G J and Sharma M M (1978) "Backmixing in Gas-Liquid Reactors." AIChE Journal 24(24): 369-400.

Shaikh A A G and Sivaram S (1996) "Organic carbonates." Chemical Reviews 96(3): 951-976.

Shelden R and Stringaro J-P (1995) Vorrichtung zur Durchführung katalysierter Reaktionen (EP 0396650B1). European Union.

Sheldon R A and van Bekkum H (2001) Fine Chemicals through Heterogeneous Catalysis. Weinheim, VCH.

Shoemaker J D and Jones J, E.M. (1987) "Cumene by Catalytic Distillation." Hydrocarbon Processing(June): 57-58.

Smith L A J (1984) Catalytic distillation structure. U.S.A.

Song W, Venimadhavan G, Manning J M, Malone M F and Doherty M F (1998) "Measurement of Residue Curve Maps and Heterogeneous Kinetics in Methyl Acetate Synthesis." Ind. End. Chem. Res. 37(5): 1917-1928.

Sorel E (1893) La rectification de l'alcool. Paris, Gauthier-Villars et fils.

SRI (1999) Chemical Economics Handbook, SRI International.

Steinigeweg S (2003) Zur Entwicklung von Reaktivrektifikationsprozessen am Beispiel gleichgewichtslimitierter Reaktionen. Oldenburg, Car-von-Ossietzky-Universität.

Steinigeweg S and Gmehling J (2002) "n-Butyl Acetate Synthesis via Reactive Distillation: Thermodynamic Aspects, Reaction Kinetics, Pilot-Plant Experiments, and Simulation Studies." Ind.Eng. Chem. Res. 41: 5483-5490.

Steinigeweg S and Gmehling J (2003) "Transesterification processes by combination of reactive distillation and pervaporation." Chemical Engineering and Processing 43(3): 447-456.

Stewart W E and Prober R (1964) "Matrix Calculation of Multicomponent Mass Transfer in Isothermal Systems." Ind. Eng. Chem. Fundam. 4: 224.

Stichlmair J and Frey T (1998) "Prozesse der Reaktivdestillation." Chemie Ingenieur Technik 70(12): 1507-1516.

Stitt E H (2001) "Reactive Distillation - A Panacea or a Solution Looking for a Problem? A Case Study Based Evaluation." Chem. Ing. Tech. 73(6): 767.

Stitt E H (2003) Multifunctional Reactors? Up to a Point Lord Copper. International Symposium on Multifunctional Reactors (ISMR-3) / Colloquium on Chemical Reaction Engineering (CCRE-18), Bath, U.K.

Sulzer (2000) Produktinformation KATAPAK.

Taylor R and Krishna R (1993) Multicomponent Mass Transfer. New York, John Wiley & Sons, Inc.

Taylor R and Krishna R (2000) "Modelling reactive distillation." Chem. Eng. Sci.(55): 5183-5229.

Taylor R, Krishna R and Kooijman H A (2003) "Real World Modeling of Distillation." Chemical Engineering Progress 99(7): 28-39.

Thiel C (1997) Modellbildung, Simulation, Design und experimentelle Validierung von heterogen katalysierten Reaktivdestillationsprozessen zur Synthese der Kraftstoffether MTBE, ETBE und TAME. Fakultät für Bergbau, Hüttenwesen und Maschinenwesen. Clausthal-Zellerfeld, Technische Universität Clausthal: 121.

Tlatlik S (2004) Beitrag zur Prozesssynthese integrierter Reaktions- und Trennoperationen. Department of Biochemical and Chemical Engineering. Dortmund, University of Dortmund.

Toor H L (1964) "Prediction of Efficiencies and Mass Transfer on a Stage with Multicomponent Systems." AIChE J. 10: 545.

Toor H L (1964) "Solution of the Linearized Equations of Multicomponent Mass Transfer." AICHE Journal 10: 545-547.

Towler G T and Frey S J (2001) Reactive Distillation. Reactive Separation Processes. Kulprathipanja S. London, Taylor and Francis.

Tundo P (2001) "New developments in dimethyl carbonate chemistry." Pure and Applied Chemistry 73(7): 1117-1124.

Tundo P and Selva M (2002) "The chemistry of dimethyl carbonate." Accounts of Chemical Research 35(9): 706-716.

Ullmann (1985) Ullmann's Encyclopedia of Industrial Chemistry, 5th edition edition. Weinheim, Wiley-VCH.

van Baten J M and Krishna R (2002) "Gas and Liquid Phase Mass Transfer within KATAPAK-S Structures studied using CFD Simulations." Chem. Eng. Sci. 57: 1531-1536.

van Hasselt B W, Calis H P A, Sie S T and van den Bleek C M (1999) "Liquid hold-up in the three-levels-of-porosity reactor." Chem. Ing. Sci. 54: 1405-1411.

van Swaaij W P M, Charpentier J C and Villermaux J (1969) "Residence Time Distribution inthe Liquid Phase of Trickle Flow in Packed Columns." Chem. Eng. Sci. 24: 1083-1095.

Wang S-J, Wong D S H and Lee E-K (2003) "Control of a Reactive Distillation Column in the Kinetic Regime for the Synthesis of n-Butyl Acetate." Ind. Eng. Chem. Res. 42: 5182-5194.

Weitkamp J, Hunger M and Rymsa U (2001) "Base catalysis on microporous and mesoporous materials: recent progress and perspectives." Microporous and Mesoporous Materials 48(1-3): 255-270.

Wesselingh J A (1990) Mass Transfer. Chichester, West Sussex (England), Ellis Horwood.

Wesselingh J A and Krishna R (1990) Mass Transfer, Ellis Horwood.

Wu K-C and Lin C-T (1999) Catalytic processes for the preparation of acetic esters. U.S. 5,998,658.

Yeoman N, Pinaire R, Ulowetz M A, Nace T P and Furese D A (1994) Method and Apparatus for Concurrent Reaction with Distillation. WO 94/08679.

Yuxiang Z and Xien X (1992) "Study on catalytic distillation processes - pt. II: simulation of catalytic distillation processes (quasi-homogeneous and rate-based model)." Trans. IChemE 70: 465-470.

Zheng Y, Flora T.T. Ng and Rempel G L (2001) "Catalitic Distillation: A Three-Phase Nonequilibrium Model for the Simulation of the Aldol Condensation of Acetone." Ind.Eng. Chem. Res. 40: 5342-5349.

Zheng Y, Rempel G L and Ng F T T (2003) "Modelling of the Catalytic Distillation Process for the Synthesis of Ethyl Cellosolve Using a Three-Phase Nonequilibrium Model." Int. J. Chem. Reac. Eng. 1: 15.

4 Reactive gas adsorption

Sven Reßler, Martin P. Elsner, Christoph Dittrich, David W. Agar, Simone Geisler, Olaf Hinrichsen

4.1 Introduction

Adsorptive reactors represent a special case of multifunctional reactors where the processes which are affected by the chemical reaction system are coupled with an adsorptive separation functionality. This approach can additionally be extended to gas-solid reactions as a selective method for the manipulation of the concentration profile of a particular species.

In this way, one or more components can be adsorptively removed from or desorptively supplied to the reaction mixture, thus affecting the kinetics and thermodynamics of the reaction system. One common application of the adsorptive reactor concept is the enhancement of the conversion in a simple reversible reaction to overcome thermodynamic limitations by the selective removal of a (by-)product.

Other concepts for selectivity control in complex reaction schemes offer the opportunity to favor desired reactions over the unwanted ones by a suitable adsorptive manipulation of the concentration profiles.

A survey of the chemical reactions investigated in gas-phase adsorptive reactors is given in Table 4.1. Most previous work has exploited adsorptive reaction processes as a means to enhance equilibrium conversions by the uptake of one of the products according to Le Chatelier's principle. Furthermore, simple substances such as ammonia, carbon dioxide, water or small organic molecules are the adsorptive of preference since they are stable on the adsorbent surface under the reaction conditions of catalytic gas-phase reactions.

Table 4.1: Experimentally investigated adsorptive gas phase reactor concepts

Reaction & conditions	Reactor	Regeneration	Reference
$CO(g) + 2\,H_2(g) \leftrightarrow CH_3OH$ (T = 225-250 °C; p = 50-63 bar)	fluidised bed	—	Kuczynski et al. 1987
$2\,NH_3(ads) + 2\,NO(g) + \frac{1}{2}\,O_2(g) \rightarrow$ $2\,N_2(g) + 3\,H_2O(g)$ (T = 300 °C)	fixed-bed	Reaction	Agar and Ruppel 1988
$CO(g) + \frac{1}{2}\,O_2(g) \leftrightarrow CO_2$ (T = 160 °C; p = 1-2,7 bar)	fixed-bed	Pressure swing	Vaporciyan and Kadlec 1989
$C_6H_{12}(g) \leftrightarrow C_6H_6(g) + 3\,H_2$ (T = 150-190 °C; p = 1 bar)	fixed-bed	Purge	Goto et al. 1993
$CO(g) + H_2O\ (g) \leftrightarrow \mathbf{\underline{CO_2}} + H_2(g)$ (T = 160 °C; p = 1-2,7 bar)	fixed-bed	Temperature swing	Harrison and Han 1994
$CO_2(g) + H_2(g) \leftrightarrow CO(g) + \mathbf{\underline{H_2O}}$ (T = 250 °C; p = 4,8 bar)	fixed-bed	Pressure swing	Carvill et al. 1996
$CH_4(g) + 2\,H_2O\ (g) \leftrightarrow \mathbf{\underline{CO_2}} + 4\,H_2(g)$ (T = 650 °C; p = 20 bar)	fluidised bed	—	Kurdyumov et al. 1996
$\mathbf{\underline{VOC}}^{1}\ (ads) + O_2(g) \rightarrow$ $m\,CO_2(g) + n\,H_2O(g)$ (T = 150-500 °C; p = 1 bar)	fixed-bed	Reaction	Zagoruiko et al 1996, Salden and Eigenberger 2001
$CH_4(g) + 2\,H_2O\ (g) \leftrightarrow \mathbf{\underline{CO_2}} + 4\,H_2(g)$ (T = 450-750 °C; p = 15 bar)	fixed-bed	Temperature swing	Balasubramanian et al. 1999
$CH_4(g) + 2\,H_2O\ (g) \leftrightarrow \mathbf{\underline{CO_2}} + 4\,H_2(g)$ (T = 428-468 °C; p = 3,1-7,2 bar)	fixed-bed	Pressure swing	Hufton et al.1999, Ding and Alpay 2000b
$NH_3(g) + CO\ (g) \leftrightarrow \mathbf{\underline{HCN}} + H_2O(g)$ (T = 350-450 °C; p = 1-10 bar)	fixed-bed	Temperature swing	Dittrich 2002, Elsner et al. 2002
$2\,H_2S(g) + SO_2\ (g) \leftrightarrow 2\,\mathbf{\underline{H_2O}} + \frac{3}{8}\,S_8(g)$ (T = 250-350 °C; p = 1 bar)	fixed-bed	Purge & Temperature swing	Elsner et al. 2002, Elsner et al. 2003, Elsner 2004
$CO(g) + H_2O\ (g) \leftrightarrow CO_2 + H_2(g)$ (T = 400 °C; p = 1-3 bar)	fixed-bed	Purge & Temperature swing	Müller and Agar 2000

---

[1] **V**olatile **O**rganic **C**ompounds

While the enhancement of equilibrium-limited reactions has been investigated experimentally in the past, with the emphasis being on CO_2 removal for hydrogen production applications, publications concerning the experimental validation of the adsorptive manipulation of complex reaction schemes are still few and far between.

In the work of Tonkovich dealing with the oxidative coupling of methane in a gas-phase simulated moving bed (SMB) reactor separation and reaction sections are operated at different temperatures, the desired intermediate is removed adsorptively and thus the selectivity enhanced (Tonkovich et al.1993). Two other articles discuss the selectivity enhancement of consecutive reactions, with the reactions occurring during the hydrogenation of acetylenes exemplifying such reactions (Kodde and Bliek 1997, Kodde et al. 2000).

4.1.1 Gas-phase adsorptive reactors – operation and regeneration strategies

The limited capacity of the adsorbent requires periodic regeneration to re-establish the adsorptive capability of the adsorptive reactor system. Various reactor configurations and regeneration concepts can be adopted, where, with respect to the reactor configuration, a distinction between fixed-bed operations, fluidised bed operation and their hybrids can be drawn (Figure 4.1).

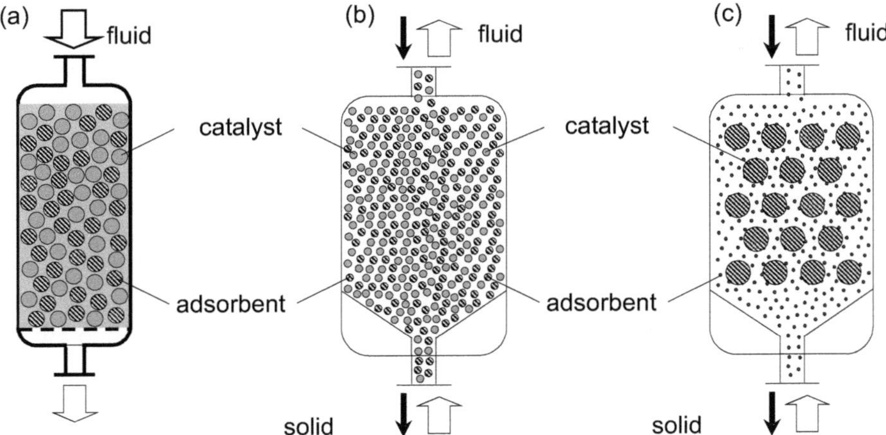

Fig. 4.1: Different reactor configurations: a) fixed-bed; b) moving bed or fluidised bed; c) trickle bed reactor

One problem in the fluidised bed configurations is the difficult solid handling, so that fixed-bed arrangements combined with regeneration cycles are the usual option of choice. In fixed-bed operation, the adsorptive reactor is operated according to a reaction - regeneration schedule, comprising cycles for reaction and regeneration and intermediate switching procedures to establish suitable pressure and temperature conditions for the main phases. In this way, the adsorbent is regenerated *in situ* under unsteady-state conditions.

Other configurations to accomplish regeneration in a quasi-steady-state manner are the rotating bed adsorber and the simulated moving bed adsorber, which are well-known from liquid chromatography. However, these chromatographic concepts have not established themselves for gas phase applications due to the large purge gas streams and the typically much lower concentration values of the species treated in gas phase compared to those treated in liquid phase. The regeneration strategies can be subdivided into pressure swing, concentration swing, temperature swing, reactive regeneration and displacement regeneration, or combinations thereof.

Pressure swing processes are based on the reduction in the adsorptive capacity at lower adsorptive partial pressure. Exploiting the compressibility properties of gas phase systems, this can be achieved by the decrease of the total pressure by decompression or evacuation, leading to rapid regeneration of the adsorbent. These processes can draw on the considerable experience available with separation technology in this area and especially lend themselves to situations in which the reaction is carried out under pressure. A critical aspect in reactor design is the pressure drop, particularly for rapid pressure swing cycles.

Elutive processes are similar to the pressure swing concept, since the desorption is accomplished by a desorptive partial pressure reduction, the difference being that the total pressure kept constant. Due to the high pressure level maintained in the elutive process, a larger amount of sweep fluid is needed compared to the pressure swing process to perform the regeneration, a major disadvantage of the elutive concept. Further processing steps are often needed to purify the large amount of the adsorptive-laden sweep fluid arising.

Temperature swing is an option for processes exhibiting much longer cycle times, due to the thermal inertia of the reactor, catalyst and adsorbent material. Another aspect is the need for a heat source within the overall process to accomplish the temperature increase of the adsorptive reactor, limiting the economic potential of the thermal swing concept for industrial purposes. Enhancement of heat transfer into the reactor can be achieved by fluidised bed operation or other innovative concepts, such as microwave (Bathen 2003) or electric heating (Judkins and Burchell 1999) of the bed.

The reactive regeneration concept cannot be generalised and its suitability has to be examined on a case-by-case basis. In some special cases, a reactive regeneration proves to be particularly appropriate, for instance for the total denitrification of flue gases or the unsteady-state Deacon process. The need for stable adsorptives to avoid selectivity losses and for reasonable adsorbent capacities tends to limit the possible applications of the reactive regeneration concept in practice.

Displacement desorption, in which the adsorbate is driven off by a more strongly adsorbing species, entails an additional regeneration cycle to desorb the expulsion agent and can thus only be justified for the recovery of a high-value adsorbed intermediates.

4.1.2 Comparison with related reactor concepts

Comparing the gas phase adsorptive reactor with other multifunctional concepts, similarities in both the overall behavior and the mathematical models become apparent, as illustrated in Figure 4.2.

The most similar type of reactor, also adsorptive in nature, is the liquid phase chromatographic reactor, in which the concentration profiles are similarly manipulated by adsorption. The differences between these two systems are due to the different magnitudes of the properties appearing in the balance equations which describe the reactor mathematically. Thus there are significant differences in the adsorptive capacity of the components with respect to their concentrations in the fluid phase and also in the selectivity of the adsorption process. The relevance of heat effects also differs, since the heat capacity of liquids is much higher than that of gases, often demanding the simultaneous solution of a heat balance for the gas phase reactor, while liquid systems can be regarded as isothermal. Additional possibilities for regeneration are available in gaseous systems as a consequence of this low heat capacity and the compressibility of the gases. Liquid phase chromatographic reactors are not treated further here, since they are the subject of another chapter in this book.

Exploiting regenerative storage effects in a similar manner to the adsorptive reactor, reverse flow reactors widely used for the catalytic combustion of lean gases, take up the heat that is released during the combustion process within the central section of a fixed-bed catalytic reactor. Due to the non-linearity of the temperature dependency of the reaction kinetics and the "autocatalytic" phenomena arising in exothermic systems, hysteresis effects are more common in these reactors than in adsorptive ones, where the emphasis is on regenerative mass storage.

As a third related multifunctional reactor concept, membrane reactors should be mentioned. By the selective diffusive removal of particular species from a reaction mixture, the course of the reactions occurring can be manipulated. The membrane can be regarded as an "adsorbent" with infinite capacity, allowing these reactors to operate continuously.

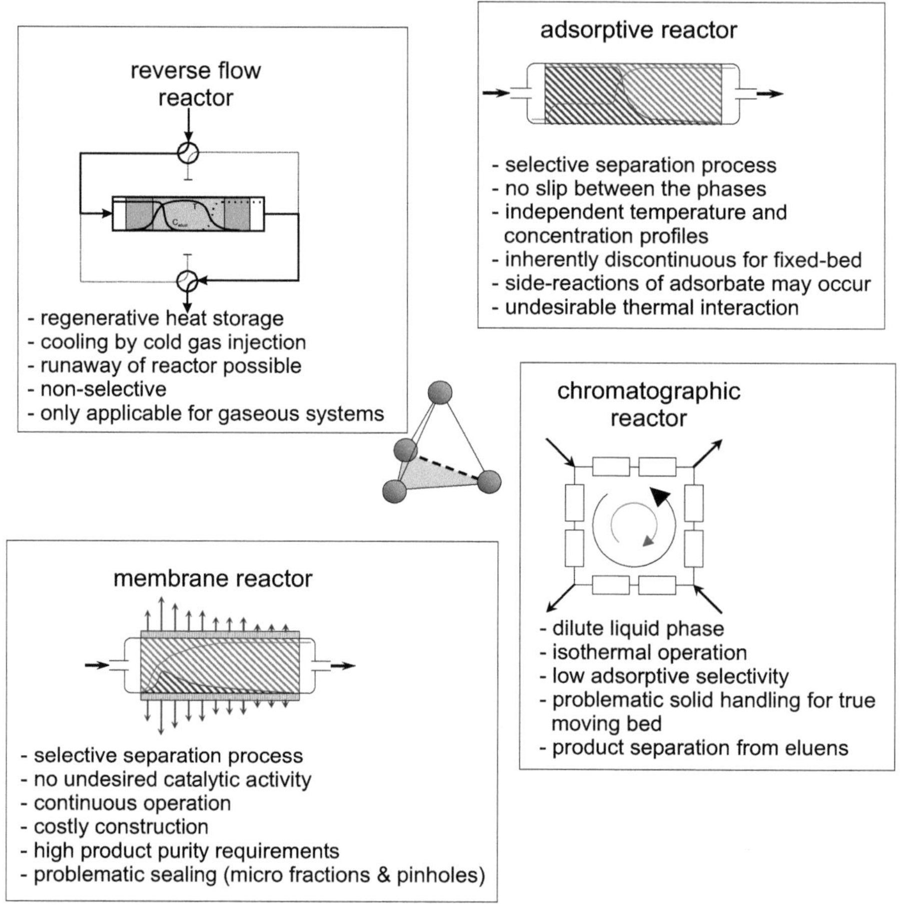

Fig. 4.2: Comparison of adsorptive gas phase reactors with related multifunctional reactor concepts

4.2 Modeling of gas-phase adsorptive reactors

Besides experiments, numerical simulations of adsorptive reactors are helpful for the preliminary assessment of their behaviour to ascertain suitable operating and design parameters. In the following section, the model equations usually employed for adsorptive fixed-bed reactors are presented and features of their implementation discussed.

4.2.1 Model equations

For the balancing of an individual species i within the fluid bulk phase of the adsorptive reactor, the following differential equation can be derived for a one-dimensional treatment of the reactor system shown in Figure 4.3.

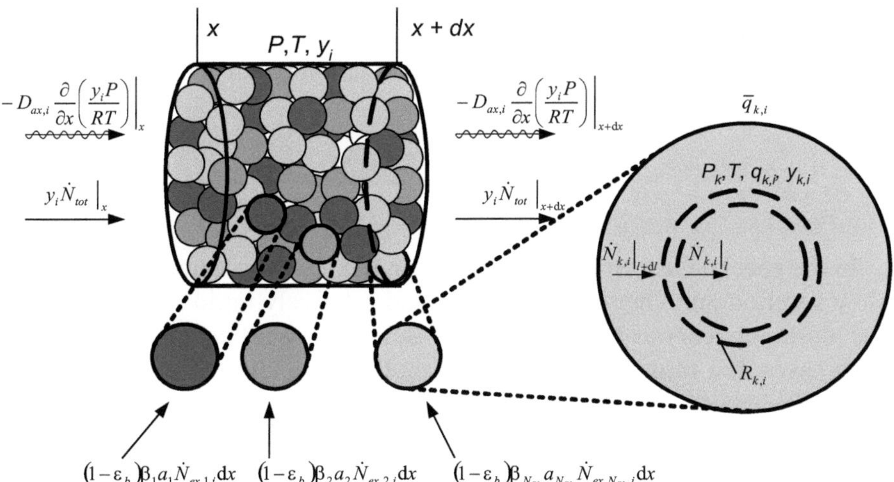

Fig. 4.3: Mass balances in adsorptive reactors

The main transport mechanism is convective flow with slight diffusive backmixing effects being taken into account by a dispersion term.

$$\varepsilon_b \frac{\partial}{\partial t}\left(\frac{y_i P}{RT}\right) = -\frac{\partial}{\partial x}\left(\dot{N}_{tot} y_i\right) + D_{ax,i} \frac{P}{RT} \frac{\partial^2 y_i}{\partial x^2}$$
$$+ \left(1-\varepsilon_b\right)\sum_{k=1}^{N_{Ph}} \beta_k a_k \dot{N}_{ex,k,i}$$

$$(4.1)$$

with the boundary conditions, depending on the flow direction, given by:

for $x = 0$:

$$\dot{N}_{tot} y_i = \dot{N}_{tot} y_{in,i} - D_{ax,i} \frac{\partial y_i}{\partial x} \quad \wedge \quad \dot{N}_{tot} > 0$$

$$\frac{\partial^2 y_i}{\partial x^2} = 0 \quad \wedge \quad \dot{N}_{tot} \leq 0$$

for $x = L_c$:

$$\frac{\partial^2 y_i}{\partial x^2} = 0 \quad \wedge \quad \dot{N}_{tot} > 0$$

$$\dot{N}_{tot} y_i = \dot{N}_{tot} y_{in,i} - D_{ax,i} \frac{\partial y_i}{\partial x} \quad \wedge \quad \dot{N}_{tot} \leq 0$$

For the total molar flux, the balance equations for each species i are summed, a step which is essential in systems exhibiting a considerable change of the molar flux due to chemical reaction or sorption effects.

$$\varepsilon_b \frac{\partial}{\partial t}\left(\frac{P}{RT}\right) = -\frac{\partial \dot{N}_{tot}}{\partial x} + (1-\varepsilon)\sum_{i=1}^{N_c}\sum_{p=1}^{N_{Ph}}\beta_p a_k \dot{N}_{ex,k,i} \tag{4.2}$$

In the special case when the particles in the fixed-bed are very small, for slow sorption and chemical processes and relatively rapid mass transfer, a pseudo-homogeneous treatment of the solid phase k becomes possible. In such cases, the molar fractions in the bulk phase and the solid phase are identical.

$$\left(\frac{\partial q_{k,i}}{\partial t} + \varepsilon_{p,k}\frac{\partial}{\partial t}\left(\frac{y_i P}{RT}\right)\right) = -a_k \dot{N}_{ex,k,i} + R_{k,i} \tag{4.3}$$

By the simple addition of this equation to the balance for the bulk phase and by eliminating the exchanged molar flux $\dot{N}_{ex,k,i}$, a total balance for component i, which comprises both the bulk phase and the solid phase, is obtained. In practice, the assumption of pseudo-homogeneity is often not justified, so that additional balance equations for the solid phases have to be solved. Balancing the component i in a differential volume within the solid phase yields:

$$\varepsilon_{p,k} \frac{\partial}{\partial t}\left(\frac{y_{k,i}P_k}{RT}\right) + \frac{\partial q_{k,i}}{\partial t} = -\frac{1}{l^m}\frac{\partial}{\partial l}\left(l^m \dot{N}_{k,i}\right) + R_{k,i} \qquad (4.4)$$

where the parameter m depends on the geometry of the particle ($m = 0$ for a slab, $m = 1$ for a cylinder and $m = 2$ for a sphere).

A very general approach to determining the molar flux is the dusty gas model, which is based on the application of the Maxwell-Stefan-Equation to the gas mixture and the solid matrix, with the solid matrix being treated as an immobile pseudo-species (Jackson 1977, Hite and Jackson 1977, Mason and Malinauskas 1983).

$$\frac{N_{k,i}}{D^e_{Kn,i}} + \sum_{s \neq i}\frac{y_s \dot{N}_{k,i} - y_i \dot{N}_{k,s}}{D^e_{is}} = -\frac{P}{RT}\frac{\partial y_i}{\partial l} - \frac{y_i}{RT}\left(1 + \frac{B_0 P}{\mu D^e_{k,i}}\right)\frac{\partial P}{\partial l} \qquad (4.5)$$

In this way, both the molecular diffusion in multicomponent mixtures, the interaction between the gas molecules and the solid pore walls (Knudsen diffusion) and viscous flow effects are taken into consideration. Additionally, surface transport may also make a considerable contribution to the overall transport within the solid particle (Krishna and Wesselingh 1997).

However, in practice efforts are usually made to develop simplified expressions instead of the generalised approach presented above, since in most cases not all of the effects covered are important enough to justify their inclusion. Additionally, it is desirable to simplify the sophisticated model as much as possible to facilitate the implementation of the equations and to reduce calculation times during the numerical treatment of the problem.

A commonly used expression to describe the mass transfer limited adsorption process within the particle is the Linear Driving Force approach (LDF), which is based on the assumption of a uniform concentration level or parabolic profile within the particle and a linear dependency of the rate of mass transfer on the driving force, which is defined as the difference of the actual loading of the particle and its potential equilibrium value (Glueckauf and Coates 1947):

$$\frac{\partial q_{p,i}}{\partial t} = k_{m,i,eff}\left(q^*_{p,i} - q_{p,i}\right) \qquad (4.6)$$

with $k_{m,i,eff} = 15\dfrac{D_{k,i}^{e}}{r_{p}^{2}}$ for spherical particles

For very rapid pressure swing processes or high mass transfer resistances, Alpay and Scott (1992) have shown that the effective mass transfer coefficient is given by:

$$k_{m,i,eff} = \frac{5.14}{\sqrt{\dfrac{r_{p}^{2}\,t_{cyc}}{D_{k,i}^{e}}}} \tag{4.7}$$

By using this simplified approach, which is also used here, it is possible to lump the concentration level within the particle, thus avoiding the need to solve large distributed model equation systems for the solid phase.

Due to the low heat capacities of the fluid in gas phase systems, heat effects caused by the release of the reaction and adsorption enthalpies during the processes often have to be considered. In the temperature regime pertinent for adsorptive reactor operation, the bulk and solid phases can be treated in a pseudohomogeneous way. The energy balance can be described as follows under non-isobaric conditions (Xiu et al. 2003)

$$\varepsilon_{t}C_{pm}\frac{P}{RT}\frac{\partial T}{\partial t} - \varepsilon_{t}\frac{\partial P}{\partial t} + \left(1-\varepsilon_{b}\right)\sum_{k=1}^{N_{p}}\beta_{k}\rho_{k}c_{p,k}\frac{\partial T}{\partial t}$$

$$= -C_{pm}\dot{N}_{tot}\frac{\partial T}{\partial x} + \lambda_{ax}\frac{\partial^{2}T}{\partial x^{2}} + \left(1-\varepsilon_{b}\right)\sum_{k=1}^{N_{Ph}}\beta_{k}\sum_{i=1}^{N_{c}}\frac{\partial q_{k,i}}{\partial t}\left(-\Delta H_{ad,k,i}\right) \tag{4.8}$$

$$+ \left(1-\varepsilon_{b}\right)\sum_{i=1}^{N_{Ph}}\beta_{k}\sum_{i=1}^{N_{c}}r_{k,j}\left(-\Delta H_{j}\right) + \frac{4U}{d}\left(T_{a}-T\right)$$

An equation which is usually of minor importance, but the relevance of which increases at higher flow velocities is the impulse balance, describing the axial pressure profile within the reactor, which affects the concentrations of the species through the compressibility of the gaseous phase (Ergun 1952, Macdonald et al. 1979, Sereno and Rodrigues 1993):

$$\frac{\partial}{\partial t}\left(\rho_g u\right) = -\frac{\partial P}{\partial x} - 180\frac{\mu_g\left[f_s(1-\varepsilon_b)\right]^2}{d_p^2\varepsilon_b^3}u$$
$$-1.75\frac{f_s(1-\varepsilon_b)\hat{\rho}_g}{d_p\varepsilon_b^3}u\cdot|u| \tag{4.9}$$

In this way a system of coupled non-linear PDE is developed, which describes the behavior of the adsorptive reactor.

4.2.2 Model implementation and numerical features

Due to their complexity, the partial differential equations that arise from the mathematical formulation of the reactive and adsorptive fixed bed processes generally have to be solved numerically. To convert the partial differential equations (PDE) to ordinary differential equations (ODE), spatial discretization methods have to be used in order to solve the equations describing the dynamic system using standard ODE-solving algorithms.

To assess the characteristics of the regenerative systems under consideration, knowledge of the cyclic steady-state is often desirable. To determine asymptotic cyclic steady-state solutions, either shooting methods or the full discretization of the PDE in space and time can be applied (Croft and Levan 1994, Salinger and Eigenberger 1996). Using the shooting methods, the system is solved dynamically, replacing the solution at the beginning of the cycle by the solution of the preceding cycle or a modification thereof, until the solution remains unchanged from cycle to cycle, indicating the attainment of the cyclic steady-state.

The second technique uses discretization in both space and time, converting the PDE into a system of algebraic equations, which can be solved by common methods such as the Newton algorithm. Using this full discretization as opposed to the shooting method, large systems of equations arise, where both calculation time and memory requirements are strongly influenced by the number of discretization intervals. As a consequence, effective discretization methods are needed, which provide high precision even when only a small number of nodes for systems with steep mobile fronts are used. This can be achieved by the local refining of the grid in regions where gradients are steep (Nowak et al. 1996) or by the use of non-linear discretisation methods, in which linear methods are combined, using a weighting factor depending on the shape of the solution profiles. If certain criteria for this weighting factor are fulfilled, the numerical oscillations arising with linear methods of a higher approximation order than one can be excluded, achieving higher accuracies than are feasible using the simple upwind-scheme (Harten 1983, Harten 1984).

The convergence behavior of dynamic simulations is thus not crucial, since the temporal progression of the solution is calculated according to a predictor-corrector-approach starting from a known initial condition. Achieving convergence for highly non-linear algebraic systems is often difficult. This complexity can be managed by a parameterization of the system and the step-by-step calculation of the solution depending on the parameter value starting from a known or trivial solution. In this way, convergence can be attained by the proximity of the predicted and the converged solutions (Seydel 1988). Additionally, this approach provides the opportunity to analyze the system with respect to multiplicity of the solutions and bifurcation phenomena.

4.3 Design principles of adsorptive reactors

Both the reaction kinetics as well as the adsorbent capacity is strongly dependent on the operational temperature.

While higher temperatures enhance the kinetics, at lower temperatures the adsorbent capacity is greater. This contradictory temperature dependency of the reactive and adsorptive processes leads to the existence of an operating window in which an adsorptive enhancement of the reaction system is possible (Schembecker and Tlatlik 2003).

Generally, the uptake of the adsorbent is raised by increasing the pressure due to the pressure-dependency of the adsorption isotherm until the maximum capacity of the adsorbent is attained. To a certain extent, the reaction kinetics are enhanced through the increased density of the reaction medium as well, provided that the kinetic inhibitions due to adsorption on the catalyst are not too pronounced. The maximum pressure level is thus governed by economic considerations as well as the sorptive enrichment relative to the gas phase. As a consequence, a reduction of the pressure is favorable to facilitate the desorption process.

An important design parameter is the ratio of the catalyst and adsorbent within the adsorbent bed. A compromise between catalytic activity and adsorbent capacity has to be found. Further improvements in the reactor performance can be achieved by modifying the ratio of the two solid phases according to the progress of the reaction and exploiting the potential available for concentration profile manipulation that are present in both time and space.

4.4 Conversion enhancement of equilibrium-limited reactions

4.4.1 Claus reaction

The catalytic reaction of H_2S with SO_2 in a stoichiometric ratio of 2:1 as occurs in the Claus process is an example of an equilibrium-limited reaction system, albeit one exhibiting very high conversions.
The overall reaction can be described by the equation

$$2\ H_2S + SO_2 \Leftrightarrow 3/8\ S_8 + 2H_2O \qquad\qquad \Delta H_R^\circ = -108\ \text{kJ/mol}$$

The difference between this reaction system and others reported in the literature is that conversions of about 93% are achievable for isothermal operation of gas containing 10 mol% H_2S, without additional measures. However, this is still inadequate to satisfy stringent environmental legislation. The objective is thus to enhance the conversion to values above 99.5%.

Considering the reaction equation for the Claus reaction, it becomes apparent that the equilibrium can be displaced by interstage or *in situ* sulphur removal, as used in conventional process variants or by the separation of the water formed. In this study the adsorptive *in situ* water removal was the subject of the investigation (Figure 4.4).

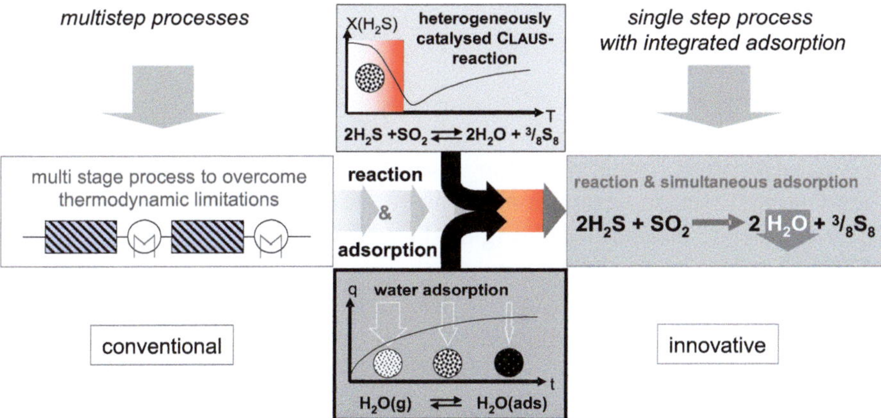

Fig. 4.4: Adsorptive enhancement of the Claus reaction by water removal

Zeolite 3A was chosen as an adsorbent for the removal of the water formed during the reaction. This adsorbent is characterised by its narrow micro pores having an effective diameter of about 0.4 nm, which makes it a suitable adsorbent for water (critical molecular diameter 0.2 nm) whilst preventing the adsorption of other components of the Claus reaction system due to their greater molecular dimensions. In this way, a high selectivity for the water adsorption is feasible. As the catalyst for the Claus reaction, a commercial γ-Al_2O_3-catalyst, doped with small amounts of Na_2O to enhance its basicity, was selected.

The reaction macro kinetics for the catalytic conversion of H_2S with SO_2 on the catalyst were determined from experiments in a differential recycle reactor of the Berty type. The experiments were conducted under steady-state conditions in the temperature range of 200 to 330°C and under atmospheric pressure with catalyst pellets of 5 mm diameter. Various models, based on information available in the literature (Dalla Lana et al. 1972, Pineda and Palacios 1995), e.g. simple power law and Hougen-Watson-approaches, were tested with respect to their ability to fit the experimental data using statistical techniques. The data could be best represented by the following empirical power law equation:

$$\eta_R = k_{1,0}e^{-\frac{E_{A1}}{RT}}P_{H_2S}^{0.95}P_{SO_2}^{0.22} - k_{2,0}e^{-\frac{E_{A2}}{RT}}P_{H_2O}^{0.99} \tag{4.10}$$

with the following parameter values:

$k_{1,0} = 5292$ mmol / (s kg mbar$^{1.17}$)
$k_{2,0} = 1.252 \cdot 10^6$ mmol / (s kg mbar$^{0.99}$)
$E_{A1} = 49.980$ kJ / mol
$E_{A2} = 86601$ kJ / mol

The adsorption isotherm for the adsorption of water on zeolite 3A under the Claus reaction conditions was determined, with an empirical Freundlich-type isotherm yielding the best fit:

$$q_{H_2O} = a_0 \exp\left(-\frac{\Delta H_{ads}}{RT}\right) \cdot \left(\frac{P_{H_2O}}{RT}\right)^{0.75} \tag{4.11}$$

with the parameter values:

$$a_0 = 0.65 \ (mol/m^3)^{0.25}$$
$$\Delta H_{ads} = 30.5 \ kJ/mol$$

The mass transfer resistances arising could be described by an LDF-appoach, the transport parameter for the mass transfer being estimated from breakthrough experiments.

Using the parameter values given in a numerical model it could be shown that there is an optimum composition of the fixed bed with respect to the utilization of the adsorbent, corresponding to the cycle time, for which a conversion of 99.5% can be maintained. Under isothermal conditions and temperatures of around 250°C, the optimal volumetric adsorbent fraction was found to be 43% for a given space time of 100 $Nm^3/(m^3 \ h)$ for the fixed-bed.

Based on the results of such orientation simulations, an adsorptive catalytic fixed-bed reactor was constructed and experiments carried out with integrated catalytic and adsorptive functionalities. The reactor was filled with a spatially homogeneous mixture of the catalyst and the adsorbent in the volumetric fraction mentioned above (uniform particle diameter of 5 mm). The reactive cycle was studied by heating up the fixed-bed to the initial temperature (230-280°C) desired with preheated N_2 sweep gas, before switching to a gas mixture consisting of a stoichiometric feed (5-10% H_2S, SO_2 in the ratio of $y(H_2S)/y(SO_2) = 2:1$). The catalytic fixed-bed had a length of 60 cm or 83 cm, while the total volumetric flow of the feed was in the range of 15-25 l_N/min. To compare the performance of the adsorptive reactor with the conventional fixed-bed reactor, the adsorbent particles were replaced by inert particles in additional control experiments.

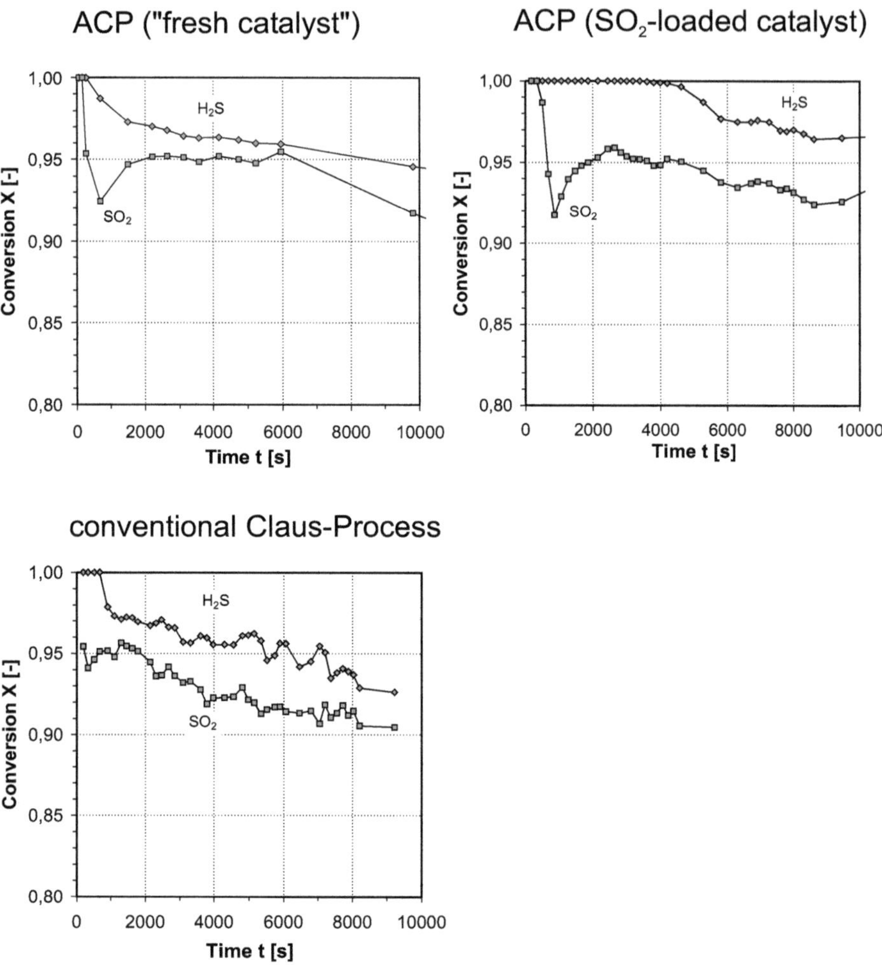

Fig. 4.5: H_2S and SO_2 conversions during adsorptive operation and in conventional Claus process

The diagrams presented illustrate that at the start of the experiments the apparent conversions of the reactants is higher than the equilibrium value for the Claus reaction under the prevailing conditions, which is approximately 90%. It is thus apparent that the thermodynamically limited Claus reaction can be enhanced to a certain extent by adsorptive operation. The diagram depicting the behavior of the conventional reactor, which only contains catalyst and inert material, is remarkable in that the time to achieve a steady-state is much higher than might be expected from the residence times of a few seconds. This observation leads to the conclusion that there are significant storage effects for the reactants on the catalyst, resulting from adsorption or chemical reaction, which also has to be taken into account.

Additionally it can be seen from the experimental results that there is a divergence between the conversions of H_2S and SO_2. This is an unexpected finding, since the consumption of the two educts should be dictated by the ratio of their stoichiometric coefficients, leading to identical conversions for an inlet ratio of $y(H_2S)/y(SO_2) = 2:1$. An explanation for the behavior observed is that different sorptive properties for each of the reactants are occurring on the catalyst, which leads to a distortion of the original stoichiometric ratio and thus to a distinct development for the conversion of each species.

It is known that H_2S weakly adsorbs on the catalyst (Deo et al. 1971, Saur et al. 1981), but not according to a strictly physisorptive mechanism (Datta and Cavell 1985a). The adsorptive can interact with the adsorbent at Al^+-sites and dissociate, abstracting water, if vicinal hydroxyl groups and O-atoms are present. The adsorptive parameters for the interaction of H_2S with the alumina-catalyst were determined by fitting simulated breakthrough and temperature profiles to experimental data, from experiments using the bench-scale reactor as an adsorption column. The isotherm parameters could be best fitted by a Langmuir isotherm, with the adsorption kinetics being described by an LDF-model (Elsner 2004).

On the other hand, the interaction between SO_2 and the catalyst is dominated by a chemisorptive process. The SO_2 sorption on the solid is due to the amphoteric character of the support, and a slow and largely irreversible process. Aluminium sulfite $Al_2(SO_3)_3$, is formed as a product of this gas-solid reaction (Karge and Dalla Lana 1985). The small delay in the breakthrough profile indicates a slight physisorptive contribution to the overall sorption process, in addition to the dominant chemisorption.

All in all, the rate of SO_2-sorption on the γ-alumina catalyst and also the overall SO_2-uptake was increased at higher temperatures, as a consequence of the higher rates of the gas-solid reaction and the transport process within the catalyst particles. The activation energy of the chemical reaction was estimated to be around 62 kJ/mol in the temperature range of 230-300°C. For the parameter fit, a shrinking core model was used for the mathematical description of the gas-solid reaction system.

An additional aspect of the SO_2-sorption on the catalyst is its influence on the Claus reaction kinetics (Datta and Carvell 1985b). It has been reported in the literature that the sorption of SO_2 on the catalyst causes an activation of the Claus catalysis. Following the addition of H_2S to a SO_2-laden catalyst, the physisorbed species reacts first, and at temperatures above 200°C the chemisorbed species, which are bound in a sulphitic form, also begin to react with the H_2S. As a result, the catalytic Claus reaction exhibits its greatest effectiveness at temperatures of about 250°C.

Regarding the results from the additional experiments, the shape of the conversion profiles shown in Figure 4.5 describing the performance of the adsorptive catalytic reactor and the conventional Claus process under unsteady state conditions can be explained as follows. The apparent SO_2-conversions are higher in all the three cases, which is due to the weak but rapid H_2S sorption on the catalyst surface, while the stronger chemisorption of SO_2 is slower, leading in general to an excess of SO_2 in the gas phase and thus to an incomplete conversion for this reactant as opposed to the H_2S.

The main difference in the reactor performance observed in the cases with a fresh and an aged catalyst is the much longer delay of H_2S breakthrough for the aged catalyst. A possible explanation for this behaviour is the higher catalyst activity, caused by the chemisorbed sulfitic species and the presence of excess SO_2 in the reactor, which is a consequence of the chemisorbed SO_2 from the previous experimental runs. In this manner, all the H_2S is consumed at the beginning of the experiment, while the excess SO_2 leaves the reactor outlet. As an additional effect, a minimum in the SO_2-conversion development is visible in all the cases. This effect is presumably due to the lack of the reaction partner H_2S, which is removed by physisorption and chemical reaction with the chemisorbed sulfite.

To recapitulate the insights obtained from experimental and modeling studies, the conclusion is that the conversion in an adsorptive Claus reactor can be enhanced beyond the thermodynamic restrictions encountered in the conventional single stage steady-state Claus process. However, the two components H_2S and SO_2 do not react according to their reaction stoichiometry, as a result of the qualitatively and quantitatively different sorptive interactions with the catalyst. The consequence of this finding is the need to operate the catalyst under steady-state conditions, while for dynamic operation a fixed-bed configuration is required for the regeneration of the adsorptive functionality. This implies that one should partially deintegrate the catalytic and the adsorptive functionalities (Figure 4.6).

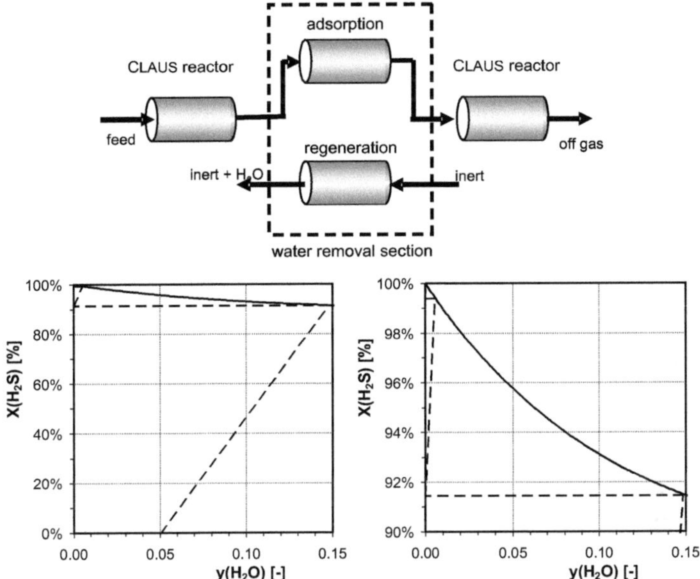

Fig. 4.6: Deintegrated concept for the adsorptive enhancement of the Claus reaction

In this deintegrated arrangement there are alternating reactive and adsorptive sections. The reactors can be operated at steady-state, while the adsorber sections are run under cyclic conditions. Since the adsorbent is inert with respect to the adsorption of the reactants H_2S and SO_2, the feed of the reaction sections can be maintained at the constant stoichiometric ratio of 2:1 required for conversions of 99.5% or more. The advantage over traditional industrial processes with interstage sulfur removal by condensation is a greater thermal efficiency, since the need for cooling and reheating of the process gases is obviated. By the application of reduced pressures for the regeneration of the adsorbent, the desulfurized off-gas from the process can be employed as a sweep gas.

4.4.2 HCN-synthesis from CO and NH_3

The BASF-formamide-process is a commercial synthesis of HCN based on carbon monoxide as the the C-source. In this process, HCN is synthesized in three steps, *via* the intermediates methyl formate and formamide (Figure 4.7)

$$CO + CH_3OH \Leftrightarrow HCOOCH_3 \qquad \Delta H_R^\circ = -21.9 \text{ kJ/mol}$$
$$HCOOCH_3 + NH_3 \Leftrightarrow HCONH_2 + CH_3OH \qquad \Delta H_R^\circ = -75.4 \text{ kJ/mol}$$
$$HCONH_2 \Leftrightarrow HCN + H_2O \qquad \Delta H_R^\circ = +79.5 \text{ kJ/mol}$$

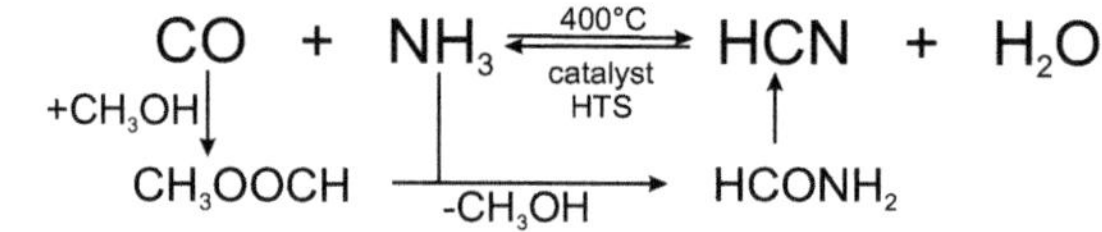

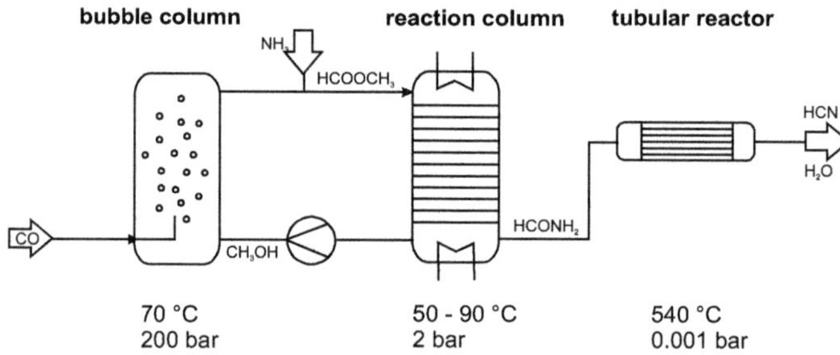

Fig. 4.7: Direct and indirect route to HCN from CO and NH_3

To attain favorable equilibria in the individual steps, suitable pressure and temperature conditions are chosen. Overall, the formation of HCN can be described by the equation

$$CO + NH_3 \Leftrightarrow HCN + H_2O$$

which is thermodynamically very unfavorable exhibiting equilibrium conversions of less than 3%. Considering this single reaction equation more closely, it becomes obvious that an enhancement of this reaction cannot be achieved by pressure variation, and the temperature window dictated by the demands of the catalyst together with the low heat of reaction leave little room for improvement for the equilibirum position either. Additionally, the potential of equilibrium displacement by supplying an excess of one of the educts is very limited.

Thus an enhancement in a single step process under the prevailing reaction conditions can only be achieved by the integration of a separation function into the reaction system, to deplete one of the reaction products and so enhance the conversion. The higher temperatures required for the catalysis that are known for this reaction (>350°C) preclude the utilization of a water adsorbent or HCN-adsorbent, since these have too low a capacity under such conditions. Since the catalysts used for the production of HCN are also known to be active for the water-gas shift reaction

$$CO + H_2O \Leftrightarrow H_2 + CO_2$$

the water formed is consumed by this additional reaction, leading to the overall equation:

$$2\ CO + NH_3 \Leftrightarrow HCN + H_2 + CO_2$$

This exposes an opportunity for an equilibrium enhancement by a CO_2-removal either using a gas-solid reaction or an adsorption process.

Since the direct HCN-synthesis from ammonia and carbon monoxide is not a process employed industrially, no commercial catalysts are available. A screening of various catalysts showed that Fe_3O_4/Cr_2O_3 - systems used for the high temperature shift reaction have sufficient activity for HCN formation. The data obtained in kinetic experiments were fitted to a reversible power law expression:

$$r_m = k_r^{HCN} \sqrt{P_{NH_3}} \cdot \left(1 - \frac{P_{HCN} P_{CO_2} P_{H_2}}{K_p^{HCN} \cdot P_{CO}^2 P_{NH_3}} \right) \tag{4.12}$$

Taking gas-solid-reactions as the first option for CO_2-removal from the reaction medium, various metal-oxides can be identified as potential candidates. A suitable solid material for the CO_2-scavenging has to be functional at reaction temperatures of around 400°C, which imposes an important limitation on appropriate solid materials. The dissociation pressure of the carbonates formed during the gas-solid-reaction should be very low in the temperature window under consideration, so that oxides like MgO, whose carbonate $MgCO_3$ is not stable at temperatures above 300°C, have to be excluded. Another aspect which determines the choice of the solid reactant is the ability to decompose the carbonate for regeneration with a tolerably low temperature rise. From his point of view, substances such as CaO, which are known to form stable carbonates that only decompose at temperatures above 700°C might be possible candidates if only the CO_2-removal is considered in isolation, but if the regeneration is taken into account, the high energy consumption for the large temperature rise necessary will render the adsorptive reactor concept uneconomic. A side-reaction, the formation of Calcium cyanamide $CaCN_2$, (Jensen et al. 1997) which consumes the HCN produced and lowers the capacity of the solid, also militates against the use of CaO as the scavenging material for CO_2-depletion in this adsorptive reactor concept for the direct HCN-synthesis.

A material which fulfills the requirements of stability of the product formed and exhibits suitable regeneration characteristics upon only a moderate temperature rise is lithium zirconate $(Li_2O)_nZrO_2$ (with n = 1, 2 or 4) which reacts with CO_2 according to the equation:

$$n\ CO_2 + (Li_2O)_nZrO_2 \Rightarrow n\ Li_2CO_3 + ZrO_2$$

with CO_2 being stored as a carbonate within the solid matrix. The capacity of the solid and the kinetics of CO_2-uptake were determined experimentally for different forms of the lithium zirconate and at various CO_2-concentrations. The measurements showed that for n = 4, the capacity is up to 10 wt-% in a concentrated CO_2-atmosphere, which corresponds to the capacity of typical technical adsorbents. The experiments demonstrated that the zirconates with a lower Li_2/Zr-ratio (n = 1) have a significantly lower capacity (up to 5 wt-%) compared to those with n = 4.

The kinetics of CO_2-uptake were initially faster for the zirconate with a high lithium content, but decreased below the levels for the zirconate with the lower Li content as the time proceeded. A variation of the CO_2-concentration in the gas phase indicated that the initial sorption-rate is more or less independent of the CO_2-concentration. On the other hand, a clear dependency becomes visible after an initial time period.

By immobilizing the lithium zirconate on an inert SiC-carrier, the sorption kinetics could be slightly enhanced compared to the kinetics on the massive particles. These results point to a strong influence of mass transfer resistances on the overall sorption kinetics. However, the time needed to achieve significant amounts of CO_2-uptake, let alone the equilibrium values, were of the order of minutes to hours, which can be explained by the slow phase separation process of lithium carbonate and zirconium oxide during the gas-solid-reaction.

Since the sorption rates were unreasonably slow in comparison to the kinetics of the HCN-synthesis reaction, adsorption was investigated as an alternative option to the gas-solid-reaction for CO_2-uptake. Suitable adsorbents for high temperature adsorption of CO_2 from the gas-phase, which have already been employed in adsorptive reactor concepts to enhance hydrogen generation from methane, are based on hydrotalcite $(Al_2Mg_6(OH)_{16}CO_3 \cdot 4H_2O)$. To extend the operating range from 300°C (Yong et al. 2001) up to 500°C, the basic hydrotalcite material is doped with K_2O to increase the basicity of the adsorbent (Mayorga et al. 1997). Being modified in this way, CO_2-loadings of the order of about 2 wt-% can be achieved, even in the presence of large amounts of water vapour. The equilibrium capacity can be described by a Langmuir isotherm and the adsorption kinetics by an LDF approach given by Ding and Alpay (Ding and Alpay 2000a).

As an additional unwanted phenomenon, the co-adsorption of HCN on the adsorbent was observed, which, although enhancing the conversion, reduces the capacity of the adsorbent and retains the product desired within the reactor, necessitating an additional separation step.

Unfortunately, apart from the HCN-synthesis reaction, the CO-decomposition to C and CO_2 according to the Boudouard-reaction

$$CO \Leftrightarrow C + CO_2$$

also takes place, which could be described by a model taking contributions of the reactor wall and inert particles as well as the heterogeneous catalyst into account, whereby the catalytic contribution dominates:

$$r_u = \left(k_r^{C,\mathrm{hom}} + k_r^{C,het} \rho_{cat}\right)\left(P_{CO}^2 - \frac{P_{CO_2} P^\circ}{K_P^C}\right) \tag{4.13}$$

Due to the occurrence of this unwanted side reaction, which is just as enhanced by the CO_2-removal as the main reaction, the selectivity towards the desired product HCN is significantly diminished. Besides the sorption capacity being cut by HCN coadsorption, the CO decomposition leads to carbon deposition, which is a major obstacle as far as the realization of an adsorptive enhancement of the direct HCN-synthesis from CO and NH_3 is concerned.

The conclusion from our research is that the potential for a sorptive enhancement of the direct HCN-synthesis can be classified as very low. The development of specially tailored catalysts for the desired HCN-synthesis which suppresses the unwanted Boudouard reaction could open up new possibilities.

4.4.3 Water-gas shift reaction

Against the backdrop of increased interest in the field of hydrogen production for fuel cell applications, the plethora of publications on the adsorptive enhancement of the steam reforming and water-gas shift reaction:

$$CH_4 + H_2O \Leftrightarrow 3\ H_2 + CO \qquad \Delta H_R° = +\ 206\ kJ/mol$$
$$CO + H_2O \Leftrightarrow H_2 + CO_2 \qquad \Delta H_R° = -\ 41\ kJ/mol$$

is hardly surprising.

In these adsorptive reactor concepts, the aim is to displace the equilibrium by the removal of CO_2 and thus enhance the thermodynamically limited combined reaction system. In this way, an increased hydrogen yield can be achieved in the water-gas shift reaction. As a further benefit, additional processes to remove CO from the hydrogen-rich gas to avoid serious catalyst poisoning in the fuel cell can be omitted.

As suitable scavengers for the CO_2-removal, the materials that have already be presented in the preceding section for the adsorptively enhanced HCN-synthesis from ammonia and carbon monoxide can also be adapted to the water-gas shift reaction system. For the agents discussed above it can be said that there is a fundamental contradiction between the requirements for a high adsorbent capacity and fast kinetics of CO_2-uptake, which are both prerequisites for a compact reactor system with reasonably long cycle times.

While lithium zirconate fulfills the criterion of high capacity, the kinetics for CO_2-sorption are poor. The adsorbent based on hydrotalcite, which we synthesised ourselves, was characterised by a low capacity, albeit with much faster sorption kinetics. As a consequence, additional research must be devoted to the development of adsorbents that fulfill both the kinetic and capacitive requirements.

4.5 Yield and selectivity enhancement for complex reaction schemes

As already mentioned in the literature review, experimental investigations concerning the adsorptive enhancement of reaction schemes with multiple reactions are still few and far between. It was thus decided to try and identify suitable reaction systems to demonstrate the application of sorptive reactors.

As an appropriate reaction system for this purpose, an adsorptive reactor concept for the direct synthesis of dimethyl ether (DME) from synthesis gas was chosen. As a second example of gas-solid reactions, the oxidative dehydrogenation of ethylbenzene to styrene is also considered in this section.

4.5.1 Direct synthesis of DME from synthesis gas

Dimethyl ether, which is used as an aerosol propellant and as an intermediate chemical product, has also been proposed as an alternative to diesel due to its advantageous combustion characteristics (Sorenson 2001). Dimethyl ether can be produced by the dehydration of methanol on solid acid catalysts like γ-alumina, H-ZSM-5, amorphous silica-alumina or titania modified zirconia (Spivey 1991, Xu et al. 1997).

An attractive alternative to the two-step synthesis of DME from methanol, which first has to be produced from synthesis gas, is the combined direct DME-production from syngas in a single stage reactor. This combined synthesis allows one to exploit the synergies arising when the reactions occur simultaneously. The reactions that take place in such a reactor can be described by the following equations:

$$
\begin{array}{llll}
CO + 2\,H_2 & \Leftrightarrow & CH_3OH & \text{(I)} \\
CO_2 + 3\,H_2 & \Leftrightarrow & CH_3OH + H_2O & \text{(II)} \\
CO + H_2O & \Leftrightarrow & CO_2 + H_2 & \text{(III)} \\
2\,CH_3OH & \Leftrightarrow & CH_3OCH_3 + H_2O & \text{(IV)}
\end{array}
$$

In this system, reactions (I)-(III) are catalyzed on $Cu/ZnO/Al_2O_3$ catalysts for the hydrogenation of the carbon oxides, while reaction (IV) takes place on the solid acid catalyst, so that the fixed-bed consists of an mixture of two catalysts or a catalyst combining both catalytic activities.

As far as the methanol forming reactions are concerned, it is known that the primary carbon source for the hydrogenation reaction is carbon dioxide, which is generated from carbon monoxide by the comparatively fast water-gas shift reaction, while the direct formation of methanol from carbon monoxide is almost negligible in comparison (Sahibzada et al. 1998). For this reason, a simplified consecutive reaction scheme can be assumed as follows:

$$
CO \quad \Leftrightarrow \quad CO_2 \quad \Leftrightarrow \quad CH_3OH
$$

As a consequence, a maximum in conversion is achieved for methanol reactors with a low CO_2-content in the synthesis gas, which can be explained by the promoting effect of the water formed during the CO_2-hydrogenation and consumed by the water-gas shift reaction, which in turn creates fresh CO_2 and by the formation of highly active $CuZnO_x$ species at a certain CO/CO_2-ratio (Kurtz 2004).

In order to integrate the dehydration functionality into the reactor, various catalysts have been discussed. While γ-alumina is known to be prone to a strong inhibition in the presence of water in the system, zeolitic systems are characterized by a stronger resistance to water. Another important aspect is a slow and reversible deactivation effect due to the deposition of carbon on the catalyst, which has been reported for H-ZSM-5 systems (Jun et al. 2003).

It has been demonstrated that the total equilibrium yields of oxygenates can be improved by the simultaneous operation of the methanol- and DME-forming reactions (Shen et al. 2000), driven by the removal of methanol through the consecutive dehydration reaction. Apart from the thermodynamic enhancement, the performance of the system is improved under operating conditions which are governed by kinetic limitations far from equilibrium.

The synergy observed is thus comprised of several effects within the reaction system (Peng et al. 1999), as shown in Figure 4.8. The first effect is the removal of methanol by the dehydration to DME, by which the equilibrium limitations of the methanol forming reactions can be overcome. The water that is formed during that reaction and which strongly inhibits both methanol and DME forming reactions by competitive adsorption on the catalyst is consumed by the water-gas shift reaction. The hydrogen that is formed by this shift leads to a further enhancement of the methanol-forming hydrogenation reaction.

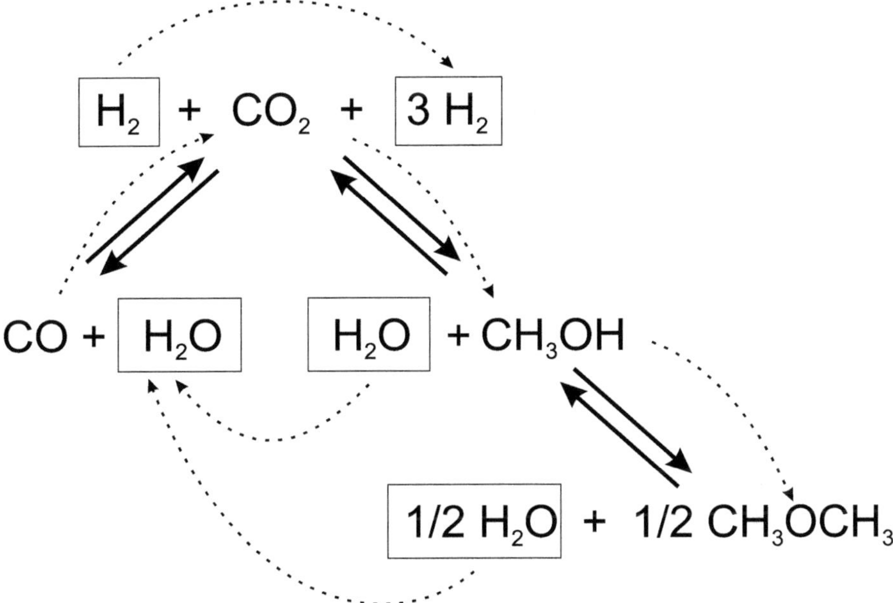

Fig. 4.8: Synergy in the direct DME-synthesis reaction scheme

Since the water gas shift reaction is rapid compared to the other reactions, the water content of the system and thus the degree of inhibition by competitive adsorption on the catalyst is a function of the ratio

$$\frac{[H_2][CO_2]}{[CO]}$$

which follows from the law of mass action for the water-gas shift reaction (Peng et al 1999). So it can be explained why the CO- and H_2-rich as well as CO_2-depleted systems exhibit a high synergy between the two reactions due to the low water concentration in the kinetically controlled regime. A consequence of this is a high degree of synergy in coal-derived syngas, while in the case of a natural-gas-derived syngas the synergetic effect is expected to be lower.

One possibility of overcoming this shortcoming would be the removal of water to enhance the rate of the dehydration reaction, thus improving the removal of methanol in kinetically limited systems as well as shifting the equilibria towards the desired oxygenated products, if a sufficient high residence time is provided. The positive effect of water removal from a slurry system for the direct production of DME from syngas was experimentally demonstrated by Kim et al. (Kim et al. 2001) by adding $MgSO_4$ as a dehydrating agent to the reaction system.

The objective of our study was to evaluate the potential of an adsorptive water removal in H_2-rich gas phase systems, using mixed fixed-beds of the catalyst systems $Cu/ZnO/Al_2O_3$ as methanol synthesis catalyst and γ-alumina as dehydration catalyst and to demonstrate the benefits of water removal from a system both kinetically and thermodynamically inhibited by this species.

As adsorbent zeolite 3A, already employed for the enhancement of the Claus reaction, was used due to its high selectivity for water adsorption as a result of the small micropore dimensions. In experimental investigations, the two catalysts and the adsorbent were physically mixed in a fixed-bed within a temperature regulated tubular reactor. The active section was enclosed within inert packing end zones to maintain well-defined temperature conditions.

In the experiments presented here, the volumetric fraction of the three solid phases was selected as follows:

$Cu/ZnO/Al_2O_3$:	25 %
γ-Al_2O_3:	25 %
Zeolite 3A:	50%

The other parameters used in the experiments are listed in Table 4.2.

Table 4.2: Parameters in integrated adsorptive reactor experiments

length of active bed	400 mm
tube diameter	10 mm
initial reaction temperature	250°C
throughput of gas mixture	625 ml_N/min
GHSV	1194 h^{-1}
Pressure	15 bar, 25 bar
Gas composition	H_2: 80%
	CO: 18%, 20%
	CO_2: 0.5%, 0%

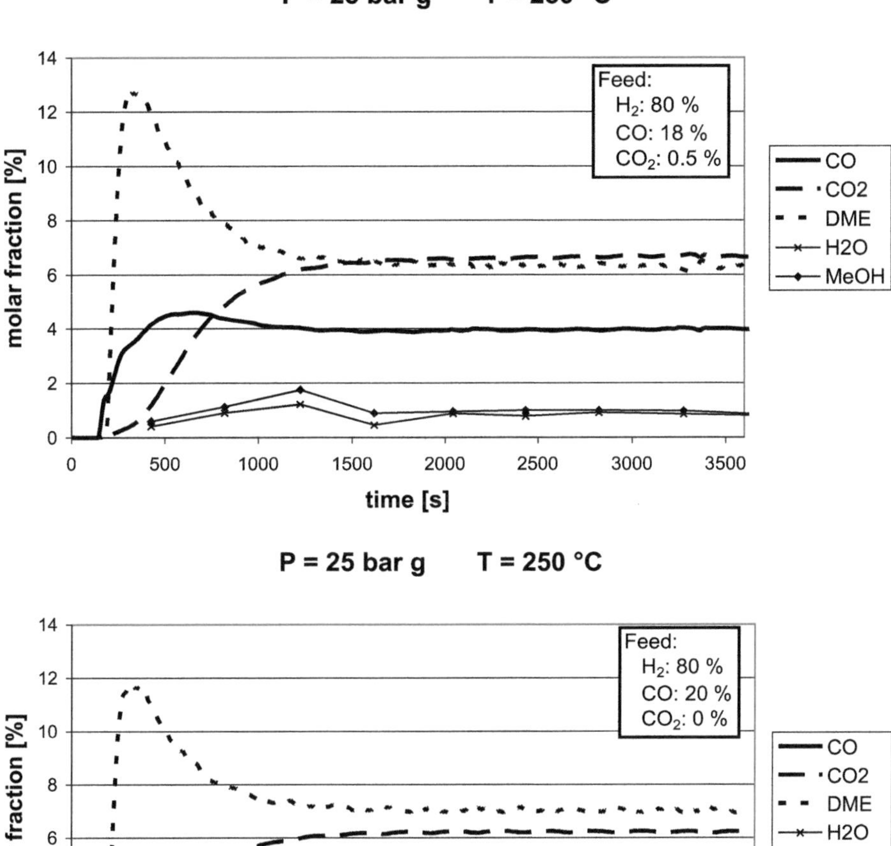

Fig. 4.9: Concentration profiles at the outlet of the adsorptive DME-reactor for a pressure of 25 bar g

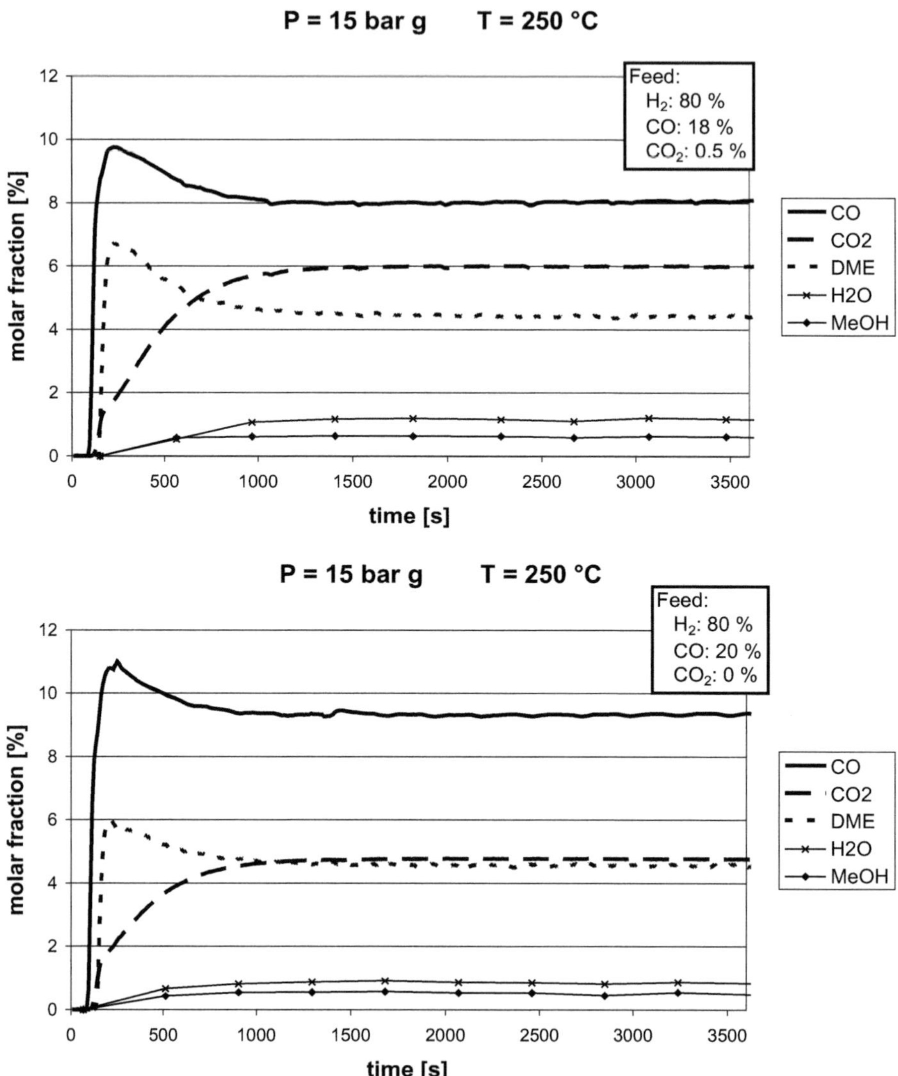

Fig. 4.10: Concentration profiles at the outlet of the adsorptive DME-reactor for a pressure of 15 bar g

Following the regeneration of the reactor with pure hydrogen under atmospheric pressure, the synthesis gas was switched to the reactor inlet. From the molar fractions that were measured at the reactor outlet during these experiments, it can be seen that DME-formation is clearly enhanced for a few minutes while CO_2 formation is suppressed (Figure 4.9 and Figure 4.10).

The concentrations of CO_2 and water are correlated by virtue of the fast water-gas shift reaction. If CO_2 is also present in the feed, a small increase in the DME-maximum is visible, while the DME-level is decreased at steady-state. It can thus be concluded that the low CO_2 content enhances the reaction process in a kinetic manner. This finding is consistent with simulation studies that have shown that the water removal plays a conflicting role.

On the one hand, the methanol and DME forming reactions are enhanced, but on the other the water-gas shift reaction is suppressed, leading to a rise in the CO-level (Reßler and Agar 2005). In this way it has been shown that the extent of regeneration influences the CO_2-level and thus the reaction kinetics in the adsorptive/reactive cycle, and is therefore an important design parameter for this adsorptive reactor concept.

4.5.2 Oxidative dehydrogenation of ethylbenzene to styrene

In case of heterogeneously catalyzed reactions performed in fixed-bed reactors, the optimal reaction conditions in terms of parameters such as temperature, pressure, residence time, and feed gas composition dominate the product selectivity and yield. With respect to oxidation catalysis, however, the redox state of the catalyst itself also exerts a powerful influence on the performance of the reactor.

Most catalytic oxidation reactions follow a two-step Mars-van Krevelen type mechanism due to the dual functions of the mixed metal oxide catalyst (Mars and van Krevelen 1954). The selective product formation occurs in the first step through participation of oxygen from the catalyst lattice with simultaneous reduction of the catalyst (reactive functionality of the catalyst). In a subsequent step, the catalyst is deoxidized using the gas-phase oxygen feed (adsorptive functionality of the catalyst). It has been shown that the products that are formed selectively during the first step can strongly depend on the surface redox state of the catalyst and the presence of loosely-bound oxygen (Bielanski and Haber 1991; Zanthoff et al. 1999). In a steady-state mode, where the hydrocarbon and oxygen are fed simultaneously, the undesirable total oxidation can hardly be suppressed, due to the presence of loosely-bound oxygen species (Zanthoff et al. 1999; Pantazidis et al. 1998). Based on a deeper understanding of the intrinsic kinetics on a microscale level, the aim is to combine catalysis and reaction engineering to come up with improved catalysts in oxidation catalysis. This approach can be demonstrated by the example of the unsteady-state operation of the oxidative dehydrogenation of ethylbenzene to styrene.

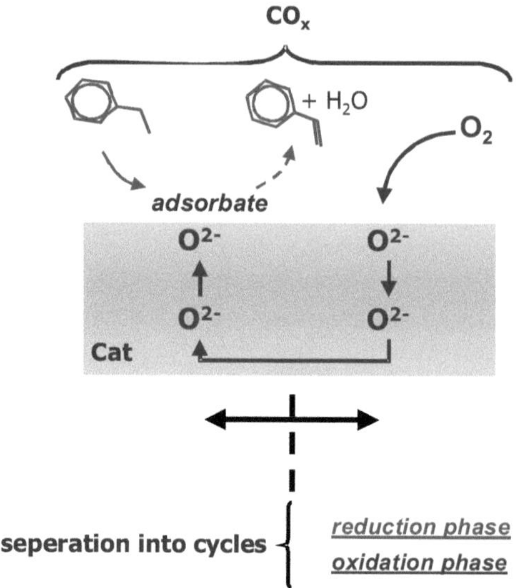

Fig. 4.11: Mars-van-Krevelen type mechanism.

The industrial large-scale production of styrene, which is usually performed *via* dehydrogenation of the basic chemical ethylbenzene, is still a challenge with respect to the improvement of the overall yield and selectivity. Today's process is based on the non-oxidative (direct) dehydrogenation over potassium doped iron-oxide based catalysts. The reaction, however,

$$C_6H_5CH_2CH_3 \Leftrightarrow C_6H_5CH=CH_2 + H_2 \qquad \Delta H_R = +125 \text{ kJ mol}^{-1}$$

is strongly limited by equilibrium constraints at temperatures below 900 K and pressures slightly above atmosphere. Due to the endothermicity of the reaction and severe coking under reaction conditions, the industrial process is operated under excess of water vapour. Problems and drawbacks of today's process are summarized in the following references (Cavani and Trifirò 1995; Watzenberger et al. 1999). In order to overcome some of the limitations occuring, the process can be run by co-feeding oxygen into the reaction mixture, leading to the overall exothermic reaction:

$$C_6H_5CH_2CH_3 + 1/2 O_2 \Leftrightarrow C_6H_5CH=CH_2 + H_2O \quad \Delta H_R° = -122 \text{ kJ mol}^{-1}$$

Multi-component oxide catalysts such as V-Mg-O catalysts are employed at a lowered reaction temperature of 750 K. Since the selectivity is limited by the undesirable total oxidation in the co-feed mode, alternative processes such as the concept of the reverse flow reactor (see section 4.1.2) and the unsteady-state ODEB (redox operation) have been proposed (c.f. Ref. (Cavani and Trifirò 1995; and the references therein).

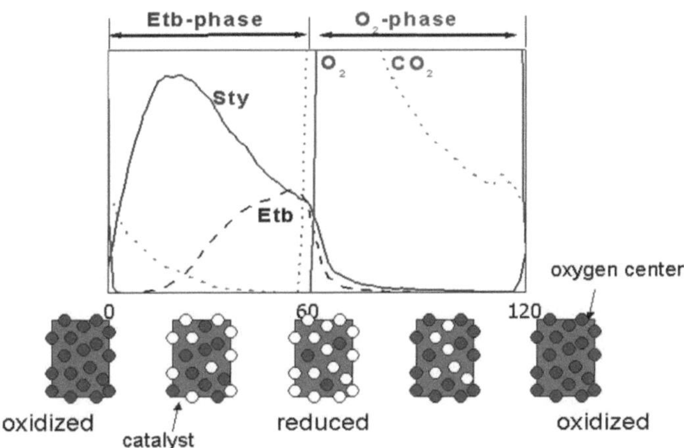

Fig. 4.12: Redox operation in unsteady-state ODEB. The change of the state of the catalyst is indicated schematically.

Catalytic results for the ODEB have been obtained in a mini fixed-bed reactor by sequentially and repeatedly passing ethylbenzene (reduction step) and oxygen (oxidation step) over the V-Mg-O catalyst, which changes the oxidation state as schematically illustrated in Figure 4.12. In order to understand the dual catalysts with different catalytic functions at a micro-scale level, the following approach was used: the results from a catalyst screening and characterization study of V-Mg-O catalysts published in (Geisler et al. 2003) was used as starting point for a detailed kinetic study. The unsteady-state operation of the ODEB provided a valuable tool for the determination of kinetic data, leading to overall conversion plots and an overall selectivity-conversion diagram depicted in Figure 4.13.

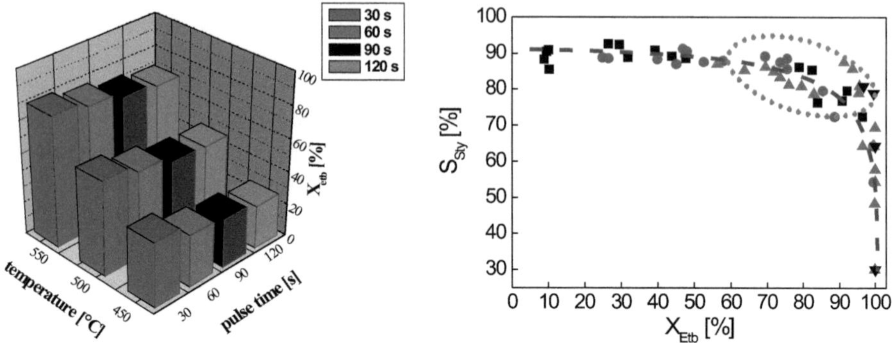

Fig. 4.13: left: Experimental results for the unsteady-state ODEB in a reduction-step; right: Activity-selectivity relationship in unsteady-state ODEB.

The best results, e.g. for an unpromoted V-Mg-O catalyst (20 wt. % V_2O_5) were obtained at a total conversion of 80 % and an overall selectivity of 85 % as indicated by the oval in Figure 4.13 (right). By changing process parameters such as reaction temperature, feed gas composition and the cycle time between the reduction and the oxidation phase, it was possible to optimize the redox behavior of the catalysts in terms of its overall selectivity and conversion. The influence of the temperature and the cycle time within a reduction step on the conversion is demonstrated in Figure 4.13 (left). Unfortunately, the kinetics of the elementary reaction steps taking place on a microscale level, i.e. the reaction steps during the reduction phase as well as the adsorption of the gas-phase oxygen followed by insertion into the lattice, are accompanied by severe coking.

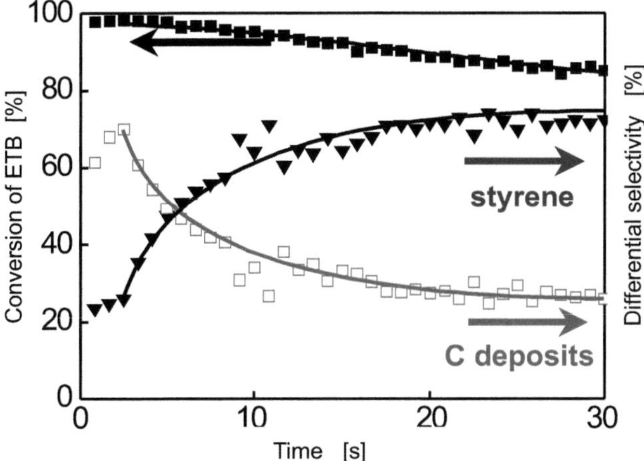

Fig. 4.14: Conversion of ethylbenzene and differential selectivity versus time in the reduction step.

However, the unsteady-state operation allows the direct observation of *in situ* generated C deposits (coke plus C-containing intermediates) as indicated in the trace concentrations in Figure 4.14 during the reduction step. Here, coking occurs while in the oxidation phase the C deposits are burnt off. Furthermore, an increase of styrene formation and a slight decrease in conversion of ethylbenzene are observed. Hence, based on the results obtained in the literature and on our own studies it can be safely concluded that the reaction proceeds via a Mars-van-Krevelen type mechanism, leading to the basic structure of the qualitative reaction network presented in Figure 4.15.

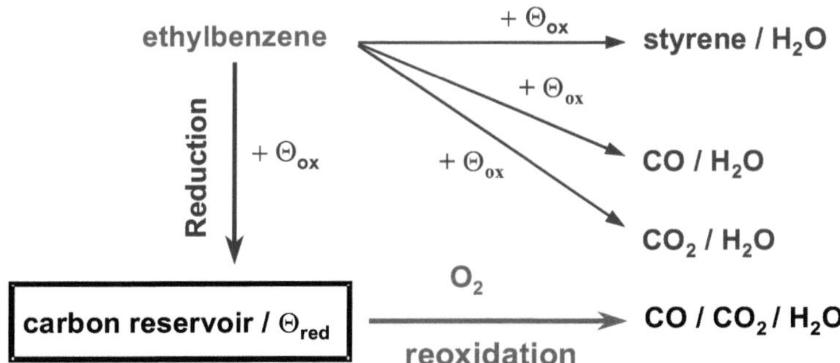

Fig. 4.15 Proposed reaction mechanism and transfer into a model for the unsteady-state ODEB.

Compared to the direct dehydrogenation of ethylbenzene to form styrene, the enhanced catalytic performance of the ODEB can be explained by the fact that the catalyst redox state can be adjusted to its optimum, minimizing the presence of non-selective short-lived oxygen species on the catalyst surface, which are responsible for the total oxidation. Industrial unsteady-state process operation is realized either by temporal separation, i.e. switching between reduction and oxidation phases, or spatial separation applying the riser-regenerator concept.

4.6 Conclusions

The case studies presented in the preceding sections illustrate that the integration of an adsorbent or other solid scavenging agent to a catalyst does not always lead to a clear improvement in the performance of the multifunctional reactor system. One feature that can be learnt from the adsorptive Claus reactor concept is the need to consider regenerative effects that are caused by the catalyst. In the case of the Claus process, this effect led to a fatal distortion of the required stoichiometric ratio of the reactants, so that a deintegration of the functionalities to circumvent the unsteady-state operation of the catalyst is a better option in this instance.

The problems that arise in the high temperature HCN-synthesis reaction during the sorptive operation mode due to side-reactions become more significant at higher temperatures. In this case, the coke-forming Boudouard reaction was also enhanced by the CO_2-removal, and proved to be the main obstacle for this adsorptive reactor concept. Additionally, it was shown by the analysis of the potential of an adsorptive enhancement of this reaction system as well as for the water-gas shift reaction that the sorbent has to be characterized by both high capacities and rapid kinetics.

Another feature of adsorptive fixed-bed reactors which are operated dynamically is a volume large enough to provide sufficient space for the propagation of well-defined adsorption fronts. The adsorbent already loaded behaves as "inert" and the zones with saturated adsorbent do not contribute to the reactor performance if the removal of a species is desired, thus resulting in high reactor residence times. Considering more complex systems with multiple reaction paths, precise tuning of the adsorbent with the reaction system becomes even more crucial.

The analysis of direct synthesis of DME from synthesis gas has shown that an enhancement of DME at the expense of CO_2 could be achieved by water removal. As an additional effect, the enhancement of the retroshift reaction was observed as well, showing that the operation conditions have to be adjusted carefully to facilitate the desired and to suppress the unwanted effects by the adsorptive functionality.

The complexity of adsorptive reactors in conjunction with the low added value of simple gaseous products still impedes the commercial realization of such reactor systems. An exception is the niche application of simple adsorptive enichment of pollutants at low concentrations, which are subsequently destroyed in oxidative regeneration processes.

4.7 Notation

Latin symbols

a_0	isotherm parameter	see equation
a_k	volumetric surface of phase k	$[m^2/m^3]$
B_0	Darcy permeability	$[m^2]$
c_p	mass-specific heat capacity	$[J/(kg\,K)]$
C_{pm}	molar heat capacity	$[J/(mol\,K)]$
d	reactor diameter	$[m]$
D_{is}^e	Maxwell-Stefan diffusivity of species i in species s	$[m^2/s]$
$D_{k,i}^e$	effective diffusion of component i within the particle phase k	$[m^2/s]$
$D_{Kn,i}^e$	effective Knudsen diffusion coefficient for component i	$[m^2/s]$
$D_{ax,i}$	axial dispersion coefficient	$[m^2/s]$
E_A	activation energy	$[J/(mol\,K)]$
f_s	particle shape factor	$[-]$
ΔH_{ad}	heat of adsorption	$[J/mol]$
ΔH_R	heat of reaction	$[J/mol]$
k_0	kinetic constant (see kinetic expressions)	varies
$k_{m,i,eff}$	effective mass transfer coefficient	$[1/s]$
l	length coordinate within a particle	$[m]$
L_c	length of column section c	$[m]$

N_c	number of components	[-]
$\dot{N}_{ex,k,i}$	molar flux of component i exchanged between the bulk phase and the solid phase k	[mol/(m^2 s)]
$\dot{N}_{k,i}$	molar flux of species i within particle k	[mol/(m^2 s)]
N_{Ph}	number of solid phases	[-]
$\dot{N}_{tot}$	total molar flux in axial direction	[mol/s
P	total pressure	[Pa]
$q_{k,i}$	loading of species i on the solid adsorbent k	[mol/m^3]
r	reaction rate	[mol/(m^3 s)]
r_p	particle radius	
R	ideal gas constant	[J/(mol K)]
$R_{k,i}$	molar rate of change of component i in phase k by chemical reaction	[mol/(s m^3)]
t	time coordinate	[s]
t_{cyc}	cycle time	[s]
T	temperature	[K]
u	superficial gas velocity	[m/s]
U	overall heat transfer coefficient	[W m^2 / K]
x	axial coordinate in the fixed-bed	[m]
y	molar fraction of species i	[-]

Greek symbols

β	volumetric fraction of solid	[-]
ε_b	bed porosity	[-]
η	effectiveness factor	[-]
λ_{ax}	axial bed thermal conductivity	[W/(m K)]
μ	dynamic viscosity of the gas mixture	[Pa s]
ρ	mass density	[kg/m^3]
$\hat{\rho}$	molar density	[m^3/mol]

Indices

$\circ$	standard conditions
ad	adsorption

ax	axial
b	bed
cat	catalyst
cyc	cycle
eff	effective
g	gas phase
i	related to species *i*
k	related to phase *k*
p	particle
R	reaction
t	total

4.8 Literature

Agar DW, Ruppel W (1988) Multifunktionale Reaktoren für die heterogene Katalyse. Chem.-Ing.-Tech. 60:731-741

Alpay E, Scott DM (1992) The linear driving force model for fast-cycle adsorption and desorption in a spherical particle. Chem. Eng. Sci. 47:499-502

Balasubramanian B, Lopez Ortiz A, Kaytakoglu S, Harrison DP (1999) Hydrogen from methane in a single-step process. Chem. Eng. Sci. 54:3543-3552

Bathen D (2003) Physical waves in adsorption technology – an overview. Separation and Purification Technology 33, 163-177

Bielanski A, Haber J (1991) Oxygen in catalysis. Marcel Dekker New York

Cavani F, Trifirò F (1995) Alternative processes for the production of styrene. Appl Catal A: Gen 133:219-239

Carvill BT, Hufton RJ, Anand M, Sircar S (1996) Sorption-enhanced reaction process, AIChE Journal 42:2765-2772

Croft DT, Levan MD (1994) Periodic states of adsorption cycles – I. Direct Determination and stability. Chem Enf. Sci. 49:1821-1829

Dalla Lana IG, Mc Gregor DE, Liu CL, Cormode AE (1972) A Kinetic Study of the Catalytic Reaction of H_2S and SO_2 to Elemental sulfur, 2nd Int. Symp. React. Eng., Elsevier Amsterdam Vol. B2:9-18

Datta A, Cavell RG (1985a) Claus Catalysis. 2. An FTIR Study of the Adsorption of H_2S on the alumina Catalyst. J. Phys Chem. 89:450-454

Datta A, Cavell RG (1985b) Claus Catalysis. 3. An FTIR Study of the Sequential Adsorption of H_2S and SO_2 on the Alumina Catalyst, J. Phys Chem. 89:454-457

Deo AV, Dalla Lana IG, Habgood HW (1971) Infrared studies of adsorption and surface reactions of hydrogen sulfide and sulfur dioxide on some aluminas and zeolites. J. Catal. 21:270

Ding Y, Alpay E (2000a), Equilibria and kinetics of CO_2 adsorption on hydrotalcite adsorbent. Chem. Eng. Sci. 55:3461-3474

Ding Y, Alpay E (2000b), Adsorption-enhanced steam reforming. Chem. Eng. Sci. 56:3929-3940

Dittrich C (2002) Bewertung eines adsorptiven Reaktors für die direkte Cyanwasserstoffsynthese aus Ammoniak und Kohlenmonoxid. PhD Thesis, University of Dortmund, Germany

Elsner MP, Dittrich C, Agar DW (2002) Adsorptive reactors for enhancing equilibrium gas-phase reactions – two case studies. Chem. Eng. Sci. 57:1607-1619

Elsner MP, Menge M, Müller C, Agar DW (2003) The Claus Process: teaching an old dog new tricks, Catalysis Today 79-80:487-494

Elsner MP (2004) Experimentelle und modellbasierte Studien zur Bewertung des adsorptiven Reaktorkonzeptes am Beispiel der Claus-Reaktion. PhD Thesis, University of Dortmund, Germany

Ergun S (1952) Fluid flow through packed columns. Chem. Engng Prog. 48:89-94

Geisler S, Vauthey I, Farussng D, Zanthoff H, Muhler M. (2003) Advances in catalyst development for oxidative ethylbenzene dehydrogenation. Catal Today 81:413-424

Glueckauf E, Coates JI (1947) Theory of chromatograhy (Part IV). J. Chem Soc. 1315-1321

Goto S, Tagawa T, Oomiya T (1993) Dehydrogenation of cyclohexane in a PSA reaction using hydrogen storage alloy. Kagaku-kogaku-ronbunshu (Chem. Eng. Essays) 19:978-983

Harrison DP, Han C (1994) Simultaneous shift reaction and carbon dioxide separation for the direct production of hydrogen. Chem. Eng. Sci. 49:5875-5883

Harten A (1983) High Resolution Schemes for Hyperbolic Conservation laws. J. Comput. Phys. 49:357-393

Harten A (1984) On a Class of High Resolution Total-Variation-Stable Finite-Difference Schemes. SIAM J. Numer. Anal. 21:1-23

Hite RH, Jackson R (1977) Pressure gradients in porous catalyst pellets in the intermediate diffusion regime. Chem. Eng. Sci. 32:703-709

Hufton JR, Mayorga S, Sircar S (1999) Sorption-enhanced reaction process for hydrogen production. AIChE Journal 45:248-256

Jackson R (1977) Transport in Porous Catalysts, Elsevier, Amsterdam

Jensen A, Johnsson JE, Dam-Johansen K (1997) Catalytic and gas-solid reactions involving HCN over limestone. AIChE Journal 43:3070-3084

Judkins RR, Burchell TD (1999) US Patent 5972077 from 26.10.1999

Jun KW, Lee HS, Roh HS, Park SE (2003) Highly Water-Enhanced H-ZSM-5 Catalysts for Dehydration of Methanol to Dimethyl Ether. Bull. Korean Chem. Soc. 24:106-108

Karge HG, Dalla Lana IG (1985) Infrared studies of SO_2 Adsorption on a Claus Catalyst by Selective Poisoning of Sites. J. Phys. Chem. 88:1538-1543

Kim HJ, Jung H, Lee KY (2001) Effect of Water on Liquid Phase DME Synthesis from Syngas over Hybrid Catalysts Composed of $Cu/ZnO/Al_2O_3$ and γ-Al_2O_3. Korean J. Chem. Eng. 18:838-841

Kodde AJ, Bliek A (1997) Selectivity enhancement in consecutive reactions using the pressure swing reactor. Stud. Surf. Sci. Cat. 109:419-428

Kodde AJ, Fokma YS, Bliek A (2000) Selectivity Effects on Series Reactions by Reactant Storage and PSA Operation. AIChE Journal 46:2295-2304

Krishna R, Wesselingh JA (1997) The Maxwell-Stefan approach to mass transfer. Chem. Eng. Sci. 52:861:911

Kuczynski M, Oyevaar, MH, Pieters RT, Westerterp KR (1987) Methanol synthesis in a countercurrent gas-solid trickle flow reactor. An experimental study. Chem Eng. Sci. 42:1887-1898

Kurdyumov SS, Brun-Tsekhovoi AR, Rozenthal AL (1996) Steam conversion of methane in the presence of a carbon dioxide acceptor. Petr. Chem. 36:139-143

Kurtz M (2004) Die Bedeutung der Globalkinetik für die Entwicklung Cu-basierter Methanolkatalysatoren, PhD Tesis, Univeristy of Bochum, Germany

Macdonald IF, Sayed MS, Mow K, Dullien FAL (1979) Flow through porous media – the Ergun equation revisited. Ind. Engng Chem. Fundam. 18:199-207

Mars P, van Krevelen DW (1954) Oxidations carried out by means of vanadium oxide catalysts. Chem Eng Sci Suppl 3:41-59

Mason EA, Malinauskas AP (1983) Gas Transport in Porous Media: The Dusty-Gas Model. Elsevier, Amsterdam

Mayorga SG, Hufton JR, Sircar S, Gaffney TR (1997) Sorption enhanced reaction process for production of hydrogen – Phase I final report. U.S. Dep. of Energy: #DE-FC36-95G010059

Müller C, Agar DW (2000) Einstufige Kohlenmonoxidkonvertierung zur Wasserstofferzeugung für mobile Anwendungen, Chem.-Ing.-Tech. 72:982-983

Nowak U, Frauhammer J, Nieken U, Eigenberger G (1996) A fully adaptative algorithm for parabolic partial differential equations in one space dimension. Comp. Chem. Eng. 20:547-561

Pantazidis A, Buchholz SA, Zanthoff HW, Schuurman Y, Mirodatos C (1998) A TAP reactor investigation of the oxidative dehydrogenation of propane over V-Mg-O catalyst. Catal Today 40:207-214

Peng XD, Toseland BA, Tijm PJA (1999) Kinetic understanding of the chemical synergy under LPDME^{TM} conditions - once-through applications. Chem. Engn. Sci. 54:2787-2792

Pineda M, Palacios JM (1995). The performance of a γ-Al_2O_3 catalyst for the Claus reaction at low temperature in a fixed bed reactor. Applied Catalysis A: General 136:81-86

Reßler S, Agar DW (2005) Enhancement of the Syngas-to-Dimethyl Ether Process by Adsorptive Water Removal, International Symposium on Multifunctional Reactors (ISMR-4), 15.-18.06.2005, Portoroz

Sahibzada M, Metcalfe IS, Chadwick D (1998) Methanol Synthesis from $CO/CO_2/H_2$ over $Cu/ZnO/Al_2O_3$ at Differential and Finite Conversions. J. Catal. 174:111-118

Salden A, Eigenberger G (2001) Multifunctional adsorber / reactor concept for waste-air purification. Chem. Eng. Sci. 56:1605-1611

Salinger AG, Eigenberger G (1996) The direct calculation of periodic states of the reverse flow reactor – I. Methodology and propane combustion results. Chem. Eng. Sci. 51:4903-4913

Saur O, Cherrau T, Lamote J, Travert J Lavalley JC (1981) Comparative adsorption of H_2S, CH_3SH and $(CH_3)_2S$ on alumina – structure of species and adsorption sites. J. Chem. Soc. Faraday Trans. I 77 (Part 2):427-437

Schembecker G, Tlatlik S (2003) Process Synthesis for Reactive Separations. Chem. Eng. Proc. 42:179:189

Sereno C, Rodrigues AE (1993). Can steady-state momentum equations be used in modeling pressurization of adsorption beds? Gas Sep. Pur. 7:167-174

Seydel R (1988) From Bifurcation to Chaos. Elsevier, New York

Shen WJ, Jun KW, Choi HS, Lee KW (2000) Thermodynamic Investigation of Methanol and Dimethyl Ether Synthesis from CO_2 Hydrogenation. Korean J. Chem. Eng. 17:210-216

Sorenson,SC (2001) Dimethyl Ether in Diesel Engines: Progress and Perspectives. Trans. ASME 123:652-658

Spivey JJ (1991) Review: Dehydration catalysts for the methanol/dimethyl ether reaction. Chem. Eng. Comm. 110:123-142

Tonkovich AL, Carr RW, Aris R (1993) Enhanced C2 yields from methane oxidative coupling by means of a separative chemical reactor. Science 262:221-223

Vaporciyan GG, Kadlec RH (1989) Periodic separating reactors: Experiments and theory. AIChE Journal 35:1334-1343

Watzenberger O, Ströfer E, Anderlohr A (1999) Unsteady-state oxidative dehydrogenation of ethylbenzene to form styrene. Chem Eng. Technol. 21:659-662

Xiu GH, Li P, Rodrigues AE (2003) New generalized strategy for improving sorption-enhanced reaction process. Chem. Eng. Sci. 58:3425-3437

Xu M, Lunsford JH, Goodman DW, Bhattacharyya A. (1997) Synthesis of dimethyl ether (DME) from methanol over solid-acid catalysts. Appl. Catal. A 149:289-301

Yong Z, Mata V, Rodrigues AE (2001) Adsorption of carbon dioxide onto hydrotalcite-like compounds (HTlc) at high temperatures. Ind. Eng. Chem. Res. 40:204-209

Zagoruiko AN, Kostenko OV, Noskov AS (1996) Development of the adsorption-catalytic reverse process for incineration of volatile organic compounds in diluted waste gases. Chem. Eng. Sci. 51:2989-2994

Zanthoff HW, Buchholz SA, Pantazidis A, Mirodatos C (1999) Selective and non-selective oxygen species determining the product selectivity in the oxidative conversion of propane over vanadium mixed oxide catalysts. Chem. Eng. Sci. 54:4397-4405

5 Reactive liquid chromatography

Thomas Borren, Jörg Fricke, Henner Schmidt-Traub

5.1 Introduction

Chromatography is a highly selective separation technique which is mainly used for the separation of components employed as fine chemicals, pharmaceuticals, food additives and biological products. The coupling of chemical or biochemical reactions and chromatographic separations leads to an integrated process for the production of high purity products. Within a chromatographic reactor the conversion of the reactants as well as the separation of the components takes place simultaneously. Therefore reversible reactions can overcome the limitation of the conversion ruled by a chemical equilibrium.

The reaction within integrated processes can be catalyzed homogeneously or heterogeneously. In case of a homogeneous catalysis the separation of the catalyst from the products has to be taken into account. Heterogeneously catalyzed reactions occur more often. In special cases, such as esterifications, the same ion exchange resin can act as catalyst for the reaction as well as adsorbent for the separation (Mazzotti et al. 1996). In general different materials are employed. This is why the establishment of a reliable packing is a major task. Different wetting properties, densities and particle sizes lead to a non uniform distribution of the species within the packed bed. The partition of the bed into several alternate sections of adsorbent and catalyst is a possibility to overcome these problems (Meurer et al. 1997).

This chapter focuses on the design and applications of preparative chromatographic reactors in the liquid phase. In contrast to analytical applications it is the aim of preparative or product chromatography to recover the desired product (Ditz 2005). Carr and Dandekar (2002) recently discussed applications of chromatographic reactors in the gas phase.

5.2 Process concepts

Reactive chromatographic processes are strongly influenced by the type of reaction and the elution order of the components. Most published investigations are restricted to equilibrium-limited reactions of the type $A \leftrightarrows B$, $A \leftrightarrows B + C$ or $A + B \leftrightarrows C + D$. Formation of by-products normally leads to reduced product purities. Therefore parallel or consecutive reactions are not favorable. The influence of consecutive reactions has been studied e.g. by Michel et al. (2003) for the application of electrochemical micro reactors within SMB processes.

The elution order is an important constraint for chromatographic reactors. In case the educts elute in the middle of the chromatogram the products can be obtained with high purity. If the reactant is the strongest or the weakest adsorbed component only the weakest or strongest adsorbed product respectively can be withdrawn with high purity and the intermediate eluting component is withdrawn together with the reactant (Grüner and Kienle 2004). Normally a reactive separation aims at the production of a single high-value product. The following sections briefly introduce reactive process concepts and provide examples for their application.

5.2.1 Chromatographic batch reactor

The basic principles of chromatographic batch reactors are discussed for a single reactive column and the reaction $A \leftrightarrows B + C$ as shown in Figure 5.1. The reactant A is solved in the desorbent and injected as a sharp pulse into the column packed with the stationary phase. At the surface of the catalyst the reaction occurs and the products B and C are formed. The components interact with the surface of the adsorbent and due to different affinities to the stationary phase their propagation velocities through the bed are different. Hence the products are separated and the driving force for the forward reaction is enhanced while the backward reaction is suppressed. Therefore chemical equilibrium can be overcome and high purity products are withdrawn at the column outlet.

This principle can be transferred to every multi-component reaction system because multiple fractions can be collected at the outlet. For binary reactions of type $A \leftrightarrows B$ the batch reactor cannot be applied because the backward reaction cannot be suppressed by separating the products. Consequently it is not possible to overcome the equilibrium and to obtain the pure product only.

Like chromatographic batch separations the batch reactor is suffering from the fact that only a small part of the whole bed is actually in contact

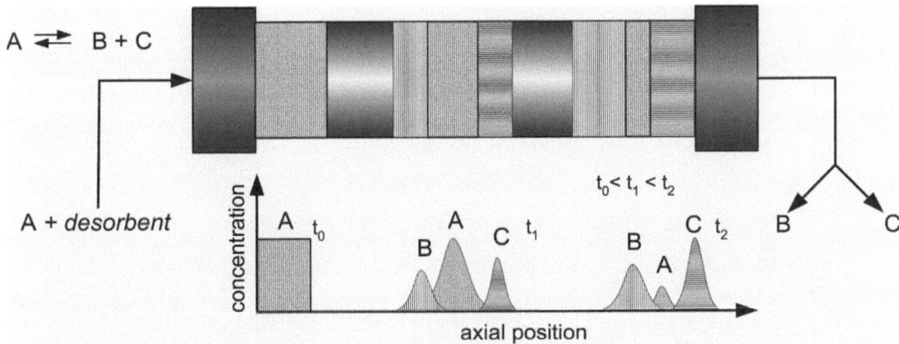

Fig. 5.1: Chromatographic batch reactor (reprinted with permission from Schmidt-Traub 2005)

with the components and therefore the productivity of the process is rather low. Improved operation concepts like closed-loop recycling and steady-state recycling chromatography can reduce these drawbacks of batch reactors. Another problem of batch operation is that the high desorbent consumption and the resulting product dilution is not avoidable.

Several applications of the batch reactor have been investigated in the laboratory scale. These applications include esterifications (Mazzotti et al. 1997) and transesterifications (Sardin et al. 1993), hydrolysis reactions (Falk and Seidel-Morgenstern 2002), saccharifications (Sarmidi and Barker 1993a) as well as the preparation of unstable reagents (Coca et al. 1993).

5.2.2 Continuous annular reactor

A continuous annular reactor can be considered as parallel arrangement of several batch reactors. Instead of individual tubes the homogeneous mixture of adsorbent and catalyst is packed into an annular gap which rotates around its central axis during operation.

The feed components enter the gap at a certain point at the top of the bed while eluent is fed into the packing over the remaining area. Similar to batch reactors reaction occurs within the bed and the components are separated at the same time due to different adsorption behavior. The stronger adsorbed components have a higher residence time and leave the reactor at the latest. At the outlet of the annulus the components are collected separately at different outlet ports around the circumference according to their elution order.

Due to the above-mentioned similarities with batch reactors, annular reactors have comparable properties but enable a continuous operation.

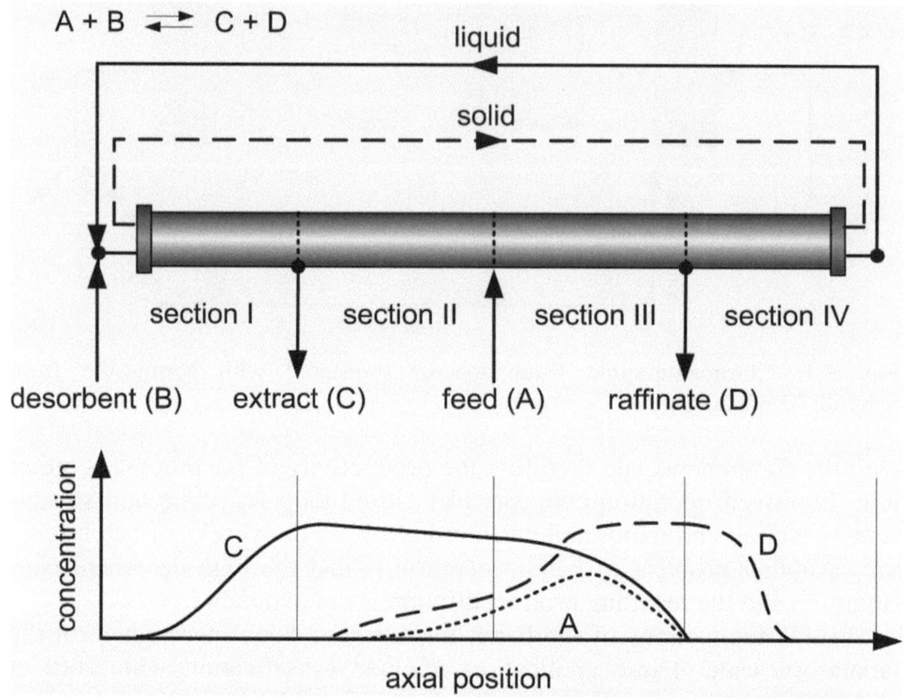

Fig. 5.2: True Moving Bed reactor (reproduced from Schmidt-Traub 2005)

The application of this process is often restricted by the packing quality within the gap as well as the limited volume flow due to the maximum allowable pressure difference at the sealing.

Since the first description by Martin (1949) saccharifications (Sarmidi and Barker 1993b), hydrolysis reactions (Carr 1993) and redox-reactions (Herbsthofer et al. 2002) have been investigated.

5.2.3 Counter-current flow reactors

True moving bed reactor

A counter-current flow of liquid and solid phase enables a more efficient continuous process operation. The True Moving Bed Reactor (TMBR) supposes an ideal counter-current movement within a single column.

As shown in Figure 5.2 the process is subdivided by external inlet and outlet streams into four sections. Considering a reaction of type $A + B \leftrightarrows C + D$ the educt component A is fed into the middle of the column and distributed by the solid and the liquid stream into the sections II and III. The second component B may also enter the process as solute of the feed stream or be the solvent itself. C and D are the reaction products. While the more strongly adsorbed component C is transported with the solid phase in the direction of the extract line, the less strongly adsorbed component D moves with the liquid phase to the raffinate outlet. Reaction and separation occur simultaneously in sections II and III. In front of section I fresh eluent is added to desorb all components from the solid phase. The regenerated adsorbent enters section IV and has to clean the liquid phase by adsorbing the solutes. Therefore sections I and IV are called regeneration sections. If the second reactant B is used as desorbent this component is present in large excess within the process and the reaction order is reduced to a reaction of type $A \leftrightarrows C + D$.

After start-up the process reaches a steady state and if the right flow rates are chosen in all process sections total conversion of the reactant A and complete separation of the products C and D can be achieved. Therefore successful process operation strongly depends on the selection of the operating conditions.

The operating conditions of a TMBR are defined by the liquid flow rates in each section $\dot{V}_{j,TMBR}$ and the volumetric flow rate of the adsorbent $\dot{V}_{ads}$. By introducing dimensionless flow rate ratios m_j as the ratio of the liquid to the solid flow rate

$$m_{j,TMBR} = \frac{\dot{V}_{j,TMBR}}{\dot{V}_{ads}} \tag{5.1}$$

the degrees of operational freedom can be reduced to four. These parameters can be estimated by the so called "triangle theory" which is based on a simple equilibrium model. Storti et al. (1993) developed this method for chromatographic separation processes for the first time. A basic result of this model says that, e.g. in case of linear adsorption isotherms, the adsorbed component moves in the direction of the solid flow if the Henry coefficient H_i is smaller than the dimensionless flow rate ratio. If the Henry coefficient is larger than the flow rate ratio the component moves in the direction of the liquid flow.

Based on the assumption of linear isotherms, ideal plug flow through the packed bed and total conversion of the reactants the following constraints for the dimensionless flow rate ratios in the different sections can be derived for total separation of the products C and D:

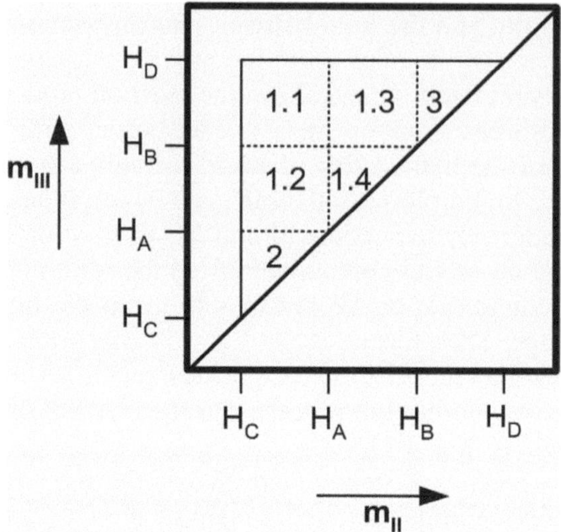

Fig. 5.3: Triangle diagram with different separation regions for reaction $A + B \leftrightarrows C + D$ and linear adsorption isotherm

$$m_I \geq H_C, \; m_I \geq H_D \text{ (section I)} \tag{5.2}$$

$$H_D \geq m_{II} \geq H_C \text{ (section II)} \tag{5.3}$$

$$H_D \geq m_{III} \geq H_C \text{ (section III)} \tag{5.4}$$

$$m_{IV} \leq H_C, \; m_{IV} \leq H_D \text{ (section IV)} \tag{5.5}$$

These constraints can be applied to estimate an initial operating point of the TMBR. Best performance regarding productivity is achieved if the feed flow rate is as high as possible and therefore the difference between m_{III} and m_{II} as large as possible. To reduce the eluent consumption m_I should be as small as possible and m_{IV} as large as possible.

All potential operating points for section II and III lie in the triangle area within the m_{II}, m_{III}-plane as shown in Figure 5.3. For a reaction of type $A + B \leftrightarrows C + D$ the region of total separation of both products is divided by the Henry coefficients of the components into six subsections. Further details of the inner structure of this triangle are explained in section 5.5.

Operating points outside the triangle do not fulfill the separation constraints and yield impure extract or raffinate.

Compared to reactive batch chromatography the TMBR enables higher productivities and lower eluent consumption. Additionally the concentration of the withdrawn products is higher and consequently the workup of the products is simplified. However, instead of several fractions achieved in a batch reactor the TMBR is limited to the withdrawal of two fractions only.

A practical implementation of the TMBR concept is difficult. Generally the movement of the solid leads to an inhomogeneous particle distribution over the cross section, back mixing of the solid and abrasion of the particles.

Simulated moving bed reactor

The practical design of a counter-current flow is realized by a series of packed bed columns as shown in Figure 5.4 (Broughton and Gerhold 1961). The counter-current movement of the liquid and solid phases is simulated by periodically shifting the external streams in the direction of the liquid flow. Hence the process is called Simulated Moving Bed Reactor (SMBR). Due to the switching of the ports after each shifting period the process finally reaches a cyclic steady state.

In case of an increasing number of columns (e.g. more than three columns per section) with a small bed length as well as short shifting periods the behavior and the properties of the SMBR are becoming very similar to the TMBR.

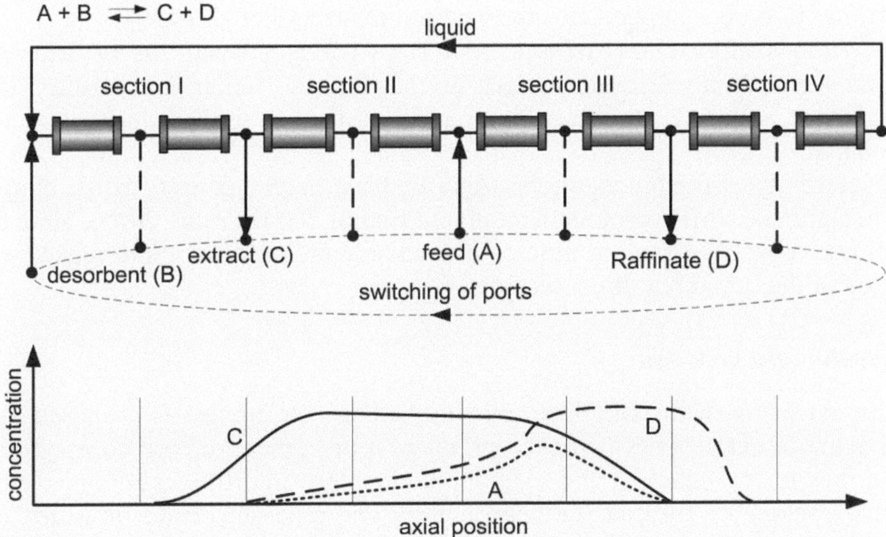

Fig. 5.4: Simulated Moving Bed reactor (reproduced from Schmidt-Traub 2005)

Hence, the TMBR model can be used for a first estimate of the operating conditions. Due to the simulated movement of the stationary phase the shifting period, the column volume V_c and the void fraction ε of the solid have to be taken into account to define the dimensionless flow rate ratios:

$$m_{j,SMBR} = \frac{\dot{V}_{j,SMBR}\; t_{shift} - (1-\varepsilon)V_c}{(1-\varepsilon)V_c} \qquad (5.6)$$

Special types of reactions may afford modifications of the process configuration. The following review briefly describes the investigations of chromatographic SMB reactors over the last decade. More studies are summarized in the literature (Fricke et al. 2003).

The most common process layout comprises four sections and a closed loop with direct recycle of the liquid leaving section IV (Figure 5.4). This configuration is often applied for esterifications (A + B ⇆ C + D). In most cases one of the reactants is simultaneously the desorbent in order to inhibit separation of the reactants and to shift the chemical equilibrium (Migliorini et al. 1998; Lode et al. 2003a; Ströhlein et al. 2004). Other investigations of esterifications have been carried out without direct recycle of the liquid leaving section IV (Kawase et al. 1999a; Kawase et al. 2001). The major advantage of this open-loop configuration is a reduced process complexity and a more robust operation. An open-loop process set-up without direct regeneration of liquid and therefore saving process section IV has also been applied to esterifications (Mazzotti et al. 1996).

Various bioreactions of type A ⇆ B + C, like biosynthesis of dextran and sucrose conversion have been investigated by Barker et al. (1992). In this case a closed loop set-up with a pulsed feed is applied during operation.

Recently also isomerizations (A ⇆ B) have been analyzed within chromatographic SMB reactors (Toumi and Engell 2004; Fricke 2005). Here a closed-loop three-section process is predominantly used without regeneration of the liquid but direct recycle.

Hashimoto process

An extension of the simulated moving bed reactor process is proposed by Hashimoto et al. (1983) for the production of higher fructose corn syrup. In this process reaction and separation take place in different columns and the reactors are located stationary within one process section. In order to implement this, the reactors are shifted together with the external streams in the direction of the liquid flow. This induces a carry-over of liquid which is contained in the reactor and causes the characteristic concentration profile shown in Figure 5.5.

A reactor with a high liquid phase concentration is shifted at the end of a shifting interval in front of a column containing significantly lower concentrations. Hence a discontinuity in the axial concentration profile arises which is balanced by adsorption and hydrodynamic effects over the shifting interval.

Applications of the Hashimoto process as a four-section process with raffinate withdrawal are discussed by Borren and Fricke (2005). A three-section process with direct recycle of the liquid phase for the production of fructose corn syrup is investigated by Hashimoto et al. (1993) and Zhang et al. (2004). Another modification of the Hashimoto process is used for the production of pure enantiomers from racemic mixtures (Borren and Schmidt-Traub 2005).

5.2.4 Degree of process integration

For optimal process synthesis different degrees of integration of chromatographic separation and chemical reaction have to be considered. The simulated moving bed reactor and the Hashimoto process have been discussed in the previous section. Additionally, a sequential process of a reactor followed by simulated moving bed separation has to be taken into account.

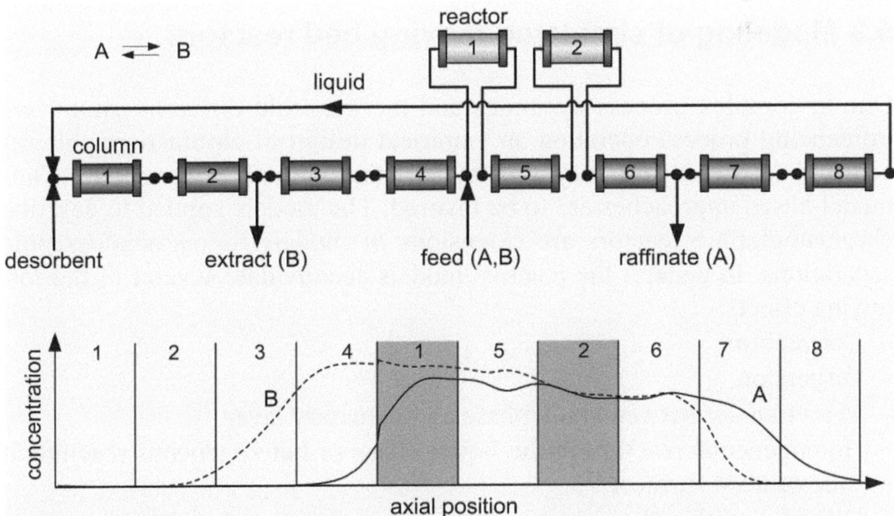

Fig. 5.5: Hashimoto process (reproduced from Schmidt-Traub 2005)

A continuously operated reactive chromatographic process involves mass transport by the liquid as well as the adsorbent, which contributes to the separation. An additional phase is the catalyst which generates the products. A total integration of reaction and separation is achieved within the simulated moving bed reactor where adsorbent and catalyst particles are either totally mixed or identical. In this case operating conditions have to be chosen which allow sufficient high reaction rates as well as efficient separation. Such a total integration is not suitable for reactions of type A $\leftrightarrows$ B because the reverse reaction cannot be reduced by a separation of different products. But a selective intensification of the reaction in one direction is possible within a partial deintegrated process like the Hashimoto process. A segregation of the functionalities also has the advantage of being able to choose different operating conditions for separation and reaction. Nevertheless a suitable solvent system is necessary and the increasing complexity of the technical realization has to be taken into account. An independent selection of operating conditions as well as the solvent is possible in case of a sequential process. Therefore the deintegrated process is used as a benchmark for the evaluation of the performance of integrated processes.

The degree of integration increases from the sequential process to the simulated moving bed reactor while the degrees of operational freedom decrease in reverse order. Therefore, the selection of a suitable process design and its operating conditions is a major task.

5.3 Modeling of simulated moving bed reactors

Due to complex process dynamics and the manifold different parameters influencing process operation an empirical design of chromatographic reactors is difficult and may result in a non-optimal performance. Thus model-based approaches are to be favored. The models applied to describe chromatographic reactors are extensions of models for chromatographic separations. In general the reactive models accomodate several of the following effects:
- Convection,
- Dispersion,
- Mass transfer between bulk phase and boundary layer,
- Homogeneous reaction in the liquid phase or heterogeneous reaction at the catalyst surface,
- Diffusion within the particle pores,
- Surface diffusion within the particle,

– Adsorption equilibrium or adsorption kinetics.

Different kinds of modeling approaches for chromatographic separations as well as chromatographic reactors have been developed and successfully applied over the last decades. An overview is given by Michel et al. (2005). More detailed descriptions can be found in publications of Guiochon et al. (1994) as well as Seidel-Morgenstern (1995).

In the following a modeling approach is discussed which takes into account convection, dispersion, homogeneous or heterogeneous reaction, adsorption equilibrium and an effective mass transfer coefficient which summarizes internal as well as external mass transfer resistances. This model is referenced in the literature as equilibrium transport dispersive model (Klatt 1999).

The mathematical models of simulated moving bed reactors are classified into models assuming a steady state true counter-current process or rigorous models taking into consideration process dynamics because of the shifting of the external streams.

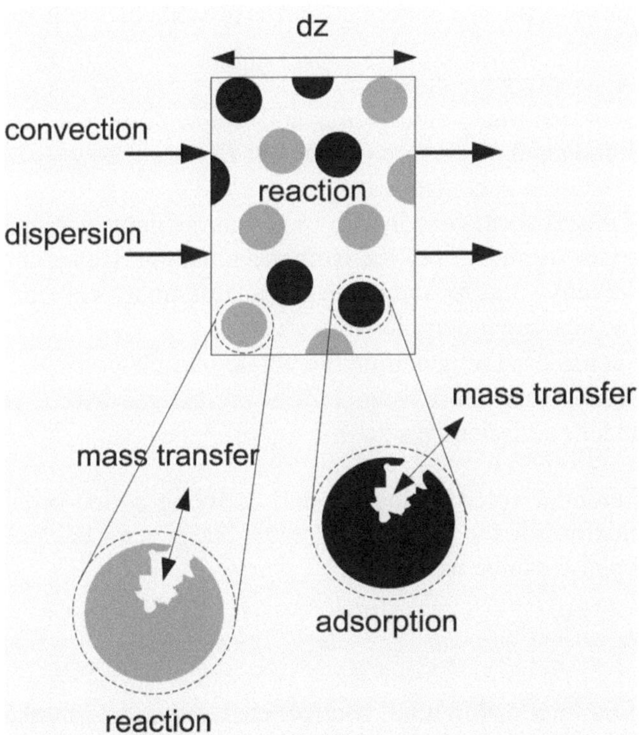

Fig. 5.6: Equilibrium transport dispersive model of a reactive column (reprinted with permission from Schmidt-Traub 2005)

5.3.1 Rigorous models

The model of a single reactive chromatographic column is the essential element of every simulated moving bed reactor. Hence the modeling approach for a single reactive column is derived first and, thereafter, the additional models necessary for the description of the more complex continuous processes are discussed.

Modeling of reactive chromatographic columns

The model of a reactive chromatographic column containing a homogeneous mixture of adsorbent and catalyst is based on the mass balances for the mobile and stationary phase in a differential volume element as depicted in Figure 5.6. For the equilibrium transport dispersive model, the following assumptions are made:
- The solid phase consists of a homogeneous mixture with uniform distribution of adsorbent and catalyst. The catalyst fraction is given by X_{cat},
- Adsorbent as well as catalyst are spherical particles with constant diameters d_p,
- Fluid density ρ and viscosity ν are constant,
- Radial distributions of internal velocity u_{int} and concentration c_i are summarized within the axial dispersion coefficient D_{ax},
- Isothermal process operation is considered,
- The influence of the eluent on the adsorption of the components is taken into account by certain parameters of the isotherm equation. Hence the isotherm equation is only valid for the given chromatographic system.
- Convection within the adsorbent particles is neglected,
- All components are considered to penetrate the whole particle pores,
- Internal and external mass transfer resistances are summarized within an effective mass transfer coefficient k_{eff}.

By balancing a differential volume element and applying a first-order Taylor series approximation for the outgoing streams the general balance equation for the liquid phase can be written as:

$$\frac{\partial m_{acc,i}}{\partial t} = -\frac{\partial \dot{m}_{conv,i}}{\partial x} dx - \frac{\partial \dot{m}_{disp,i}}{\partial x} dx - \dot{m}_{mt,ads,i} - \dot{m}_{mt,cat,i} + \dot{m}_{reac,hom,i} \qquad (5.7)$$

The accumulation within a differential volume element is determined by:

$$m_{acc,i} = \varepsilon\, c_i\, A_c\, dx \qquad (5.8)$$

Convection can be calculated by:

$$\dot{m}_{conv,i} = \varepsilon\, A_c\, u_{int}\, c_i \tag{5.9}$$

The dispersion is described by:

$$\dot{m}_{disp,i} = -D_{ax}\, \varepsilon\, A_c\, \frac{\partial c_i}{\partial x} \tag{5.10}$$

Mass transfer to the adsorbent and the catalyst can be determined by:

$$\dot{m}_{mt,ads,i} = \left(1 - X_{cat}\right)\left(1 - \varepsilon\right) A_c\, dx\, \frac{\partial q_{ads,i}}{\partial t} \tag{5.11}$$

$$\dot{m}_{mt,cat,i} = X_{cat}\left(1 - \varepsilon\right) A_c\, dx\, \frac{\partial q_{cat,i}}{\partial t} \tag{5.12}$$

The homogeneous reaction rate is calculated by:

$$\dot{m}_{rec,hom,i} = \varepsilon\, A_c\, dx\, \upsilon_i\, k_{hom}\, \Phi_{hom}\left(c_i^*\right) \tag{5.13}$$

Axial dispersion eq. (5.10) is defined in analogy to Fick's first law. The dispersion coefficient D_{ax} represents the deviations of fluid dynamics from ideal plug flow and depends on the packing quality. A calculation is possible by the correlation proposed by Chung and Wen (1968):

$$D_{ax} = \frac{u_{int} d_{p,ads}}{\dfrac{0.2}{\varepsilon} + \dfrac{0.011}{\varepsilon}\left(\varepsilon\,\mathrm{Re}\right)^{0.48}} \tag{5.14}$$

Introducing eq. (5.8) to (5.13) into the general mass balance the differential balance for the liquid phase can be written as:

$$\begin{aligned}
\frac{\partial c_i}{\partial t} = &-u_{int}\frac{\partial c_i}{\partial x} + D_{ax}\frac{\partial^2 c_i}{\partial x^2} - \left(1 - X_{cat}\right)\frac{1-\varepsilon}{\varepsilon}\frac{\partial q_{ads,i}}{\partial t} \\
&- X_{cat}\frac{1-\varepsilon}{\varepsilon}\frac{\partial q_{cat,i}}{\partial t} + \upsilon_{i,r}\, k_{hom}\, \Phi_{hom}\left(c_i^*\right)
\end{aligned} \tag{5.15}$$

By changing the catalyst fraction X_{cat} this model can be applied to describe a chromatographic separation ($X_{cat} = 0$), an integrated chromatographic reactor ($0 \leq X_{cat} \leq 1$) as well as a fixed bed reactor ($X_{cat} = 1$).

A linear driving force approach is used to describe the mass transfer between bulk phase and the pore concentration within the different particles. Therefore the balance for the stationary catalyst phase can be written as:

$$\frac{\partial q_{cat,i}}{\partial t} = k_{eff,cat,i} \frac{6}{d_{p,cat}} \left(c_i - c_{p,cat,i} \right) + \upsilon_i k_{het} \Phi_{het} \left(c^*_{p,cat,i} \right) \qquad (5.16)$$

and for the stationary adsorbent phase:

$$\frac{\partial q_{ads,i}}{\partial t} = k_{eff,ads,i} \frac{6}{d_{p,ads}} \left(c_i - c_{p,ads,i} \right) \qquad (5.17)$$

Regarding a porous adsorbent particle, the overall adsorbent loading is the sum of the pore concentration and the solid loading:

$$q_{ads,i} = \varepsilon_p c_{p,ads,i} + \left(1 - \varepsilon_p \right) q^*_{ads,i} \qquad (5.18)$$

The solid load is a function of the pore concentrations of all components and can be described by the isotherm equation:

$$q^*_{ads,i} = f \left(c_{p,ads,1}, \ldots, c_{p,ads,N} \right) \qquad (5.19)$$

To solve the partial differential equations of the column model, a set of initial and boundary conditions is necessary. As initial condition zero concentrations and loadings are assumed:

$$c_i \left(t = 0, x \right) = 0 \qquad (5.20)$$

$$c_{p,ads \, / \, cat,i} \left(t = 0, x \right) = 0 \qquad (5.21)$$

$$q_{ads \, / \, cat,i} \left(t = 0, x \right) = 0 \qquad (5.22)$$

Mass transfer in a chromatographic column is normally governed by convection. Thus the boundary condition proposed by Dünnebier (2000) can be applied at the column inlet:

$$c_i \left(t, x = 0 \right) = c_{in,i} \left(t \right) \qquad (5.23)$$

At the column outlet the closed boundary condition proposed by Danckwerts (1953) is used:

$$\frac{\partial c_i \left(t, x = L_c \right)}{\partial x} = 0 \qquad (5.24)$$

Due to the large number of influencing parameters the transfer into dimensionless equations is recommended to analyze the process performance. The feed concentration c_{feed} is selected as reference for the concentration as well as the loading:

$$c_i = c_{feed} C_i \qquad (5.25)$$

$$c_{p,ads/cat,i} = c_{feed} C_{p,ads/cat,i} \qquad (5.26)$$

$$q_{ads/cat,i} = c_{feed} Q_{ads/cat,i} \qquad (5.27)$$

By using the column length L_c the dimensionless coordinate Z is obtained:

$$x = L_c Z \qquad (5.28)$$

The residence time of the liquid phase is the reference for the dimensionless time:

$$t = \frac{L_c}{u_{int}} \Theta \qquad (5.29)$$

Introducing the eq. (5.16), (5.17) and (5.25) to (5.29) into the differential mass balance (5.15) leads to the following dimensionless mass balance:

$$
\begin{aligned}
\frac{\partial C_i}{\partial \Theta} = & -\frac{\partial C_i}{\partial Z} + \frac{D_{ax}}{u_{int} L_c} \frac{\partial^2 C_i}{\partial X^2} \\
& - (1 - X_{cat}) \frac{1-\varepsilon}{\varepsilon} k_{eff,ads,i} \frac{6}{d_{p,ads}} \frac{L_c}{u_{int}} \left(C_i - C_{p,ads,i} \right) \\
& - X_{cat} \frac{1-\varepsilon}{\varepsilon} k_{eff,cat,i} \frac{6}{d_{p,cat}} \frac{L_c}{u_{int}} \left(C_i - C_{p,cat,i} \right) \\
& + \upsilon_i k_{hom} \frac{L_c}{u_{int}} \Phi_{hom} \left(C_i^* \right)
\end{aligned}
\qquad (5.30)
$$

In the dimensionless mass balance equation the following dimensionless parameters can be identified:

Phase ratio:

$$\varepsilon_v = \frac{1-\varepsilon}{\varepsilon} \qquad (5.31)$$

Bodenstein number:

$$Bo = \frac{u_{int} L_c}{D_{ax}} \tag{5.32}$$

Effective Stanton number:

$$St_{eff,ads\,/\,cat,i} = k_{eff,ads\,/\,cat,i} \frac{6}{d_{p,ads\,/\,cat}} \frac{L_c}{u_{int}} \tag{5.33}$$

Damkoehler number:

$$Da = \frac{k_{reac} L_c}{u_{int}} \tag{5.34}$$

This finally leads to the dimensionless partial differential equation for the liquid phase:

$$
\begin{aligned}
\frac{\partial C_i}{\partial \Theta} = & -\frac{\partial C_i}{\partial Z} + \frac{1}{Bo} \frac{\partial^2 C_i}{\partial X^2} \\
& -\left(1 - X_{cat}\right) \varepsilon_v St_{eff,ads,i} \left(C_i - C_{p,ads,i}\right) \\
& - X_{cat} \varepsilon_v St_{eff,cat,i} \left(C_i - C_{p,cat,i}\right) \\
& + \upsilon_i Da_{hom} \Phi_{hom} \left(C_i^*\right)
\end{aligned}
\tag{5.35}
$$

and for the two solid phases:

$$\frac{\partial Q_{cat,i}}{\partial \Theta} = St_{eff,cat,i} \left(C_i - C_{p,cat,i}\right) + \upsilon_i Da_{het} \Phi \left(C_{p,cat}^*\right) \tag{5.36}$$

$$\frac{\partial Q_{ads,i}}{\partial \Theta} = St_{eff,ads,i} \left(C_i - C_{p,ads,i}\right) \tag{5.37}$$

Modeling of complex flowsheets

In order to obtain reasonable agreement between experimental results and process simulation extra column effects have to be taken into account. These effects are caused by the chromatographic plant itself and include the influence of piping, valves, pumps, detectors as well as further equipment and result in dead times and back mixing. The effects are summarized and mathematically described by pipe and tank models (Epping 2005).

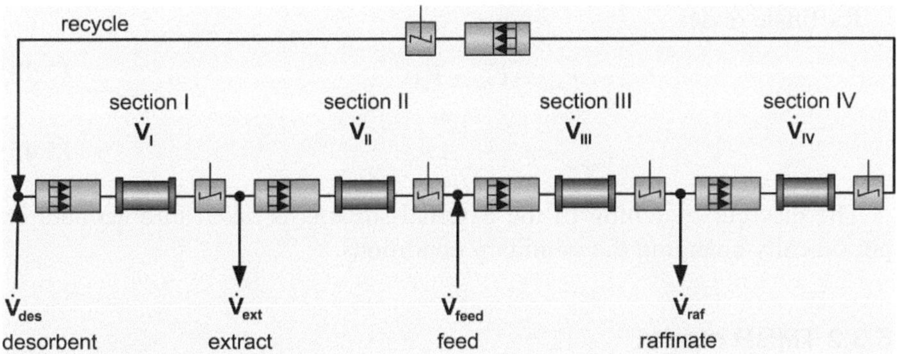

Fig. 5.7: Simulation flowsheet of the SMBR process (reprinted with permission from Schmidt-Traub 2005)

An SMBR plant consists of several columns connected by piping and maybe by additional reactors. All single parts of the plant are represented by own models which are implemented into a modular simulation tool. In a flowsheeting approach the boundary conditions of these single models are connected by node balances and the material balances. Figure 5.7 gives an example of the simulation flowsheet of the SMBR process.

The node balances also connect the internal process streams within the single sections with the external in- and outlet streams:

Desorbent node:

$$\dot{V}_{des} = \dot{V}_I - \dot{V}_{IV} \qquad (5.38)$$

$$c_{des,i}\,\dot{V}_{des} = c_{I,in,i}\,\dot{V}_I - c_{IV,out,i}\,\dot{V}_{IV} \qquad (5.39)$$

Extract node:

$$\dot{V}_{ext} = \dot{V}_{II} - \dot{V}_I \qquad (5.40)$$

$$c_{ext,i} = c_{II,in,i} = c_{I,out,i} \qquad (5.41)$$

Feed node:

$$\dot{V}_{feed} = \dot{V}_{III} - \dot{V}_{II} \qquad (5.42)$$

$$c_{feed,i}\,\dot{V}_{feed} = c_{III,in,i}\,\dot{V}_{III} - c_{II,out,i}\,\dot{V}_{II} \qquad (5.43)$$

Raffinate node:

$$\dot{V}_{raf} = \dot{V}_{III} - \dot{V}_{IV} \tag{5.44}$$

$$c_{raf,i} = c_{III,out,i} = c_{IV,in,i} \tag{5.45}$$

The discrete switching of the external streams is taken into account by periodically changing the boundary conditions.

5.3.2 TMBR model

As already mentioned, it is also possible to describe the simulated moving bed reactor by a steady state true counter-current flow of solid and liquid phase. Using the same assumptions of the equilibrium transport dispersive model, the mass balance for the liquid phase in section j can be written as:

$$\begin{aligned}
\frac{\partial c_i}{\partial t} = {}^{-}\!\!\Big/_{+}\, u_{int,j}\,\frac{\partial c_{i,j}}{\partial x} + D_{ax}\frac{\partial^2 c_{i,j}}{\partial x^2} \\
- \left(1 - X_{cat}\right)\frac{1-\varepsilon}{\varepsilon}k_{eff,ads,i}\,\frac{6}{d_{p,ads}}\frac{L_c}{u_{int}}\left(c_{i,j} - c_{p,ads,i,j}\right) \\
- X_{cat}\frac{1-\varepsilon}{\varepsilon}k_{eff,cat,i}\,\frac{6}{d_{p,cat}}\left(c_{i,j} - c_{p,cat,i,j}\right) \\
+ \upsilon_i k_{hom}\Phi_{hom}\left(c_{i,j}^*\right)
\end{aligned} \tag{5.46}$$

$$\text{for } j = I,II \,/\, III,IV$$

Please note, that the term representing convection has different signs due to the origin of the *x-axis* located in the feed port, as shown in Figure 5.8. The model is used to describe the real counter-current flow of the two phases, but non-idealities of the solid movement are neglected. Hence the balances of the solid phases can be written as:

$$\begin{aligned}
\varepsilon_{p,ads}\frac{\partial c_{p,ads,i,j}}{\partial t} + \left(1 - \varepsilon_{p,ads}\right)\frac{\partial q_{ads,i,j}}{\partial t} = \\
{}^{+}\!\!\Big/_{-}\, u_{ads}\left[\varepsilon_{p,ads}\frac{\partial c_{p,ads,i,j}}{\partial x} + \left(1 - \varepsilon_{p,ads}\right)\frac{\partial q_{ads,i,j}}{\partial x}\right] \\
+ k_{eff,ads,i}\,\frac{6}{d_{p,ads}}\left(c_{i,j} - c_{p,ads,i,j}\right)
\end{aligned} \tag{5.47}$$

$$\text{for } j = I,II \,/\, III,IV$$

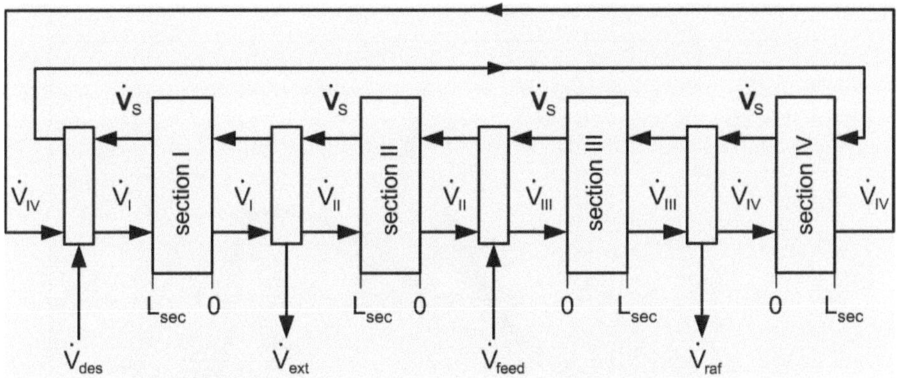

Fig. 5.8: TMBR model (reprinted with permission from Schmidt-Traub 2005)

$$\varepsilon_{p,cat}\frac{\partial c_{p,cat,i,j}}{\partial t}+\left(1-\varepsilon_{p,cat}\right)\frac{\partial q_{cat,i,j}}{\partial t}=$$

$$+\!\!\!\diagup\!\!\!_{-}\,u_{ads}\left[\varepsilon_{p,cat}\frac{\partial c_{p,cat,i,j}}{\partial x}+\left(1-\varepsilon_{p,cat}\right)\frac{\partial q_{cat,i,j}}{\partial x}\right]$$

$$+k_{eff,cat,i}\frac{6}{d_{p,cat}}\left(c_{i,j}-c_{p,cat,i,j}\right)$$

$$+\upsilon_i k_{het}\Phi_{het}\left(c^*_{p,cat,i,j}\right)$$

$$\text{for } j = \text{I,II / III,IV}$$

$$(5.48)$$

The relationship between the internal and external streams of the TMBR is described by the same total mass balance eq. (5.38), (5.40), (5.42) and (5.44) as for the SMBR. Component balances, however, also have to consider the solid streams.

Desorbent node:

$$c_{i,des}\dot{V}_{des}=c_{i,I}^{(L_{sec})}\dot{V}_I-c_{i,IV}^{(L_{sec})}\dot{V}_{IV}+\left(q_{ads,i,IV}^{(L_{sec})}-q_{ads,i,I}^{(L_{sec})}\right)\left(1-X_{cat}\right)\dot{V}_{solid}$$

$$+\left(q_{cat,i,IV}^{(L_{sec})}-q_{cat,i,I}^{(L_{sec})}\right)X_{cat}\dot{V}_{solid}$$

$$(5.49)$$

Extract draw-off node:

$$c_{i,ext}\dot{V}_{ext}=c_{i,I}^{(0)}\dot{V}_I-c_{i,II}^{(L_{sec})}\dot{V}_{II}+\left(q_{ads,i,II}^{(L_{sec})}-q_{ads,i,I}^{(0)}\right)\left(1-X_{cat}\right)\dot{V}_{solid}$$

$$+\left(q_{cat,i,II}^{(L_{sec})}-q_{cat,i,I}^{(0)}\right)X_{cat}\dot{V}_{solid}$$

$$(5.50)$$

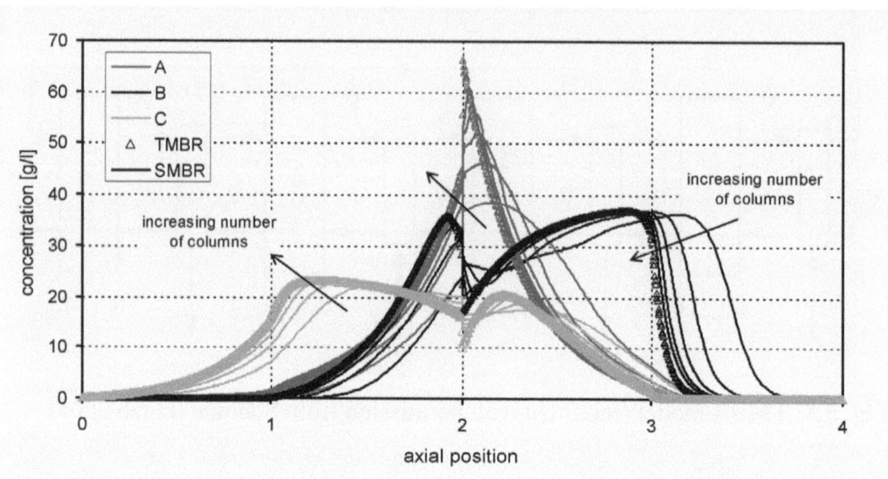

Fig. 5.9: Comparison of the axial concentration profiles of a TMBR and SMBR's with 8 to 128 columns for a split reaction

Feed node:

$$c_{i,feed}\dot{V}_{feed} = c_{i,III}^{(0)}\dot{V}_{III} - c_{i,II}^{(0)}\dot{V}_{II} + \left(q_{ads,i,II}^{(0)} - q_{ads,i,III}^{(0)}\right)\left(1 - X_{cat}\right)\dot{V}_{solid}$$
$$+ \left(q_{cat,i,II}^{(0)} - q_{cat,i,III}^{(0)}\right)X_{cat}\dot{V}_{solid} \tag{5.51}$$

Raffinate draw-off node:

$$c_{i,raf}\dot{V}_{raf} = c_{i,III}^{(L_{sec})}\dot{V}_{III} - c_{i,IV}^{(0)}\dot{V}_{IV} + \left(q_{ads,i,IV}^{(0)} - q_{ads,i,III}^{(L_{sec})}\right)\left(1 - X_{cat}\right)\dot{V}_{solid}$$
$$+ \left(q_{cat,i,IV}^{(0)} - q_{cat,i,III}^{(L_{sec})}\right)X_{cat}\dot{V}_{solid} \tag{5.52}$$

The model described here can be reduced by neglecting dispersion, mass transfer resistances and by disregarding the catalyst phase. In this case an analytical solution for steady state is possible for the mass balance equations as discussed in section 5.5.

5.3.3 Comparison of TMBR and SMBR

Concentration profiles calculated with a TMB model are in good agreement with results obtained by the rigorous SMB model if more than three columns per section are used (Ruthven and Ching 1989). But due to high investment costs SMB plants often consist of two or even fewer columns in the different sections. This explains why the real behavior can differ significantly from the performance predicted by the TMB model (Strube et al. 1998).

The same conclusions are valid for chromatographic reactors. As shown in Figure 5.9, the differences in the concentration profiles of a TMBR and a SMBR increase drastically with a decreasing number of columns within the SMBR. The total bed length is always kept constant, while the number of columns per section varies. The operating parameters are taken from an optimized TMBR process.

Additionally, differences in the concentration profiles result in wrong assumptions about the residence time of the reactants within the single sections of a TMBR and, therefore, higher conversion rates compared to a SMBR (Lode et al. 2003b). The carry-over, i.e. the transport of liquid against the direction of liquid flow induced by switching, is also not taken into account by true moving bed models. Therefore the use of a rigorous SMBR model is recommended for the detailed design as well as optimization of continuous counter-current reactors. Nevertheless, the TMBR model is a useful tool to approximate the behavior of an SMBR and to get a first estimate of the operating conditions.

5.4 Experimental model validation

The following investigations are based on the rigorous equilibrium transport dispersive model. So this section describes the determination of the model parameters for this approach. Different types of reaction as well as process set-ups are utilized to validate the model. Further examples are described by Fricke (2005).

5.4.1 Parameter determination

Chromatographic reactors are often limited by the chromatographic separation (Matsen et al. 1965). As a result of this a careful selection of the chromatographic parameters is crucial. Process design starts with the selection of the chromatographic system and the process concept. These are characterized by the following design parameters specifying the process and are subject of optimization:
– Stationary phase ($d_{p,ads}$),
– Catalyst ($d_{p,cat}$),
– Mobile phase composition,
– Chromatographic bed (length L_c and diameter d_c),

- Catalyst fraction (X_{cat}),
- Maximum allowable pressure drop (Δp),
- Temperature,
- Process-specific design parameters (e.g. segmentation).

Depending on the design parameters, the system-inherent physical and chemical parameters have to be determined. These model parameters are fixed by the following design and operating parameters:
- Plant parameters (extra plant effects),
- Packing parameters (void fraction, porosity, dispersion),
- Equilibrium isotherms,
- Mass transfer coefficients,
- Reaction kinetics.

After selection and specification of the chromatographic system as well as the process design the operating parameters have to be chosen. These parameters are also subject of optimization. In case of SMBR operation, the operating parameters are:
- Feed concentration (c_{Feed}),
- Switching time (t_{shift}),
- Dimensionless flow rate ratios (m_j).

In order to simulate reactive chromatography as exactly as possible, the model parameters have to be determined with sufficient accuracy. The method of consistent model parameters is depicted in Figure 5.10 (Altenhöner et al. 1997). In this concept the simplest parameters are determined first, and are thereafter used to determine the more complex parameters. It is of major importance to determine the model parameters for the same stationary phase and catalyst as well as the same solvent as applied later for preparative runs.

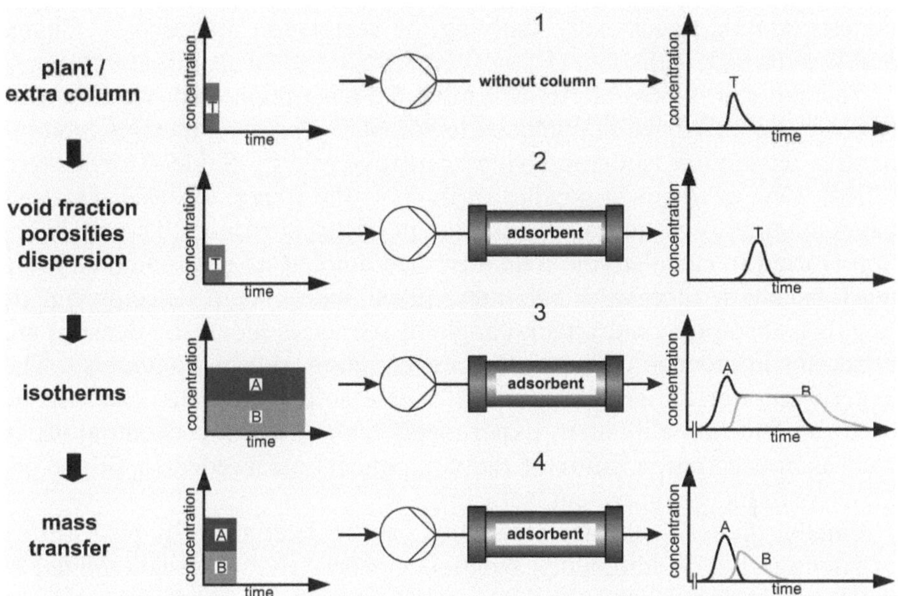

Fig. 5.10: Parameter determination concept (reproduced from Schmidt-Traub 2005)

All experiments are based on small pulses or breakthrough curves of non-reacting mixtures. The model parameters describing the reaction should be determined independently of the model parameters describing the chromatographic separation. Due to the applied flowsheeting approach, all model parameters are determined within a single column. Before operating an SMBR, the similarity of all columns has to be proven by prior experiments.

In the following the most important steps for the determination of the model parameters are summarized. More detailed information on different methods and further applicable approaches are discussed by Michel et al. (2005).

In a first step the plant behavior, e.g. the extra column effects, are characterized by pulse experiments. For that purpose a small amount of one component is injected on the plant without the column. The measured chromatogram is used to determine the dead time and the dispersion of the plant.

In a second step the column packing is characterized by tracer pulses. A tracer which cannot penetrate the pore system is used to calculate the void fraction of the packing, and a non-adsorbable tracer which penetrates the pore system of the particles is used to determine the total porosity. The axial dispersion coefficient D_{ax} can be calculated from experiments with a non-penetrating tracer or by applying the correlation proposed by Chung and Wen (eq. (5.14)).

Thereafter it is possible to determine the adsorption isotherm. The different static and dynamic methods to measure single component or competitive adsorption isotherms have been reviewed by Seidel-Morgenstern (2004). One of the most popular methods is the frontal analysis. This dynamic method is based on a breakthrough of the substances. Online detection is used to calculate the total adsorbed amount of each component by mass balances. In case of mixtures, linear independent detector signals (e.g. UV absorption and rotation angle or refractive index and density) are necessary in order to calculate the concentration of each component. The experiments for the determination of single solute isotherms can also be used for detector calibration. Experiments with different concentrations as well as concentration ratios of the components are needed to fit the parameters of multi-component isotherms.

Finally, the mass transfer coefficient can be determined by peak fitting of single pulses. Additionally, reaction kinetics have to be determined in the absence of the adsorbent. In case of homogeneous catalyzed reactions simple batch experiments are suitable to determine the required parameters. In case of heterogeneously catalyzed reactions, a parameter set similar to the parameters describing the adsorption is necessary. This implies that a packed bed of catalyst is used and additional parameters (e.g. porosity of the catalyst particle or mass transfer coefficients) have to be determined.

5.4.2 Production of β-phenethylacetate

System properties

The production of β-phenethylacetate serves to illustrate a heterogeneously catalyzed esterification reaction. β-Phenethylacetate is a flavor which finds application in the industrial production of scents and perfumes. By separation of water β-phenethyl alcohol and acetic acid are completely transformed to the ester:

$$CH_3COOH + C_6H_5CH_2CH_2OH \rightleftarrows C_6H_5CH_2CH_2COOCH_3 + H_2O \quad (5.53)$$

The acidic ion exchanger resin Amberlite$^®$ 15™ (Rohm and Haas) acts as adsorbent as well as catalyst. eq. (5.54) describes the heterogeneously catalyzed reaction:

$$r_{i,het} = \upsilon_i\, k\, \frac{c_{acid}\, c_{alcohol} - \dfrac{c_{ester}\, c_{water}}{K_{eq}}}{\left(1 + b_{alcohol}\, c_{alcohol} + b_{water}\, c_{water}\right)^2} \qquad (5.54)$$

1,4-Dioxane is used as eluent for the separation. The adsorption equilibrium of the alcohol and water are described by a Langmuirian isotherm.

$$q_i = \frac{H_i\, c_i}{1 + b_i\, c_i} \qquad (5.55)$$

The adsorption of the acid and the ester are described by a linear Henry isotherm. All process, operating and model parameters of the SMBR process are summarized in Table 5.1 (Fricke 2005).

Table 5.1: Parameters for the esterification of β-phenethyl alcohol

Parameter		Value	
Segmentation		2-2-3-1	
Column length	L_c	30	cm
Column diameter	d_c	1	cm
Flow rate section I	V_I	4.83	$cm^3\ min^{-1}$
Flow rate section II	V_{II}	1.45	$cm^3\ min^{-1}$
Flow rate section III	V_{III}	2.00	$cm^3\ min^{-1}$
Flow rate section IV	V_{IV}	0.34	$cm^3\ min^{-1}$
Shifting period	t_{shift}	45	min
Temperature	T	85	°C
Feed concentration (equimolar)	c_{feed}	5.69	$mmol\ cm^{-3}$
Total porosity	ε_t	0.36	-
Viscosity	η	$5.8\ 10^{-3}$	$g\ cm^{-1}\ s^{-1}$
Density	ρ	1.03	$g\ cm^{-3}$
Mass transfer coefficient alcohol	$k_{eff,alcohol}$	$4.39\ 10^{-4}$	$cm\ s^{-1}$
Mass transfer coefficient acid	$k_{eff,acid}$	$4.82\ 10^{-4}$	$cm\ s^{-1}$
Mass transfer coefficient ester	$k_{eff,ester}$	$4.17\ 10^{-4}$	$cm\ s^{-1}$
Mass transfer coefficient water	$k_{eff,water}$	$1.11\ 10^{-3}$	$cm\ s^{-1}$
Henry coefficient alcohol	$H_{alcohol}$	0.884	-
Langmuir coefficient alcohol	$b_{alcohol}$	24.16	$cm^3\ mol^{-1}$

Henry coefficient acid	H_{acid}	0.687	-
Henry coefficient ester	H_{ester}	0.479	-
Langmuir coefficient water	b_{water}	568.54	$cm^3\ mol^{-1}$
Equilibrium constant	K_{eq}	3.18	-
Reaction rate constant	k	0.165	$cm^3\ s^{-1}\ g^{-1}$

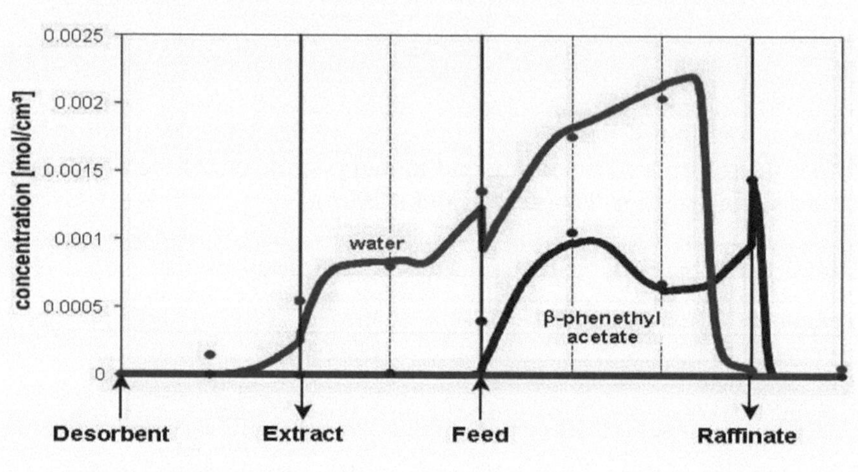

Fig. 5.11: Model verification for the production of β-phenethylacetate

SMBR experiment

An SMBR experiment with eight columns and a segmentation of 2-2-3-1 is employed for model validation. The lab scale SMB plant is operated as an open-loop configuration without direct recycle of the regenerated eluent. To ensure a sufficient regeneration of the stationary phase a high flow rate in section I is necessary. The SMBR experiments have been done in cooperation with M. Kawase, Kyoto University.

After two cycles plant operation has been stopped and samples were taken from the connecting lines of the columns and thereafter analyzed off-line to determine the content of ester and water. In Figure 5.11 the experimental axial concentration profiles of both products are compared with the concentration profiles calculated by the rigorous model. A good agreement for the profiles is obtained.

5.4.3 Thermal racemization of Troegers Base

System properties

The thermal racemization of Troegers Base exemplifies a homogeneously catalyzed reaction. Troegers Base is chosen as an example for the large group of enantiomers showing the ability to racemize. The thermal racemization is characterized by a large energy of activation. As a consequence the reaction can be neglected at lower temperatures (e.g. room temperature) and is considerably fast at elevated temperatures:

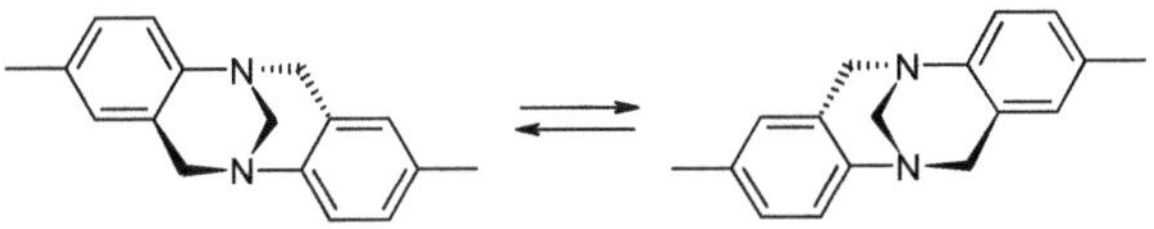

$$\tag{5.56}$$

Acidic media catalyze the reaction of Troegers Base. But the solubility of the substance in aqueous media is very low. Therefore a mixture of 2-propanole and acetic acid (1:1) serves as eluent. The temperature-dependent rate equation is given by eq. (5.57):

Table 5.2: Parameters for the racemization of Troegers Base

Parameter		Value	Units
Segmentation		2-2-2+2-2	
Column length	L_c	10	cm
Column diameter	d_c	1	cm
Reactor length	L_r	100	cm
Reactor diameter	d_r	0.53	cm
Flow rate section I	$\dot{V}_I$	6.44	cm^3 min^{-1}
Flow rate section II	$\dot{V}_{II}$	1.53	cm^3 min^{-1}
Flow rate section III	$\dot{V}_{III}$	2.75	cm^3 min^{-1}
Flow rate section IV	$\dot{V}_{IV}$	1.32	cm^3 min^{-1}
Shifting period	t_{shift}	420	s
Asynchronous switching time	t_{asyn}	415	s
Temperature reaction	T	80	°C
Temperature separation	T	25	°C
Feed concentration	c_{feed}	0.005	g cm^{-3}
Total porosity	ε_t	0.632	-
Particle porosity	ε_p	0.4	-
Viscosity at 25°C	η	2.680 10^{-2}	g cm^{-1} s^{-1}
Density at 25°C	ρ	0.845	g cm^{-3}
Viscosity at 80°C	η	6.479 10^{-3}	g cm^{-1} s^{-1}
Density at 80°C	ρ	0.788	g cm^{-3}

Mass transfer coefficient TB	k_{eff}	$4.00 \ 10^{-4}$	$cm \ s^{-1}$
Henry coefficient TB+	H_+	3.028	-
Henry coefficient TB-	H_-	10.997	-
Langmuir coefficient	b_{++}	0	$cm^3 \ g^{-1}$
Langmuir coefficient	b_{+-}	3412.72	$cm^3 \ g^{-1}$
Langmuir coefficient	b_{-+}	543.422	$cm^3 \ g^{-1}$
Langmuir coefficient	b_{--}	132.077	$cm^3 \ g^{-1}$
Activation energy	E_A	92.318	$kJ \ mol^{-1}$
Reaction rate constant	k_0	$7.432 \ 10^{10}$	s^{-1}

$$r_i = \upsilon_i \, k_0 \exp\left(-\frac{E_A}{RT}\right)\left(c_{TB+} - c_{TB-}\right) \tag{5.57}$$

The adsorption equilibrium can be illustrated by an asymmetric modified multi-component Langmuir isotherm:

$$q_i = \frac{H_i c_i}{1 + \sum_k b_{i,k} c_k} \tag{5.58}$$

The model parameters for the equilibrium transport dispersive model are summed up in Table 5.2.

Hashimoto experiment

For model validation a Hashimoto experiment with eight columns and two reactors within section III (segmentation 2-2-2+2-2) was carried out. The semi-preparative SMB plant is operated in a closed-loop arrangement. Liquid, regenerated in section IV, is directly recycled into section I by a recycle pump. To compensate the dead volume of the recycle line, an asynchronous switching of the external streams is applied and the flow rate of the extract stream is controlled by a PI controller in order to stabilize the pressure in the recycle line. The operating and design parameters are summarized in Table 5.2.

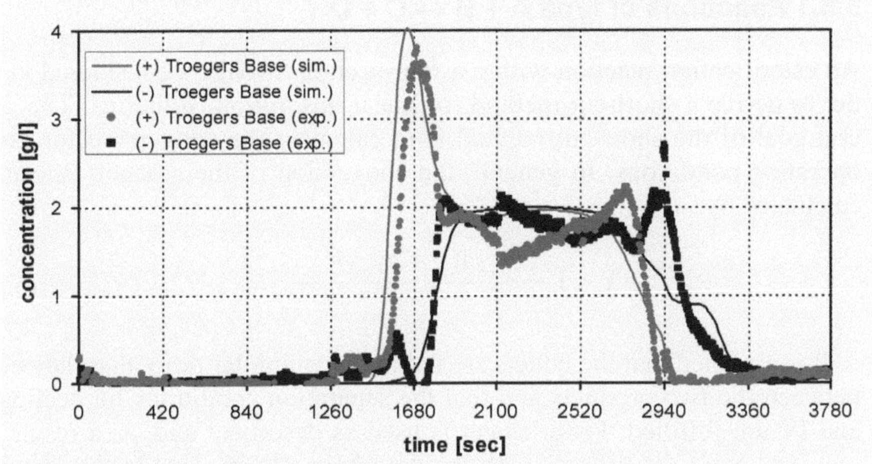

Fig. 5.12: Model verification for the racemization of Troegers Base

During the experiments temporal concentration profiles are recorded in the recycle line by an on-line detection system consisting of a polarimeter and an UV detector. Figure 5.12 pictures concentration profiles measured in the recycle line as well as the concentration profiles calculated by the rigorous model. Again there is a good agreement between experiment and simulation. Therefore the transport dispersive model appears to be validated for the theoretical simulation of the SMBR reactor.

5.5 Short-cut design methods for SMB reactors

Successful operation of counter-current chromatographic reactors strongly depends on the conversion of the educts and the separation of the products. Resulting from this the choice of the internal flow rates in the different sections with respect to the adsorption behavior of the different components is essential. This section introduces a method to estimate a priori the conversion rate of an educt on the basis of the given operating conditions. The method is based on the ideal TMBR model. The results can be used as first estimate for further detailed process optimization. This method has been derived in parallel by Lode (2002) and Fricke (2003). The authors outline this procedure in detail and present solutions for further types of reactions.

5.5.1 Reactions of type A + B ↔ C + D

An esterification reaction within a four-section SMBR is considered in order to derive a short-cut method for linear adsorption isotherms. The general goal of the short-cut method is to calculate the conversion for given operating conditions. In general, the conversion of the reactant A can be calculated by:

$$X = 1 - \frac{c_{A,ext}\,\dot{V}_{ext} + c_{A,raf}\,\dot{V}_{raf}}{c_{A,feed}\,\dot{V}_{feed}} \tag{5.59}$$

It is assumed that the educts are fed in an equimolar ratio, that they elute between the two products and that the separation conditions for sections I and IV are fulfilled. Fresh eluent is used as desorbent and, as a result, no reactants enter the process with the desorbent stream. Due to the assumption of an ideal TMB reactor educt A does not leave sections II and III and is not present in sections I and IV. Hence the inlet concentrations into sections I and IV are zero. After substituting the external streams by the component balances eq. (5.50) to (5.52) and transforming the volume flows and concentrations in their dimensionless form, the conversion rate can be calculated by:

$$X = 1 - \frac{\left[Q^1_{A,II} - m_{II}C^1_{A,II}\right] + \left[m_{III}C^1_{A,III} - Q^1_{A,III}\right]}{\left[Q^0_{A,II} - m_{II}C^0_{A,II}\right] + \left[m_{III}C^0_{A,III} - Q^0_{A,III}\right]} \tag{5.60}$$

To determine the conversion X, the concentrations at the inlet (index 0) and outlet (index 1) of reaction and separation sections II and III have to be known. Due to this the differential mass balances have to be solved. For a solid phase reaction, adsorption equilibrium without any transport hindrances and steady state the eq. (5.46) and (5.47) can be combined and reduced to:

$$\left(m_{II} - H_i\right)\frac{\partial C_{i,II}}{\partial x} + \upsilon_i\, Da_{II}\left(C_{A,II}\,C_{B,II} - \frac{C_{C,II}\,C_{D,II}}{K_{EQ}}\right) = 0 \tag{5.61}$$

$$\left(H_i - m_{III}\right)\frac{\partial C_{i,III}}{\partial x} + \upsilon_i\, Da_{III}\left(C_{A,III}\,C_{B,III} - \frac{C_{C,III}\,C_{D,III}}{K_{EQ}}\right) = 0 \tag{5.62}$$

As the mass balance equations of the four components in sections II and III depend on each other, a direct integration is not possible. Fish and Carr (1989) suggested to eliminate the reaction term in the mass balance of component B by subtraction and in the balances of components C and D by addition of the balance for component A. In the following the equations for section II are outlined in detail. The net flow rate for the educts A and B is:

$$W_{B,II} = \left(m_{II} - H_B\right)C_{B,II} - \left(m_{II} - H_A\right)C_{A,II} \tag{5.63}$$

and for educt A and products C or D respectively:

$$W_{i,II} = \left(m_{II} - H_i\right)C_{i,II} + \left(m_{II} - H_A\right)C_{A,II} \tag{5.64}$$

$$\text{for } i = C, D$$

The constant $W_{B,II}$ describes the net flow rate of component A and B in section II and is not dependent on the axial position within the section. The dimensionless concentration $C_{B,II}$ is calculated by eq. (5.63):

$$C_{B,II} = \frac{W_{B,II}}{\left(H_B - m_{II}\right)} + \frac{\left(H_A - m_{II}\right)}{\left(H_B - m_{II}\right)}C_{A,II} = K_{B,II} + M_{B,II}C_{A,II} \tag{5.65}$$

Table 5.3: Variables used to calculate the reaction region for the reaction $A + B \leftrightarrows C + D$ and linear isotherms

	Section II	Section III
$K_{B,j}$	$\dfrac{W_{B,II}}{H_B - m_{II}}$	$\dfrac{W_{B,III}}{m_{III} - H_B}$
$K_{C/D,j}$	$\dfrac{-W_{C/D,II}}{H_{C/D} - m_{II}}$	$\dfrac{-W_{C/D,III}}{m_{III} - H_{C/D}}$
$M_{i,j}$	$\dfrac{H_A - m_{II}}{H_i - m_{II}}$	$\dfrac{m_{III} - H_A}{m_{III} - H_i}$
$a_{1,j}$	$M_{B,j}K_{EQ} - M_{C,j}M_{D,j}$	
$a_{2,j}$	$K_{B,j}K_{EQ} + K_{C,j}M_{D,j} + K_{D,j}M_{C,j}$	
$a_{3,j}$	$-K_{C,j}K_{D,j}$	
$a_{4,j}$	$\dfrac{-Da_j}{\left(m_j - H_A\right)K_{EQ}}$	
Δ_j	$4a_{1,j}a_{3,j} - a_{2,1}^2$	

Similar equations can be retrieved in order to describe the net flow rates of the products in section II. After applying the same procedure to section III, a general differential mass balance equation for product A can be determined:

$$\frac{dC_{A,j}}{dZ} = a_{4,j}\left[a_{1,j}C_{A,j}^2 + a_{2,j}C_{A,j} + a_{3,j}\right] \tag{5.66}$$

All variables are summarized in Table 5.3. During integration two different cases have to be distinguished:

$$\frac{2}{\sqrt{\Delta_j}}\left[\arctan\frac{2a_{1,j}C_{A,j}^{(1)} + a_{2,j}}{\sqrt{\Delta_j}} - \arctan\frac{2a_{1,j}C_{A,j}^{(0)} + a_{2,j}}{\sqrt{\Delta_j}}\right] = a_{4,j}$$

$$\text{for } \Delta j \geq 0$$

$$\frac{1}{\sqrt{-\Delta_j}}\left[\ln\frac{2a_{1,j}C_{A,j}^{(1)} + a_{2,j} - \sqrt{-\Delta_j}}{2a_{1,j}C_{A,j}^{(1)} + a_{2,j} - \sqrt{-\Delta_j}} - \ln\frac{2a_{1,j}C_{A,j}^{(0)} + a_{2,j} - \sqrt{-\Delta_j}}{2a_{1,j}C_{A,j}^{(0)} + a_{2,j} - \sqrt{-\Delta_j}}\right] = a_{4,j}$$

$$\text{for } \Delta j < 0$$

$$\tag{5.67}$$

Each of these equations gives an analytical solution for the inlet and outlet concentrations of the reactant A in section j. In order to evaluate the equation the net flow rate $W_{i,j}$ has to be known. The net flow rates depend on the elution behavior for the given flow rate ratios m_j. Therefore the separation region has to be divided into several sub-regions as indicated in Figure 5.3. The boundary conditions for the different regions are as follows:

Boundary conditions sub-region 1

In sub-region 1 product D is the only stronger adsorbed component and thus moving with the solid. All other components are transported with the liquid in direction of the raffinate withdrawal port and do not enter section II. Hence, the concentrations of the reactants A and B at the inlet of section III are known and the net flow rate of B can be calculated by:

$$W_{B,III} = \left(H_A - m_{III}\right)C_{A,III}^{(0)} - \left(H_B - m_{III}\right)C_{B,III}^{(0)} \tag{5.68}$$

Both products are formed inside section III. As a result, the concentration of product C is neglected at the inlet and the net flow rate is determined by eq. (5.69):

$$W_{C,III} = \left(H_A - m_{III}\right)C_{A,III}^{(0)} \tag{5.69}$$

Product D propagates towards the extract port. Therefore the inlet concentration can be neglected. At the raffinate port the net flow rate is defined by:

$$W_{D,III} = \left(H_A - m_{III}\right)C_{A,III}^{(1)} \tag{5.70}$$

After the net flow rates of all components have been defined, the numerical solution of the differential eq. (5.67) is possible.

Boundary conditions sub-region 2

In sub-region 2 only the weaker adsorbed product C is moving with the liquid in direction of the raffinate port. All other components are transported with the solid in direction of the extract port and do not enter section III. Hence, the concentrations of the reactants A and B at the inlet of section II is calculated by a balance at the feed port. The net flow rate of B is then calculated by eq. (5.71):

$$W_{B,II} = \left(m_{II} - H_A\right)C_{A,II}^{(0)} - \left(m_{II} - H_B\right)C_{B,II}^{(0)} \tag{5.71}$$

Reaction occurs only in section II which means that the concentration of product C can be neglected at the extract port and the net flow rate is defined by:

$$W_{C,II} = \left(m_{II} - H_A\right)C_{A,II}^{(1)} \tag{5.72}$$

The concentration of product D at the feed port is neglected due to the propagation towards the extract port. The net flow rate is consequently as follows:

$$W_{D,II} = \left(m_{II} - H_A\right)C_{A,II}^{(0)} \tag{5.73}$$

Boundary conditions sub-region 3

Depending on the elution behavior of the reactants, four different cases have to be distinguished in sub-region 3. As in sub-region 1 the reaction in region 3.1 occurs only in section III resulting in the application of the eq. (5.68) to (5.70). Please note that the inlet concentration of reactant B is equal for both sections. In the same way the conditions for sub-region 2 (eq. (5.71) to (5.73)) can be applied to calculate the operating points in region 3.2 for equal inlet concentrations of reactant A.

In case of operating points in sub-section 3.3, reaction occurs in section II and III and the concentration of the weaker adsorbed product D cannot be neglected in section III. By combining eq. (5.70) and the mass balance at the feed port the net flow rate for component C in section III is:

$$W_{C,III} = \left(-m_{III} + H_A\right) - \left(m_{II} - H_C\right)\left(K_{C,II} - M_{C,II}\right)$$

(5.74)

and for component D in section II:

$$W_{D,II} = \left(m_{II} - H_A\right) + \left(m_{III} - H_D\right)\left(K_{B,III} - M_{B,III}\right)$$

(5.75)

If the operating points are located in sub-region 3.4, the educts propagate into different process sections due to their different adsorption behavior. This leads to the fact that operating points within this sub-region are not advisable for SMBR operation.

5.5.2 Other types of reaction

Reactions of type A $\leftrightarrows$ C + D are also of interest for application in chromatographic SMB reactors. By introducing the net flow rates for the products and thereby eliminating the reaction term a partial differential equation like eq. (5.66) can be retrieved for these types of reaction and therefore the same solution can be applied (Fricke 2005). The definitions of the introduced variables for a four-section SMBR are summarized in Table 5.4.

Further reactions of interest are of type A $\leftrightarrows$ C. After introducing the net flow rate for the product, the following partial differential equation is received:

$$\frac{dC_{A,j}}{dZ} = a_{4,j}\left[a_{2,j}C_{A,j} + a_{3,j}\right]$$

(5.76)

All introduced variables are also visualized in Table 5.4. After integration the following equation can be retrieved:

$$\frac{1}{a_{2,j}}\ln\frac{a_{2,j}C_{A,j}^{(1)} + a_{3,j}}{a_{2,j}C_{A,j}^{(0)} + a_{3,j}} = a_{4,j}$$

(5.77)

To determine the boundary conditions for the different types of reactions, similar assumptions can be made as discussed in the section before.

Table 5.4: Variables used to calculate the reaction region for different types of reaction and a linear isotherm

	A ⇆ C+D		A ⇆ C	
	Section II	Section III	Section II	Section III
$K_{C/D,j}$	$\dfrac{-W_{C/D,II}}{H_{C/D}-m_{II}}$	$\dfrac{-W_{C/D,III}}{m_{III}-H_{C/D}}$	$\dfrac{-W_{C,II}}{H_C-m_{II}}$	$\dfrac{-W_{C,III}}{m_{III}-H_C}$
$M_{i,j}$	$\dfrac{H_A-m_{II}}{H_i-m_{II}}$	$\dfrac{m_{III}-H_A}{m_{III}-H_i}$	$\dfrac{H_A-m_{II}}{H_C-m_{II}}$	$\dfrac{m_{III}-H_A}{m_{III}-H_C}$
$a_{1,j}$	$-M_{C,j}M_{D,j}$		-	
$a_{2,j}$	$K_{EQ}+K_{C,j}M_{D,j}+K_{D,j}M_{C,j}$		$K_{EQ}+M_{C,j}$	
$a_{3,j}$	$-K_{C,j}K_{D,j}$		$-K_{C,j}$	
$a_{4,j}$	$\dfrac{-Da_j}{\left(m_j-H_A\right)K_{EQ}}$		$\dfrac{-Da_j}{\left(m_j-H_A\right)K_{EQ}}$	
Δ_j	$4a_{1,j}a_{3,j}-a_{2,1}^2$		-	

5.5.3 Short-cut calculation for irreversible esterification

The short-cut method derived in the previous section can be applied for design studies of chromatographic reactors. As an example the results for an irreversible esterification (A + B → C + D) are presented here. Further examples are discussed in the literature (Lode 2002; Fricke 2005).

For linear adsorption, isotherms of the two reactants ($H_A = 0.54$ and $H_B = 0.56$) the operating area for a conversion higher than 80% is shown in Figure 5.13. The reaction region for this bimolecular reaction is formed by three intersecting ellipses. Shape and size of the ellipses depend on the adsorption behavior of all components. Depending on the feed flow rate the operating point has to be chosen within different sub-regions.

For the smallest indicated feed flow rate two maxima concerning the conversion occur. These maxima originate in different adsorption behavior. In order to achieve high conversion rates, the reaction time has to be maximized. Due to this each maximum is caused by an optimal residence time of one component.

The effect is diminishing with increasing feed flow rates until only one maximum is observed for the highest indicated flow rate.

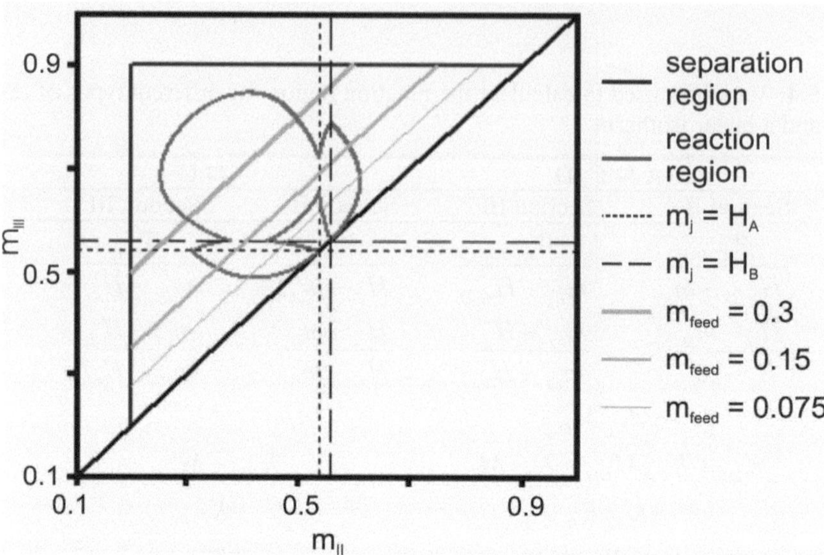

Fig. 5.13: Reaction and separation region for the irreversible esterification

5.6 Design of chromatographic reactors

The design of chromatographic reactors involves the following steps which strongly influence each other:
- Selection of the chromatographic system,
- Selection of the design parameters,
- Selection of the operating parameters,

5.6.1 Choice of the chromatographic system

The initial step to design a chromatographic reactor is the selection of the chromatographic system, i.e. a suitable adsorbent as well as a desorbent in respect of the reaction conditions. In the following section some key aspects are briefly discussed. Further and more detailed information are presented by Wewers et al. (2005).

Regarding solvent selection, the primary aim is the application of a mobile phase which exhibits a large solubility for all reactants and offers a good stability for the products. Additionally, it is recommended to use pure solvent or at least a mixture which can be easily reused for further separations. The chromatographic system is completed by the adsorbent. In this case it is important to find a stationary phase which is stable under the given process conditions, offers a suitable selectivity in combination with the mobile phase and shows a high loadability for the reactants.

In most cases the reaction requires an additional homogeneous or heterogeneous catalyst. Due to the combination of reaction and separation the operating conditions are often limited and a compromise between good selectivity at lower temperature on one side and large reaction rate at high temperature on the other side. An analysis by Matsen et al. (1965) has shown that a sufficient reaction rate is recommended and therefore the separation should be the limiting step within a chromatographic reactor.

The selection of the chromatographic system is the crucial step in the design of chromatographic reactors in respect of process economy. Suboptimal decisions during this step cannot be regained during optimization of design and operating parameters.

5.6.2 Model based optimization of design and operating parameters

The most relevant objective functions for process optimization are process economics and, finally, the price for a certain product. But in practice very often other functions are used which are of major influence on the overall objective function and are easier to determine. In case of chromatographic processes it is common practice to optimize productivities or eluent consumptions. They are unambiguous and easy to calculate as they result from mass balances. But in any case, it should be checked if they are the determining criteria for optimal economics of the current process. In an example case Susanto et al. (2005) point out how process design changes if productivity, eluent consumption or yield are used as objective function.

Optimization of the operating parameters

A systematic approach to process layout is essential in order to derive an optimal operating point. The strategy described in the following is based on the sequential optimization strategy proposed by Jupke et al. (2002) for the design of simulated moving-bed processes. Based on the validated rigorous process model, the adsorption and desorption frontiers are moved by changing the flow rate ratios within the different sections in order to remove the products with the required purity. This sequential optimization of different operating and design parameters does not necessarily yield global optima. Major advantages of this procedure are information about the quality of the determined optimum as well as the increase of procedural process understanding.

Based on initial assumptions on the process configuration and plant geometry, initial values of the design parameters, especially the flow rate ratios, are determined. For a first assumption short-cut methods can be applied (see section 5.5). Afterwards the product purities as well as the value of the objective function are determined by rigorous simulation. Further operating points within the m_{II}, m_{III}-plane are evaluated along an iso-feed line (a line parallel to the diagonal with constant feed rate) in order to determine the maximum value of the objective function which fulfils the given boundary conditions. Afterwards the feed flow rate is increased and m_{II}, m_{III} values along this new iso-feed line are evaluated. If the desired products are polluted, the responsible adsorption or desorption frontiers of the concentration profile might be shifted by changing the m_I or m_{IV} value respectively. In this case the shifting time is changed and the variation of the flow rate ratios is repeated. This optimization step offers the largest improvements of the objective function.

Optimization of the design parameters

For the optimization of the design parameters a set of dimensionless design parameters is introduced. The separation behavior can be characterized by the dimensionless total number of stages of the SMBR plant, which depends on the total chromatographic bed length and the height of an equivalent theoretical stage (*HETP*):

$$N_{i,tot} = \frac{n_c\left(1 - X_{cat}\right)L_c}{HETP_i} \tag{5.78}$$

Using the *HETP* plot (Van Deemter et al. 1956) and neglecting the axial diffusion of the solute molecule in the fluid phase, the total number of stages can be calculated by:

$$N_{i,tot} = \frac{n_c\left(1 - X_{cat}\right)L_c}{A_i + B_i\, u_{int}} \tag{5.79}$$

The reaction behavior is characterized by the dimensionless Damkoehler number for a heterogeneous or homogeneous reaction:

$$Da_{het} = \frac{k_{het} X_{cat} L_c}{u_{int}}$$

$$Da_{hom} = \frac{k_{hom} \varepsilon L_c}{u_{int}}$$

(5.80)

An optimization of the dimensionless design parameters together with the dimensionless flow rates yields not only one unique SMBR configuration but a set of different SMBR configurations with similar behavior. As for SMB plants (Susanto et al. 2005) almost identical dimensionless axial concentration profiles are retrieved for the SMBR. The final plant design is determined by considering the maximum allowable pressure drop.

5.6.3 Evaluation and application of chromatographic reactors

This section explains different applications of chromatographic SMB reactors and points out the influence of the type of reaction on process design. All results are calculated by process simulation based on the validated rigorous model as described in section 5.3.1.

Esterification

Typical reactions performed in chromatographic reactors are esterifications $(A + B \leftrightarrows C + D)$. Different adsorption behavior results in a separation of the educts and therefore a limited conversion. This problem can be solved by process or design measures.

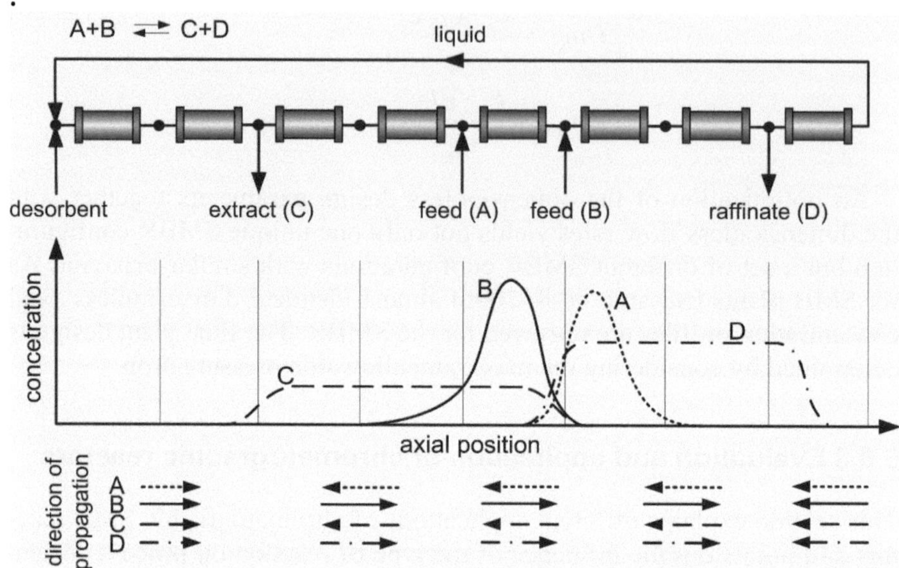

Fig. 5.14: Scheme of the 5-section SMBR (reprinted with permission from Schmidt-Traub 2005)

If separation occurs it is of advantage to use one of the educts as eluent. In this case the stoichiometry of the reaction is reduced to A ⇌ C + D and the chemical equilibrium is shifted into the direction of the products.

If separation occurs and one of the reactants cannot be used as eluent, the process design has to be adjusted by adding a fifth section in the SMBR. The stronger adsorbed reactant B is now fed to the SMBR at an additional downstream feed port as shown in Figure 5.14. Due to their adsorption behavior, the educts propagate in counter-current direction through the fifth section. Because of the increased contact time complete conversion can be achieved.

In order to assess the capabilities of the 4-section and the 5-section SMB reactor, the processes have been optimized in respect of conversion for a fixed feed rate and different separation factors. In case of the production of β-phenethyl acetate from β-phenethyl alcohol and acetic acid the separation factor of the reactants is 1.7. Within a 4-section process the maximum conversion is 92.9% with an ester purity of 88.5%. The conversion within the 5-section process reaches 87.8% and does not exceed the performance of the 4-section process, but the purity of the withdrawn ester can be increased to more than 95%.

Another esterification is the production of n-butyl methacrylate from n-butanole and methacrylic acid (Kawase et al. 1999b). Due to a separation factor of 3.1 a significant separation of alcohol and acid occurs in a 4-section SMB reactor. Therefore the conversion for the same feed rate does not exceed 65.8%. The ester is polluted by non-converted acid and the raffinate purity is only 60.8%. In case of a 5-section SMBR a conversion of 70.3% is reached and the purity increases up to 89.6%.

Isomerization

Further interesting reactions for chromatographic reactors are isomerizations (A $\leftrightarrows$ B). As an example the isomerization of glucose to fructose is discussed here. Higher fructose corn syrup is used with a fructose content of 42 and 52 wt.% (dry-base) as sweetener in the food industry. The 42 wt.% syrup is produced from glucose within an enzymatic batch reactor. Afterwards glucose is separated from fructose by SMB chromatography. The withdrawn fructose rich stream is blended with 42 wt.% syrup to produce finally the 55 wt.% syrup. Glucose, obtained as raffinate, can be reused as feed stock after reconcentration and repeated isomerization.

Besides this standard industrial process, other process concepts are applicable (Borren and Schmidt-Traub 2004). For instance, a 3-section SMB reactor with a homogeneous mixture of the immobilized enzyme and the adsorbent can be used. Due to the missing raffinate withdrawal glucose breaks through into section I and reduces the purity of fructose. Another option is the Hashimoto process with deintegrated isomerization reactors in section III. In this case the withdrawn glucose can also be reused as feed stock.

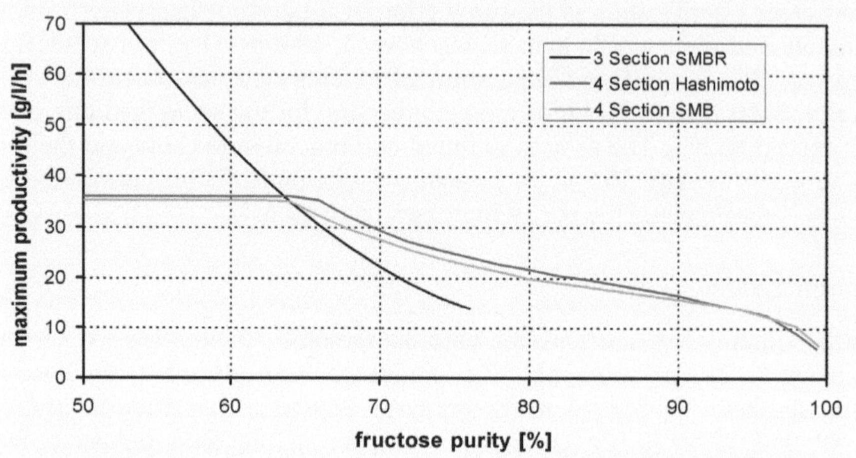

Fig. 5.15: Comparison of the process concepts (reproduced from Schmidt-Traub 2005)

Due to the industrial quality standard of fructose syrup, no constraint exists for the concentration of the fructose rich fraction. Hence, all products with fructose concentrations higher 55 wt.-% are acceptable. Figure 5.15 shows the retrieved optimal trajectories for the three process concepts: SMB, Hashimoto and SMBR. In all cases the dimensions of the SMB plant are assumed to be the same. Depending on the fructose purity, maximum productivity is achieved by different processes. In case of low purity demands of the fructose syrup an SMB reactor is advantageous. But with increasing product purity the driving force for the reverse reaction increases and, as a result, the purity strongly decreases. In the intermediate purity range the Hashimoto process reveals the best performance. But for high purity demands the maximum feed flow rate is limited. Hence the SMB process, which reaches only a slightly lower productivity compared to the Hashimoto process, is the most favorable choice for higher concentration.

Further optimization of all concepts might be possible but the results are representative for the characteristics of these processes. As a general rule the favorable degree of process integration decreases with increasing purity demands. Only for low purities the totally integrated SMB process is a promising opportunity. In the range of intermediate purity the partially deintegrated Hashimoto process enables highest productivities. Finally, for high purity demands a totally deintegrated sequential process has to be selected.

Racemization

As further example of the application of chromatographic reactors a thermal racemization is chosen. The thermal racemization of Troegers Base has already been discussed in section 5.4.4

Table 5.5: Comparison of the optimized process concepts for Troegers Base racemization

Purity [%]	Productivity [kg/l/d]	
	SMB	Hashimoto
99.9	0.5726	0.5724
99	0.5958	0.5958
95.1	0.6579	0.6581
91.7	0.7121	0.7121
84.1	0.8221	0.8221

The stronger adsorbed enantiomer is chosen as desired product of the different continuous processes. An SMBR is not applicable for this process due to the different operating conditions for reaction and separation. Therefore only a SMB process and a Hashimoto process are investigated. In both processes the weaker adsorbed enantiomer is withdrawn with the raffinate stream. In order to increase the overall yield this product is concentrated, racemized and reused as additional feed stock.

As objective function the productivity in respect of the stronger adsorbed enantiomer is chosen. The results for both processes are summarized in Table 5.5. Both processes yield almost identical results in respect of productivity. At the optimal operating point their axial concentration profiles are very similar. Differences can be determined only in process section III. The Hashimoto process establishes chemical equilibrium within the reactors and the concentrations are almost equal. But in section III of the SMB process the concentration of the stronger adsorbed enantiomer is mainly larger then the concentration of the weaker adsorbed enantiomer. Due to the strong isotherm interactions the different concentration profiles within the liquid phase result in similar loading profiles of the adsorbent. Therefore, the Hashimoto process is limited in this case by the loadability of the chiral stationary phase.

5.7 Notation

Symbols

Symbol	Description	Units
$a_{n,j}$	Variable	misc.
A_c	Column cross section	cm^2
A_i	Coefficient of the van Deemter equation	-
$b_{i,j}$	Langmuir coefficients	$cm^3\ g^{-1}$
B_i	Coefficient of the van Deemter equation	$s\ cm^{-1}$
Bo	Bodenstein number	-
c_i	Concentration in the liquid phase	$g\ ml^{-1}$
c_i^*	Molar concentration	$g\ mol^{-1}$
$c_{p,i}$	Concentration inside the particle pores	$g\ ml^{-1}$
C_i	Dimensionless concentration in the liquid phase	-
$C_{p,i}$	Dimensionless concentration inside the particle pores	-
D_{ax}	Axial dispersion coefficient	$cm^2\ s^{-1}$
d_p	Particle diameter	cm
Da	Damkoehler number	-
E_A	Activation energy	$kJ\ mol^{-1}$
H_i	Henry coefficient	-
k	Kinetic rate constant	misc.
$k_{eff,i}$	Effective mass transfer coefficient	$cm^2\ s^{-1}$
K_{eq}	Equilibrium constant	misc.
K_{EQ}	Modified equilibrium constant	misc.
$K_{i,j}$	Variable depending on the elution behavior	-
L_c	Column length	cm
m_j	Dimensionless flow rate ratio	-
m_i	Mass flow	$g\ s^{-1}$
$M_{i,j}$	Variable depending on the elution behavior	-
n_i	Number of columns	-
$N_{i,tot}$	Total number of stages	-
q_i	Solid load	$g\ ml^{-1}$
q_i^*	Total load	$g\ ml^{-1}$
Q_i	Dimensionless solid load	-
R	Gas constant	$J\ mol^{-1}\ K^{-1}$
Re	Reynolds number	-
$St_{eff,i}$	Effective Stanton number	-
t	Time	s
t_{shift}	Switching interval	s
T	Temperature	K
u	Velocity	$cm\ s^{-1}$

$\dot{V}$	Liquid phase flow rate	ml s^{-1}
$\dot{V}_{ads}$	Adsorbent flow rate	ml s^{-1}
V_c	Column volume	ml
$W_{i,j}$	Net flow rate	-
x	Axial coordinate	cm
X	Conversion	-
X_{cat}	Catalyst fraction	-
Z	Dimensionless coordinate	-

Greek Symbols

Symbol	Description	Units
ε	Void fraction	-
ε_p	Porosity of the solid phase	-
ε_v	Phase ratio	-
ρ	Density	g ml^{-1}
ν	Kinematic viscosity	cm^2 s^{-1}
ν_i	Stoichiometric coefficient	-
Φ	Kinetic law equation	-
Θ	Dimensionless time	-

Subscripts

Subscript	Description
I,II,III,IV	Process section
(0),(1)	Outlet, Inlet
acc	Accumulation
ads	Adsorbent
c	Column
cat	Catalyst
conv	Convection
des	Desorbent
disp	Dispersion
dl	Dimensionless
ext	Extract
feed	Feed
het	Heterogeneous
hom	Homogeneous
i	Component i
in	Inlet
int	Interstitial
j	Section j

mt	Mass transfer
out	Outlet
raf	Raffinate
reac	Reaction
SMBR	Simulated moving bed reactor
TMBR	True moving bed reactor

5.8 Literature

Altenhöner U, Meurer M, Strube J, Schmidt-Traub H (1997) Parameter estimation for the simulation of liquid chromatography. Journal of Chromatography A 769:59-69

Barker PE, Ganetsos G, Ajongwen J, Akintoye A (1992) Bioreaction-separation on continuous chromatographic systems. The Chemical Engineering Journal and the Biochemical Engineering Journal 50:23-28

Borren T, Fricke J (2005) Chromatographic Reactors. In: Schmidt-Traub H (ed) Preparative Chromatography of Fine Chemicals and Pharmaceutical Agents. Wiley-VCH, Weinheim

Borren T, Schmidt-Traub H (2004) Vergleich chromatographischer Reaktorkonzepte. Chemie Ingenieur Technik 76:805-814

Borren T, Schmidt-Traub H (2005) Process Intensification by integrated chromatographic processes. In: Guiochon G (ed) PREP, Philadelphia, p 62

Broughton DB, Gerhold CG (1961) Contnuous Sorption Process Employing Fixed Bed of Sorbent and Moving Inlets and Outlets. US Patent 2985586

Carr JR, Dandekar HW (2002) Adsorption with Reaction. In: Kulprathipanja S (ed) Reactive Separation Processes. Taylor & Francis, New York

Carr RW (1993) Continous Reaction Chromatography. In: Ganetsos G, Barker PE (eds) Preparative and Production Scale Chromatography. Marcel Dekker Inc., New York, pp 421-447

Chung SF, Wen CY (1968) Logitudinal Dispersion of Liquid Flowing through fixed and fluidized beds. AIChE Journal 14:857-866

Coca J, Adrio G, Jeng CY, Langer SH (1993) Gas and liquid chromatographic reactors. In: Ganetsos G, Barker PE (eds) Preparative and Production Scale Chromatography. Marcel Dekker Inc., New York, pp 449-475

Danckwerts PV (1953) Continuous flow systems - Distribution of the residence times. Chemical Engineering Science 2:1-13

Ditz R (2005) In: Schmidt-Traub H (ed) Preparative Chromatography of Fine Chemicals and Pharmaceutical Agents. Wiley-VCH, Weinheim

Dünnebier G (2000) Effektive Simulation und mathematische Optimierung chromatographischer Trennprozesse. Shaker-Verlag, Aachen

Epping A (2005) Modellierung, Auslegung und Optimierung chromatographischer Batch-Trennungen. Shaker-Verlag, Aachen

Falk T, Seidel-Morgenstern A (2002) Analysis of a discontinuously operated chromatographic reactor. Chemical Engineering Science 57:1599-1606

Fish BB, Carr RW (1989) An Experimental Study of the countercurrent moving-bed chromatographic reactor. Chemical Engineering Science 44:1773-1783

Fricke J (2005) Entwicklung einer Auslegungsmethode für chromatographische SMB-Reaktoren. Dissertation. Bio- und Chemieingenieurwesen. Universität Dortmund, Dortmund

Fricke J, Kawase M, Schmidt-Traub H (2003) Chromatograhic Reactors. In: Ullmann's Encyclopedia of Industrial Chemistry, 7th edn. Wiley-VCH, Weinheim

Fricke J, Schmidt-Traub H (2003) A new method supporting the design of simulated moving bed chromatographic reactors. Chemical Engineering and Processing 42:237-248

Grüner S, Kienle A (2004) Equilibrium theory and nonlinear waves for reactive distillation columns and chromatographic reactors. Chemical Engineering Science 59:901-918

Guiochon G, Golshan-Shirazi S, Katti AM (1994) Fundamentals of Preparative and Non-Linear Chromatography. Academic Press, Boston

Hashimoto K, Adachi S, Noujima H, Ueda Y (1983) A New Process Combining Adsorption and Enzyme Reaction for Producing Higher-Fructose Syrup. Biotechnology and Bioengineering 25:2371-2393

Hashimoto K, Adachi S, Shirai Y (1993) Development of New Bioreactors of a Simulated Moving Bed Type. In: Ganetsos G, Barker PE (eds) Preparative and Production Scale Chromatography. Marcel Dekker Inc., New York, pp 395-419

Herbsthofer R, Bart HJ, Brozio J (2002) Influence of reaction kinetics on the function of a chromatographic reactor. Chemie Ingenieur Technik 74:1006-1011

Jupke A, Epping A, Schmidt-Traub H (2002) Optimal design of batch and simulation moving bed chromatographic separation processes. Journal of Chromatography A 944:93-117

Kawase M, Inoue Y, Araki T, Hashimoto K (1999a) The simulated moving-bed reactor for production of bisphenol A. Catalysis Today 48:199-209

Kawase M, Masaki Y, Fricke J, Tanigawa T, Hashimoto K (1999b) Separation of reactants in the simulated moving bed reactor. In: APCRE Symposium, Hong Kong, pp 275 - 280

Kawase M, Pilgrim A, Araki T, Hashimoto K (2001) Lactosucrose production using a simulated moving bed reactor. Chemical Engineering Science 56:453-458

Klatt KU (1999) Modellierung und effektive numerische Simulation von chromatographischen Trennprozessen im SMB-Betrieb. Chemie Ingenieur Technik 71:555-566

Lode F (2002) A Simulated Moving Bed Reactor (SMBR) for Esterifications. Dissertation. ETH Zürich, Zürich

Lode F, Francesconi G, Mazzotti M, Morbidelli M (2003a) Synthesis of methylacetate in a simulated moving-bed reactor: Experiments and modeling. AIChE Journal 49:1516-1524

Lode F, Mazzotti M, Morbidelli M (2003b) Comparing true countercurrent and Simulated Moving-Bed chromatographic reactors. AIChE Journal 49:977-990

Ludemann-Hombourger O, Nicoud RM, Bailly M (2000) The "VARICOL" process: A new multicolumn continuous chromatographic process. Separation Science and Technology 35:1829-1862

Martin AJP (1949) Organic and Biochemical - Summarizing Paper. Discussions of the Faraday Society:332-336

Matsen JM, Harding JW, Magee EM (1965) Chemical Reaction in Chromatographic Columns. Journal of Physical Chemistry 69:522-527

Mazzotti M, Kruglov A, Neri B, Gelosa D, Morbidelli M (1996) A continuous chromatographic reactor: SMBR. Chemical Engineering Science 51:1827-1836

Mazzotti M, Neri B, Gelosa D, Morbidelli M (1997) Dynamics of a chromatographic reactor: Esterification catalyzed by acidic resins. Industrial & Engineering Chemistry Research 36:3163-3172

Meurer M, Altenhoner U, Strube J, Schmidt-Traub H (1997) Dynamic simulation of simulated moving bed chromatographic reactors. Journal of Chromatography A 769:71-79

Michel M, Epping A, Jupke A (2005) Modeling and Determination of Model Parameters. In: Schmidt-Traub H (ed) Preparative Chromatography of Fine Chemicals and Pharmaceutical Agents. Wiley-VCH, Weinheim

Michel M et al. (2003) Development of an integrated process for electrochemical reaction and chromatographic SMB-separation. Journal of Applied Electrochemistry 33:939-949

Migliorini C, Mazzotti M, Morbidelli M (1998) Analysis of simulated moving bed reactors. SPICA98

Ruthven DM, Ching CB (1989) Counter-current and simulated counter-current adsorption separation processes. Chemical Engineering Science 44:1011-1038

Sardin M, Schweich D, Villermaux J (1993) Preperative Fixed-Bed Chromatographic Reactor. In: Ganetsos G, Barker PE (eds) Preparative and Production Scale Chromatography. Marcel Dekker Inc., New York, pp 477-521

Sarmidi MR, Barker PE (1993a) Saccharification of Modified Starch to Maltose in a Continuous Rotating Annular Chromatograph (Crac). Journal of Chemical Technology and Biotechnology 57:229-235

Sarmidi MR, Barker PE (1993b) Simultaneous Biochemical Reaction and Separation in a Rotating Annular Chromatograph. Chemical Engineering Science 48:2615-2623

Seidel-Morgenstern A (1995) Mathematische Modellierung der präparativen Flüssigkeitschromatographie. Deutscher Universitätsverlag, Wiesbaden

Seidel-Morgenstern A (2004) Experimental determination of single solute and competitive adsorption isotherms. Journal of Chromatography A 1037:255-272

Storti G, Mazzotti M, Morbidelli M, Carra S (1993) Robust Design of Binary Countercurrent Adsorption Separation Processes. AIChE Journal 39:471-492

Ströhlein G, Lode F, Mazzotti M, Morbidelli M (2004) Design of stationary phase properties for optimal performance of reactive simulated-moving-bed chromatography. Chemical Engineering Science 59:4951-4956

Strube J, Haumreisser S, Schmidt-Traub H (1998) Auslegung, Betrieb und ökonomische Betrachtung chromatographischer Trennprozesse. Chemie Ingenieur Technik 70:1271-1279

Susanto A, Wekenborg K, Epping A, Jupke A (2005) Model based design and optimization. In: Schmidt-Traub H (ed) Preparative Chromatography of Fine Chemicals and Pharmaceutical Agents. Wiley-VCH, Weinheim

Toumi A, Engell S (2004) Optimization-based control of a reactive simulated moving bed process for glucose isomerization. Chemical Engineering Science 59:3777 - 3792

Van Deemter JJ, Zuiderweg FJ, Klinkenberg A (1956) Longitudinal diffusion and resistance to mass transfer as causes of nonideality in chromatography. Chemical Engineering Science 5:271-289

Wewers W, Dingenen J, Schulte M, Kinkel J (2005) Choice of Chromatographic Systems. In: Schmidt-Traub H (ed) Preparative Chromatography of Fine Chemicals and Pharmaceutical Agents. Wiley-VCH, Weinheim

Zhang Y, Hidajat K, Ray AK (2004) Optimal design and operation of SMB bioreactor: production of high fructose syrup by isomerization of glucose. Biochemical Engineering Journal 21:111-121

Zhang ZY, Mazzotti M, Morbidelli M (2003) PowerFeed operation of simulated moving bed units: changing flow-rates during the switching interval. Journal of Chromatography A 1006:87-99

6 Reactive extraction

Arno Behr, Joachim Seuster

6.1 Introduction

In homogeneous transition metal catalysis it is very important to use a process which allows the catalyst to be recycled. Reactive extraction belonging to the class of integrated reaction and separation processes may be applied for catalyst recycling and it allows also influencing the reaction selectivity. Its principle is to perform the reaction in one phase, while the second phase is catalytically inactive and contains reactants and products.

In our work on the palladium catalyzed telomerization of butadiene with ethylene glycol (Behr u. Urschey 2003, Behr et al. 2003, Behr u. Urschey 2003, Urschey 2004) we investigated different alternatives of reactive extraction processes. To get a deeper insight into the fundamentals of these processes, both thermodynamic and reactive aspects were considered, as well as the knowledge of heuristic rules.

6.2 Reactive extraction systems

The typical scheme for a reactive extraction system is illustrated in Figure 6.1. Two inmiscible phases contain the solvents and the reactants. The catalyst is solved in the stationary phase, while the product is extracted to the transport phase. It allows separating product and catalyst using a simple settler.

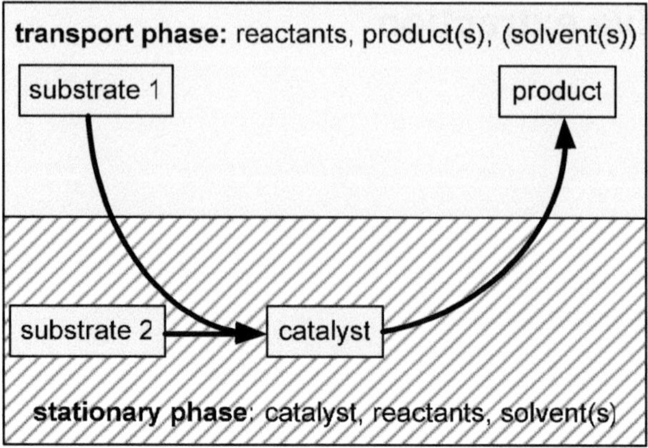

Fig. 6.1: Typical scheme for reactive extraction systems

In the simplest case there is only need for one solvent. This can be implemented in a process using only one mixer-settler unit and a following product separation. However, reactive extraction systems can become more complex, depending on the necessity of additional solvents and the distribution of the reactants and products between the phases.

Integrated reaction and separation processes have by their nature both reaction and separation aspects. Therefore, reactive extraction processes can generally be grouped into processes where either the separation or the reaction is the main focus (Kulprathipanja 2002). Therefore, the following overview of industrial processes will use this classification.

6.2.1 Separation processes

Reactive extraction is widely utilized for the purification of solvents, the extraction of products or the selective separation of physically similar components. It is applied to separate metals or pharmaceuticals, wastewater treatment or purification of organic mixtures.

An interesting example of reactive extraction is the separation of plutonium and uranium from nuclear waste using the PUREX process Swanson 1991. In this process, the selective oxidation and reduction of the plutonium is used to affect its solubility in the polar or nonpolar phase.

Another industrial example of the reactive extraction is the fuel desulforization using the UOP-Merox process (Holbrook 1997) (Figure 6.2).

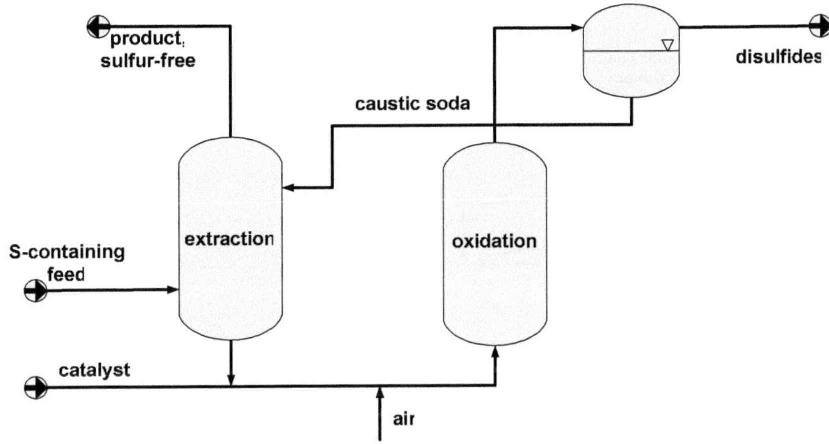

Fig. 6.2: UOP-Merox process

The feed is desulforized in a reactive extraction step using aqueous caustic soda. The aqueous solution is regenerated in a two-step reactive extraction making use of air to form disulfides which can be separated from the caustic soda solution.

Many other industrial processes adopt the reactive extraction for selective separation of components (e.g. the extraction of penicillin G (Gaidhani 2002)).

6.2.2 Synthesis processes

One of the most important processes employing reactive extraction for production purposes is the Shell-Higher-Olefin-Process (SHOP) (Vogt 2002). In this process linear α-olefins are produced, starting from a homogeneously catalyzed oligomerization step, followed by isomerisation and metathesis reactions. In the first step ethylene is oligomerized in a polar phase in which the nickel catalyst is dissolved. Figure 6.3 illustrates the oligomerization reactor of the SHOP-process.

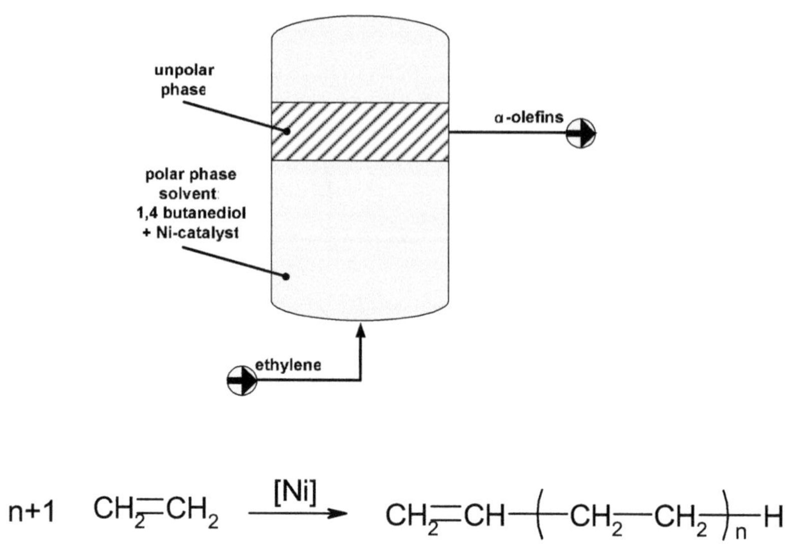

$$n+1 \quad CH_2{=}CH_2 \quad \xrightarrow{[Ni]} \quad CH_2{=}CH{-}\left(CH_2{-}CH_2\right)_n{-}H$$

Fig. 6.3: Nickel-catalyzed oligomerization of ethylene (SHOP)

The hydroformylation of propene to butyraldehyde using the Ruhrche-mie/Rhône-Poulenc-process is another industrially relevant reactive ex-traction process (Frohning et al. 2002). In this process propene is hydro-formylated using an aqueous rhodium/TPPTS catalytic system. TPPTS is the abbreviation of the phosphine ligand triphenylphosphine-trisulfonate, which has a very high solubility in the aqueous stationary phase. Com-pared to other industrial hydroformylation processes, the Ruhrche-mie/Rhône-Poulenc-process has a very high selectivity to linear aldehydes at very low catalyst losses.

An example of a desintegrated reactive extraction process is the telom-erization of butadiene with water developed by Kuraray (Yoshimura 2002). The reaction uses a palladium catalyst dissolved in sulfolane with a sequential extraction of the product.

6.3 System analysis and plant design

As a model reaction for our systematic investigations of reactive extraction processes the palladium catalyzed telomerization of 1,3-butadiene with ethylene glycol was utilized (Figure 6.4). The desired products of this reaction are the monotelomers **1**, which are of highly industrial interest since their saturated follow-up products can be used, e.g. as plasticizer alcohols for PVC.

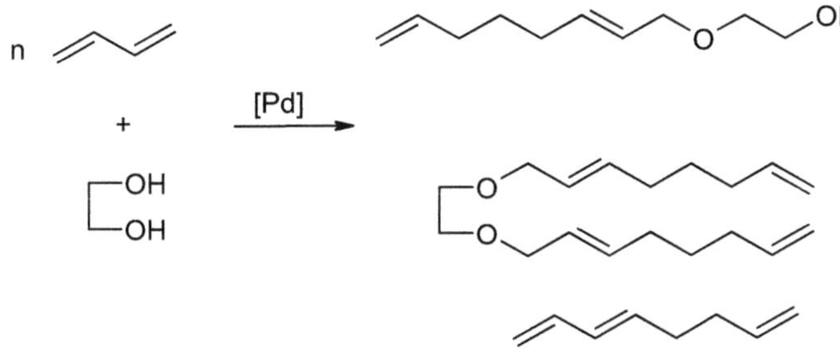

Fig. 6.4: Telomerization of butadiene with ethylene glycol

Due to the second alcohol group of the ethylene glycol, also the ditelomers **2** can be formed in a follow-up reaction. These ditelomers are di-ethers which contain no further functional groups and are therefore the non-desired products. As a suitable catalyst for the aqueous biphasic system Pd/TPPTS was used.

Compared to the mono-phase system, the biphasic system shows several advantages. Besides the possibility of catalyst recycling, the biphasic system shows a significant increase of selectivity to the desired monotelomers (Table 6.1). Due to the in-situ extraction of the monotelomers into the organic phase, the follow-up reaction to the ditelomers is suppressed.

Table 6.1: Mono- and biphasic system in batch experiments

system	t	yield product **1**	yield product **2**	selectivity[a]
	[h]	[%]	[%]	[%]
monophasic	2	60	23	72
biphasic	4	75	< 1	> 99

[a] selectivity to monotelomers related to ethylene glycol

Caused by the low butadiene concentration in the aqueous catalyst phase, the reaction rate involving the biphasic system is significantly slower than in the monophasic one. Recycle experiments on a bench scale in the laboratory showed the catalyst activity being stable over several batch-runs. The use of an additional in-situ solvent (cyclohexane) to enhance the phase separation showed no effect on the selectivity but led to lower yields.

6.3.1 Analysis of the reaction system

Due to the fact that the reaction system consists of two liquid phases, the phase and reaction thermodynamics are strongly coupled.

Batch experiments on a bench scale showed that reaction times longer than 4 hours do not lead to higher yields. After excluding other effects, e.g. catalyst deactivation, the phase thermodynamics were investigated and found to be the reason for the limitation of the reaction. Figure 6.4 shows the projection of the system in a triangular diagram of the reactants and the product . One can see that the reaction follows paths which are defined through the reaction stoichiometry. For instance, the straight line AB is the theoretical reaction path if the reaction is started with a stoichiometric ratio of two mols of butadiene and one mol of ethylene glycol. The other lines which start at other ratios of the two starting compounds take all a parallel course to the line AB, if the ternary map is mass-based. It could be experimentally shown that at the point where the reaction paths cross the binodal curve the reaction stops due to the fact that all reactants have been extracted from the water to the organic phase (Figure 6.5). Further investigations to the miscibility gap and its modelling are discussed in chapter 2.2 (Example 4) .

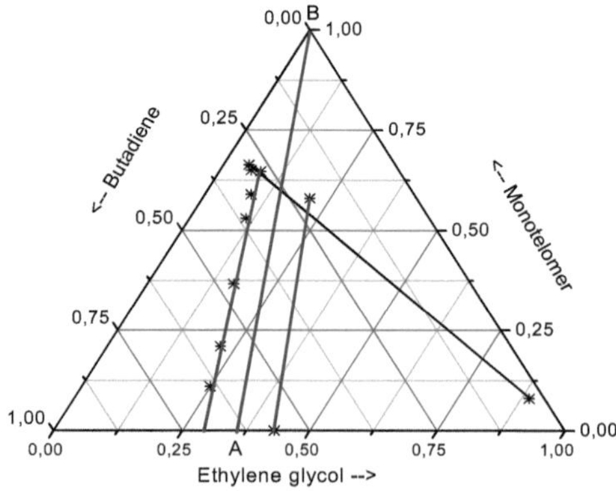

Fig. 6.5: Limitation by miscibility gap

One crucial point of the process economics is the suppression of the catalyst leaching into the organic phase. Experiments have shown that the catalyst loss into the organic phase is mainly caused by the solubility of water into the organic phase. As butadiene is the most unpolar component in the reaction system, experiments were made to investigate the influence of the amount of butadiene on the catalyst leaching. Figure 6.6 illustrates the catalyst loss related to the telomers as a function of the butadiene amount in the mixture.

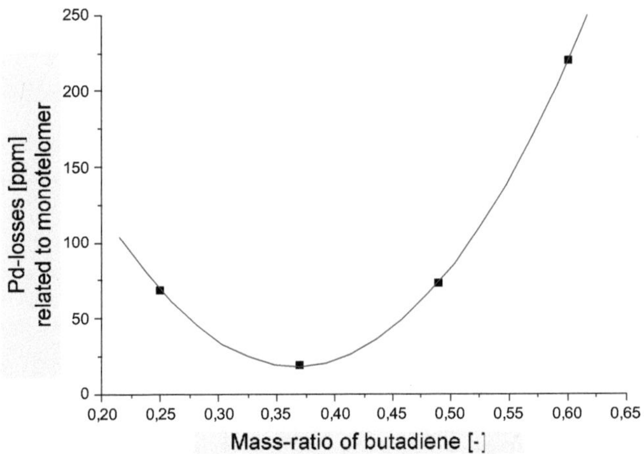

Fig. 6.6: Palladium loss versus amount of butadiene

On the left side of the diagram the effect of too low butadiene concentrations can be seen. Due to the low concentration of butadiene in the mixture, the polarity of the organic phase increases, which leads to a higher solubility of water and therefore to higher catalyst losses. On the right side of the diagram the amount of butadiene is too high so that the total amount of water increases again. In this case the high palladium-product ratio is caused by the big amount of butadiene which solves, though in low concentrations, water and catalyst. These remain in the product phase after the flash and lead to the high catalyst losses. The optimal range is between both regimes.

Also mass transfer influences biphasic catalysis significantly, because of the necessity to transport the butadiene into and the product out of the reaction phase. To optimize this mass transport experiments, were carried out varying the stirrer velocity (Figure 6.7).

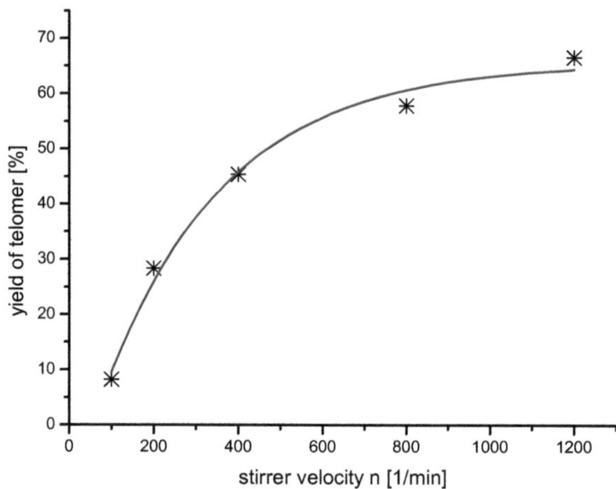

Fig. 6.7: Influence of stirrer velocity

At low stirrer velocities the mass transfer is the limiting factor for the reaction system. Stirrer velocities higher than 1200 rpm do not lead to higher product yields. Due to these experimental results, all further reactions were carried out using 1200 rpm so that mass transport limitations can be excluded.

6.4 Modelling

As discussed before, the miscibility of the reaction system water/ethylene glycol/butadiene/telomer has a strong influence on both the reaction yield and the catalyst losses. Thus knowledge of the miscibility gap is crucial for the process development. For the experimental estimation of NRTL-parameters, a technical scale reactor with a volume of one liter was used. After filling in water, ethylene glycol, butadiene and the palladium catalyst, the experiment was started. Samples were taken every half hour out of the organic phase and every hour out of the aqueous phase. Adopting these data, the parameters for the NRTL-model were fitted based on the software ASPEN PROPERTIES PLUS (Figure 6.8).

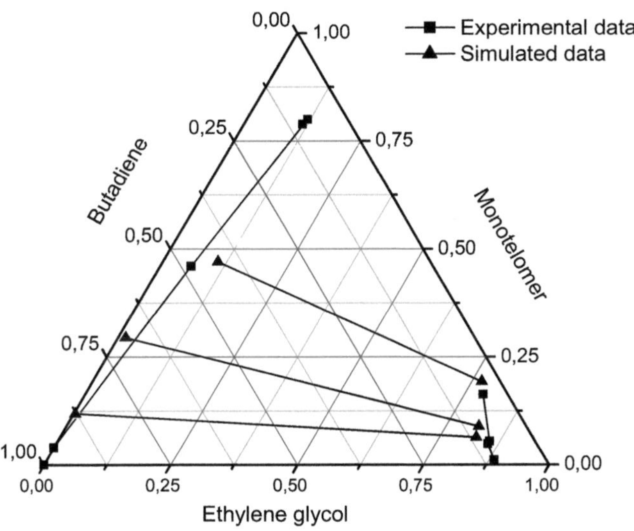

Fig. 6.8: Experimental and simulated phase equilibrium data

This NRTL model was utilized for the simulation of the process alternatives.

6.4.1 Mini-plant design

Based on the bench-scale experiments a process was designed which was realized in a continuous mini-plant. First step was the definition of targets for the process development. As mentioned before, the catalyst losses to the product phase are crucial to the process economics. For preliminary simulation studies the amount of water in the product phase was used as a target, due to the fact that water is the primary cause for the catalyst losses to the transport phase. The second target was the reaction enhancement to increase selectivity and yield to the desired monotelomers.

At first conventional process alternatives of the extraction systems were investigated. Figure 6.9 and 6.10 schematically picture all typical process alternatives, where the dark colour indicates a very polar phase and the white colour an unpolar phase containing mainly the starting compound butadiene and the products, the monotelomers.

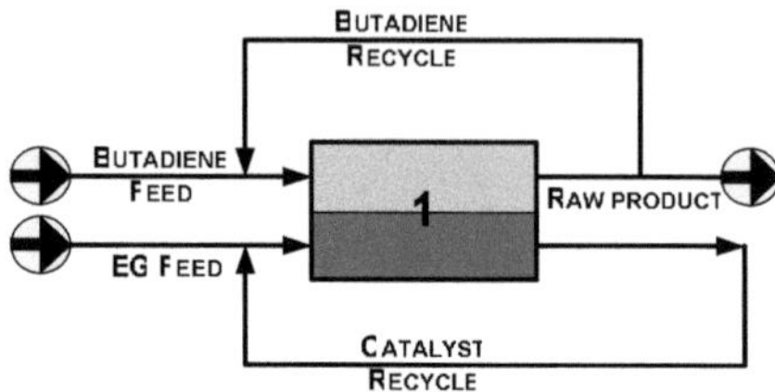

Fig. 6.9: Cocurrent process (EG = ethylene glycol)

Because in the cocurrent setup the only possibilities to influence the process are feed composition and reaction time in just one mixer-settler unit, this cocurrent setup seems not to be suitable for our reaction system.

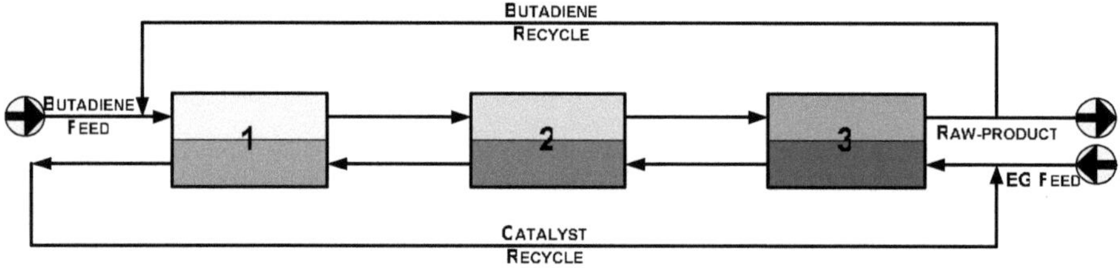

Fig. 6.10: Countercurrent process (EG = ethylene glycol)

In the countercurrent process (Figure 6.10) , the butadiene feed enters from the left side into a system of three mixers and settlers, while the ethylene glycol is fed from the right. The setup unit 3 is important for the catalyst leaching because there the product phase leaves the process. As can be seen in Figure 6.10, the difference between the polarities in unit 3 is very low, which leads to a high solubility of water in the organic phase and therefore to high catalyst losses. In units 1 and 2 a small difference between the polarities would be beneficial, because a higher solubility of butadiene in the catalyst phase enhances the reaction. However, using the countercurrent process, the difference of the polarities in units 1 and 2 is relatively high.

Therefore we developed, based on the countercurrent setup, the new "cross-countercurrent process" (Figure 6.11):

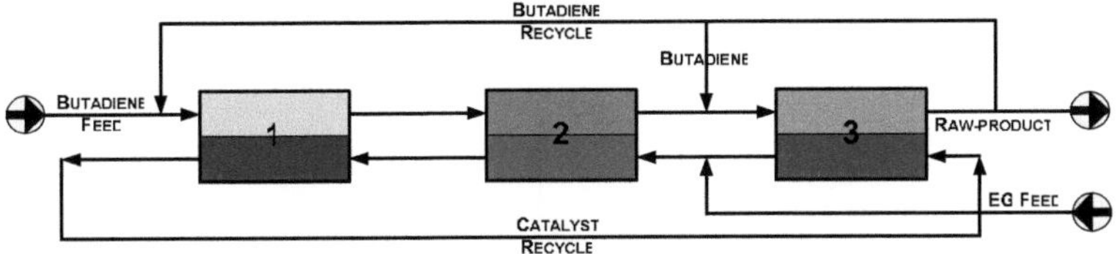

Fig. 6.11: Cross-countercurrent process (EG = ethylene glycol)

Here we assigned the diverse functionalities to the different units. In this process setup we have two butadiene feeds into the first and third stage, while the fresh ethylene glycol is only fed into the second stage. This leads to two units (1 and 3) where the focus is to improve the separation, and one unit (2) where the reactive functionality is enhanced, because of the relatively similar polarity of the two liquid phases. In the third stage, additional butadiene is fed to enable a better separation of the product phase from the aqueous catalyst phase. In the second stage, where ethylene glycol is fed, the mutual solubility of both phases is enhanced so that the reaction functionality is forced.

To verify the process alternatives, simulations were carried out and experiments were run in a continuous mini-plant. This mini-plant (Figure 6.12) consists of mainly three parts: A butadiene condenser unit (B1), three mixer-settler units (R1-R3 and S1-S3) and a butadiene flash (K1). To be variable in testing other process alternatives, the plant was built with three mixer-settler units which allowed varying the process flow.

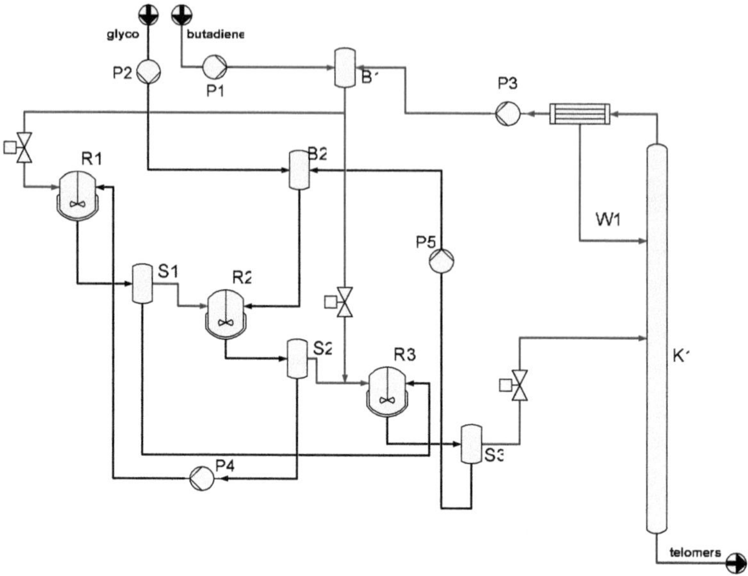

Fig. 6.12: Flowsheet of the mini-plant

The mixer-settler setup was chosen due to the fact that it can be used as a good model for integrated apparatus e.g. rotating-disc-contactors or extraction columns. The complete mixer-settler cascade can be inertized by 17 bar argon pressure to allow the natural flow and to inhibit the polymerisation of butadiene in the gaseous phase. The flash K1 works at a pressure of 1 bar.

The main targets in the development of the mixer-units were a variable residence-time and a hermetic decoupled drive to prevent a leaking of butadiene. The vessel of the mixers was built using 1.4571 steel with heating drills within the shell. The residence time can be varied by using an adjustable spillway which is pressed using a PTFE-gasket and screw-cap. The whole volume of the reactor is about 1 liter. The hermitic drive was constructed using a magnetic coupler.

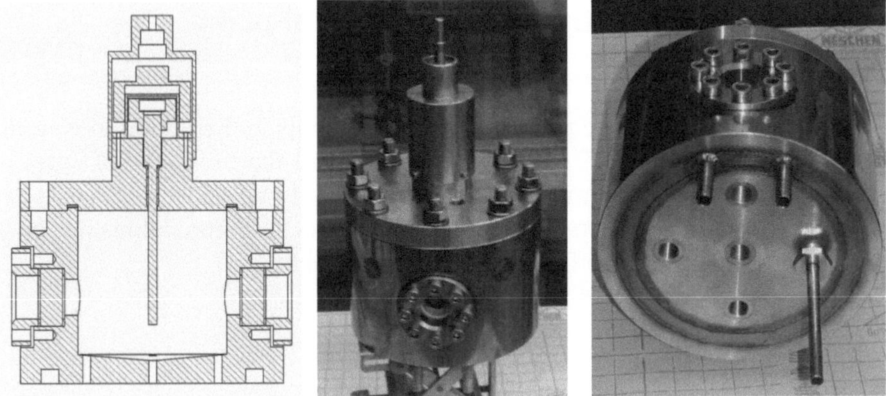

Fig. 6.13: Mixer units (profile, front view, bottom view)

Due to the relatively short time which is needed to separate the transport and the stationary phase, a simple settler geometry can be used. The settler units could be heated independently from the mixer units which allows to influence the phase equilibrium.

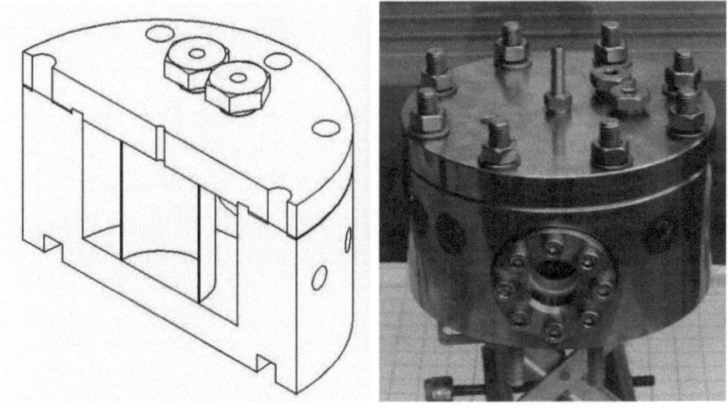

Fig. 6.14: Settler Units (isometric profile, front view)

6.5 Experiments in the continuous mini-plant

For the verification of the plant design experiments in the continuous mini-plant were run. We used the countercurrent and the cross-countercurrent setup to investigate the improvement of the newly developed reactive extraction process. To warrant the experiments to be comparable, the constraints in Table 6.2 were fixed for both experiments.

Table 6.2: Constraints for continuous experiments (Bu = butadiene, EG = ethylene glycol)

reactor residence time	pressure	temperature	stirring velocity	Bu/EG ratio
[h]	[bar]	[°C]	[rpm]	[-]
2 each unit	17	80	2000	2.5

The feed stream of butadiene in the cross-countercurrent experiments was split into two thirds for the first and one third for the last mixer, according to the preliminary experiments.

For the comparison between the continuous experiments, the mass fraction of the product after the second reactor is employed, because of the fact that this reactor is the main reactor in the cross-countercurrent experiment, according to our process design. The evaluation of the whole process is based on the total yield, selectivity and palladium losses after the flash.

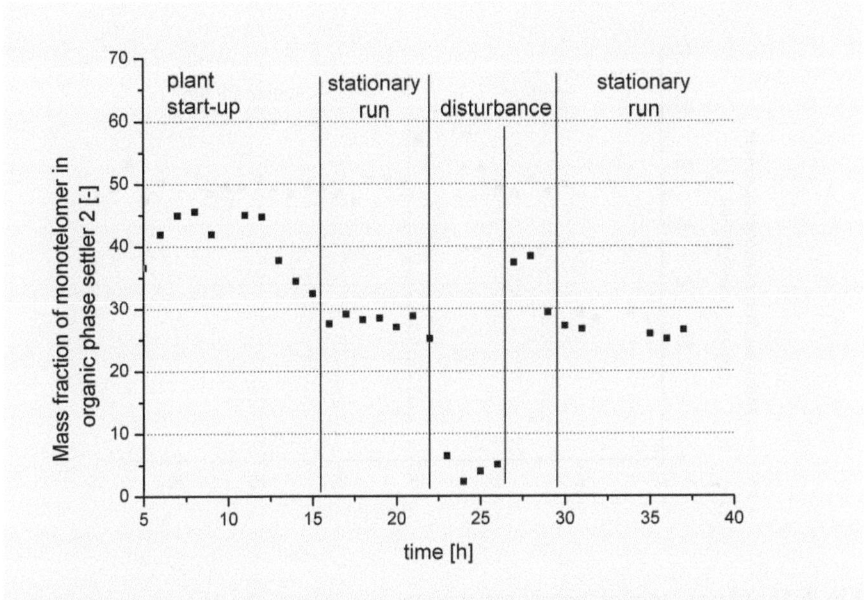

Fig. 6.15: Countercurrent experiment

In Figure 6.15 the results of the countercurrent experiment are illustrated. After 15 h the start-up phase of the plant is finished and the process runs in steady state. To evaluate the process stability, the ethylene glycol feed was then stopped for about one hour. As it can be seen in the figure, the telomer fraction drops significantly during this disturbance. After restarting the ethylene glycol feed, the process returns to its former steady state level, yielding a reaction mixture which contains about 15-17 % of monotelomers.

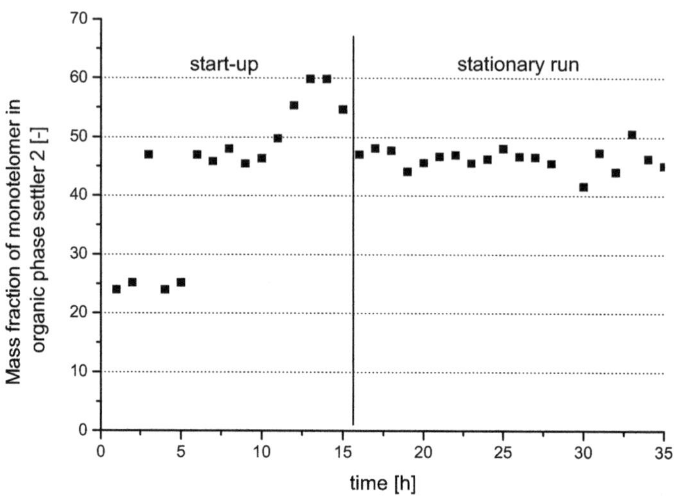

Fig. 6.16: Cross-countercurrent experiment

The results of the cross-countercurrent experiment are shown in Figure 6.16. The time for the plant start-up is approximately the same as in the countercurrent experiment (15-16h). As expected, the telomer fraction in the second mixer is significantly higher than in the countercurrent experiment (45-50 %).

In Table 6.3 the results of the mini-plant experiments are compared. By the cross-countercurrent setup, the total yield of telomers could be increased by 11 % from 68 % to 79 %. At the same time, the catalyst losses could be suppressed: Instead of a loss of 29 ppm palladium using the countercurrent setup, only 19 ppm are lost in the cross-countercurrent alternative. This is in good relation to the water content in the product phase, which could be reduced from 5 % to 4 %. The different process alternatives had no influence on the selectivity, which amounts in both setups to > 98 %.

Table 6.3: Comparison of the mini-plant experiments

			countercurrent	cross-countercurrent
	telomers	[wt%]	68	79
	butadiene	[wt%]	13	8
product phase	ethylene glycol	[wt%]	14	9
	water	[wt%]	5	4
	palladium	[ppm]	29	19
	yield	[%]	65	75
	selectivity[a]	[%]	>98	>98

[a] selectivity of monotelomers related to ethylene glycol

6.6 Conclusions

After an intensive study on the reactive extraction and a systematic investigation of different reactive extraction systems we developed a new alternative for this type of processes. In our studies we utlized the telomerization of butadiene and ethylene glycol as model system. Adopting the knowledge of our reaction system, we distributed the functionalities to different steps of the process and could improve the performance of the process. After theoretical work, experimental work on bench scale and simulation we managed to run long-term experiments in the mini-plant. We could approve the beneficial plant set-up and the stability of the catalyst system. The yield has been increased from 68 % to 79 % while the catalyst losses were reduced by 10 ppm palladium. The selectivity of our reaction system was stable at > 98 %.

For further investigations it should be possible to apply the developed system to other reaction systems, for example the telomerisation of butadiene with other diols, triols or sugar compounds.

6.7 Literature

Behr A, Urschey M, J. Mol. Catal. *A* **197** (2003) Nr. 1-2, p. 101-113
Behr A, Urschey M und Brehme VA, Green Chem. **5** (2003) p. 198-204
Behr A, Urschey M, Adv. Synth. Catal. **345** (2003) p. 1242-46
Urschey M, Dissertation, Universität Dortmund 2004
Kulprathipanja S, "Reactive Separation Process" Taylor & Francis, New York, 2002
Swanson JL, CRC Press Inc., Boca Raton, FL, III, 1991
Holbrook Dl, "Handbook of Petroleum Refining Processes", McGraw-Hill, New York, 1997
Gaidhani HK, Chem. Eng. Sci **57** (2002) 1979-84
Vogt D, in "Applied Homogeneous Catalysis with Organometallic Compounds", Wiley-VCH, Weinheim, 2nd ed., 2002
Frohning CD, Kohlpaintner CW, Bohnen HW, "Applied Homogeneous Catalysis with Organometallic Compounds", Wiley-VCH, Weinheim, 2nd ed., 2002
Yoshimura N, "Aqueous Phase Organometallic Catalysis – Concepts and Applications", Wiley-VCH, Weinheim, 2nd ed., 2002

7 Optimization and control of reactive chromatographic processes

Achim Küpper, Abdelaziz Toumi, Sebastian Engell

7.1 Introduction

This chapter deals with the rigorous optimization and the optimizing control of advanced chromatographic separation processes: simulated moving bed (*SMB*) processes, the *VARICOL* process (SMB with asynchronous port switching) and the *Hashimoto SMB* process where reactors are placed in some zones of the separation process to perform reaction and separation in an integrated fashion.

Chromatographic separation processes are based on different affinities of the involved components to a solid adsorbent that is packed in a chromatographic column. In most industrial applications, the separation is performed discontinuously in a single chromatographic column which is charged with pulses of the feed solution. The feed injections are carried through the column by pure desorbent. The more retained component travels through the column more slowly, thus leaving the column after the less adsorptive component so that pure components can be withdrawn.

In recent years, the continuous Simulated Moving Bed (SMB) process has increasingly been applied due to its advantages with respect to the utilization of the adsorbent and the consumption of solvent. The SMB process consists of several chromatographic columns which are connected in series to constitute a closed loop. An effective counter-current movement of the liquid phase and the solid phase is achieved by periodic and simultaneous switching of the inlet and outlet ports by one column in the direction of the liquid flow.

An improved process performance can be obtained by the VARICOL operation (Ludemann-Hombourger and Nicoud 2000) where the ports are switched asynchronously, leading to a better utilization of the solid bed.

A step towards process intensification that has recently gained much attention is the integration of reaction into chromatographic separations. In the simplest variant of a reactive SMB process, the columns are packed with catalyst as well as with adsorbent so that chemical reaction and adsorption take place simultaneously in all columns. For equilibrium reactions, this process can be favorable compared to sequential reaction and separation units. However, the presence of the catalyst is often superfluous, if not counterproductive in some of the zones of the SMB process. This observation led to the introduction of the *Hashimoto SMB* process (Hashimoto et al. 1983) where the functionalities of separation and reaction are performed in different columns and the reactors are fixed in the separation zones of the SMB process. For a detailed discussion of reactive chromatographic processes, see Chapter 5 .

SMB processes in general are characterized by mixed discrete and continuous dynamics, spatially distributed state variables with steep slopes, and slow and strongly nonlinear responses of the concentration profiles to changes of the operating parameters. In order to exploit the full economic potential of such processes, advanced optimization and control approaches based upon rigorous nonlinear process models employing efficient algorithms for simulation and optimization are needed. An overview of recent achievements in the optimization and control of chromatographic separations can be found in (Engell and Toumi 2005).

In the following sections, we will discuss the rigorous optimization of SMB, VARI*COL* and the Hashimoto SMB processes. A design study for the Hashimoto SMB process that focuses on the influence of the positions of the reactors and their lengths on the productivity of the process is presented. We then discuss a model-based control concept for the Hashimoto SMB process which computes operating parameters that maximize an economical objective over a finite horizon on-line while the purity requirements are considered as constraints.

7.2 The simulated moving bed process

The Simulated Moving Bed (SMB) process was developed and patented by Broughton (Broughton 1961). It is the practical implementation of the True Moving Bed (TMB) that separates components of different affinities to the adsorbent by a counter-current flow between the solid and the fluid phase. Since a movement of the solid phase is very difficult to realize, the counter-current movement is simulated by switching the inlet ports and the outlet ports in a closed loop of chromatographic columns periodically by one column in the direction of the liquid flow. The movement of the more strongly adsorbed component is dominated by the solid flow and this component is withdrawn at the raffinate port, while the less strongly adsorbed component moves with the liquid phase and is withdrawn at the extract port. After a start-up phase, a periodic steady state is reached (see Figure

7.2) where the outlet concentration profiles change dynamically within a period, but are identical (thick line) from period to period.

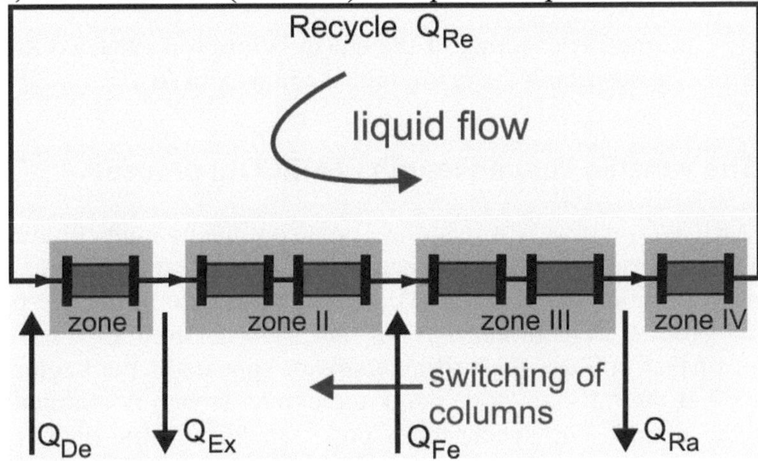

Fig. 7.1: Simulated Moving Bed principle

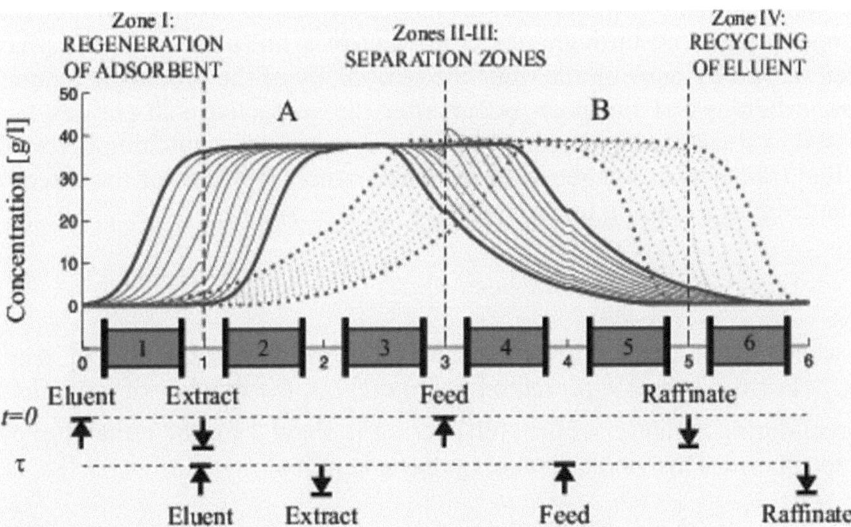

Fig. 7.2: Axial concentration profiles of the SMB process

The different zones of an SMB process are classified according to their function and their position relative to the ports:

    – Zone I: regeneration of the solid bed by desorption of the more strongly retained component (A),

   - Zone II: desorption of the less retained component (B),
   - Zone III: adsorption of the more strongly retained component (A).
   - Zone IV: cleaning of the eluent before it is recycled to zone I by adsorbing the less retained component (B).

7.2.1 The variable column length (VARICOL) process

The VARI*COL* process is a relatively new continuous multicolumn chromatographic process. In (Ludemann-Hombourger and Nicoud 2000, Ludemann-Hombourger et al. 2003), it was proposed to switch the ports of an SMB process asynchronously. As can be seen from Figure 7.2, the component that pollutes the raffinate stream appears at the beginning of the period at the raffinate port, while the extract stream is polluted at the end of the period. The synchronous port switching in the SMB process does not take advantage of this fact. The VARI*COL* operation takes advantage of the sequence in which the fronts appear at the ports by switching the ports individually by one column, leading to an improved utilization of the solid bed and a higher optimization potential in terms of eluent consumption and feed throughput. At the end of a full period, all ports have been moved by one column, thus the periodicity of the process is retained. The switchings of the ports occur after the respective sub-periods have passed as illustrated by Figure 7.3. The asynchronous switching gives rise to real (rather than integer as in the SMB process) values of the effective zone lengths according to:

$$\overline{N}_z = \sum_{k=0}^{4} N_{z,k} \Delta t_{z,k} \qquad z = 1,..,4 \quad , \tag{7.1}$$

where the index k corresponds to the respective subperiod of length $\Delta t_{z,k}$. For the example presented in Figure 7.3, the number of columns in zone I during a quarter of the (full) period is 1 and 2 for the remaining part of the period. This yields an average zone length of

$$\overline{N}_1 = 1 \cdot \tfrac{1}{4} + 2 \cdot \tfrac{3}{4} = \tfrac{7}{4} \tag{7.2}$$

columns in zone I. Analogously, for the other zones, the zone lengths are

$$\overline{N}_2 = \tfrac{6}{4}; \quad \overline{N}_3 = \tfrac{9}{4}; \quad \overline{N}_4 = \tfrac{2}{4}. \tag{7.3}$$

The sum of the average lengths of the zones is of course equal to the total number of columns.

$$\sum_{i=1}^{4} \overline{N}_i = N_{col} = 6 \qquad (7.4)$$

Hence, three zone lengths can be chosen independently during the operation of the process. The degrees of freedom of the VARICOL process are thus three sub-period lengths, the length of the full period and the flow rates in each zone. The SMB operation is a special case of the VARICOL operation where all sub-periods are equal to the full period.

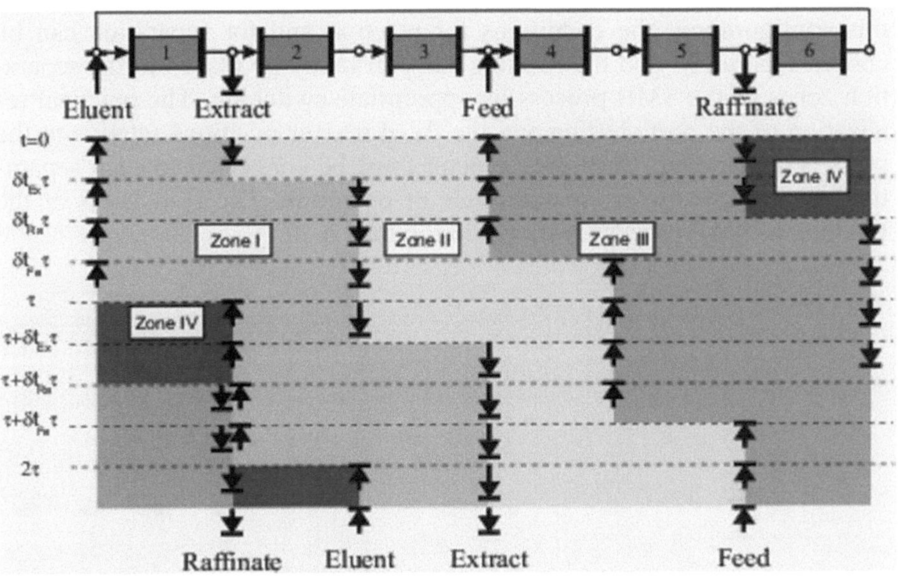

Fig. 7.3: Principle of the VARICOL process

7.3 Integration of reaction and separation – the Hashimoto SMB process

The integration of chemical reactions into chromatographic separations offers the potential to improve the conversion of equilibrium-limited reactions. By the simultaneous removal of the products, the reaction equilibrium is shifted to the side of the products. This combination of reaction and chromatographic separation can be achieved by packing the columns of the SMB process uniformly with adsorbent and catalyst, which leads to the reactive SMB (SMBR) process. The SMBR process can be advantageous in terms of higher productivity in comparison to a sequential arrangement of reaction and separation units (see Chapter 5).

However, a uniform catalyst distribution in the SMBR promotes the backward reaction near the product outlet which is detrimental for the productivity the renewal of deactivated catalyst is difficult when it is mixed with adsorbent pellets, and the same conditions must be chosen for separation and reaction which may lead to either suboptimal reaction or suboptimal separation. The Hashimoto SMB process (Hashimoto et al. 1983) overcomes the disadvantages of the SMBR by performing separation and reaction in separate units that contain only adsorbent or only catalyst. In this configuration, the conditions for reaction and for separation can be chosen separately, and the reactors can constantly be placed in the separation zones of the SMB process by appropriate switching. The practical realization of the port shifting and the fixed reactor positions relative to the ports is demanding, since each reactor must be connected to each separative column once over the full cycle of operation. The flow sheet of the Hashimoto SMB process is shown in Figure 7.4.

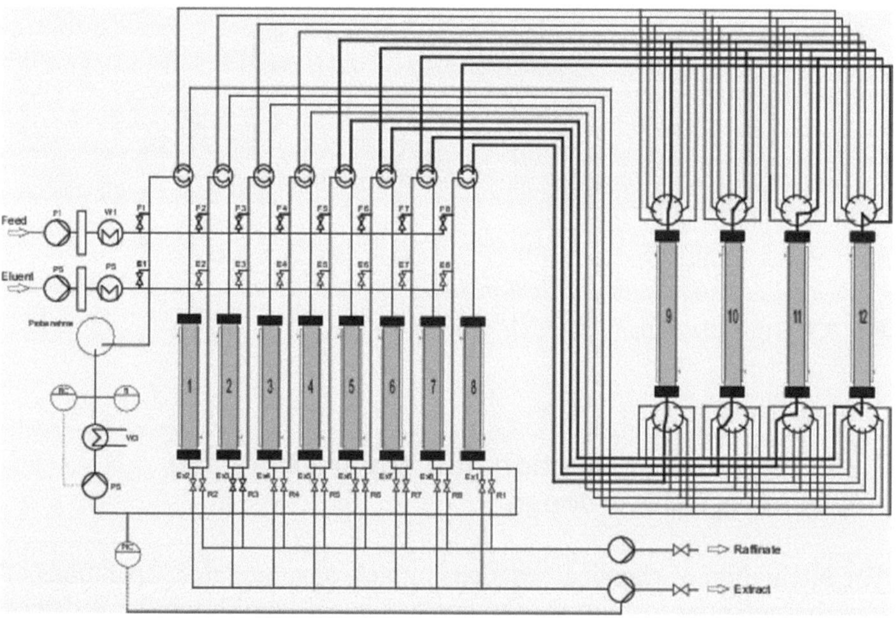

Fig. 7.4: Flow sheet of the Hashimoto SMB process

The Hashimoto SMB process can be implemented as a three-zone process or as a four-zone process. These variants are illustrated by Figure 7.5 and Figure 7.6. In the three-zone process, the feed stream is completely converted to a product stream with the required purity. The reactors and the separators are placed in alternating sequence in order to increase conversion by reaching the reactive equilibrium within the reactor and removing the product in the following separating unit. The four-zone process has an additional raffinate stream containing the educt and an additional zone IV in order to improve the regeneration of the eluent. Thereby, at the expense of an additional stream that is not the desired product and of additional columns, the process can be operated with smaller desorbent consumption or a higher feed throughput and a breakthrough of the components over the recycle stream can be prevented more easily.

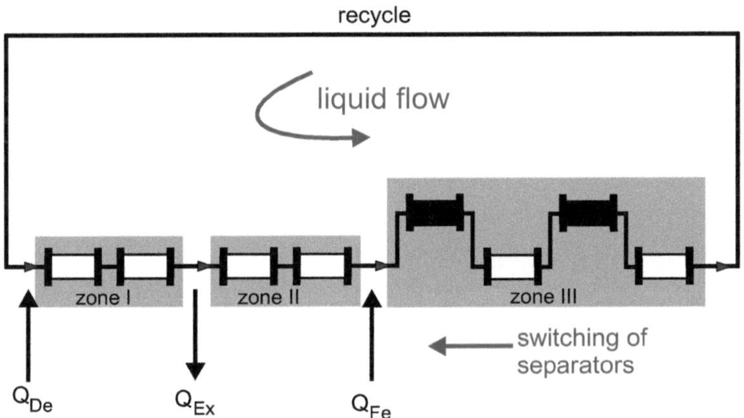

Fig.7.5: Three-zone Hashimoto SMB process

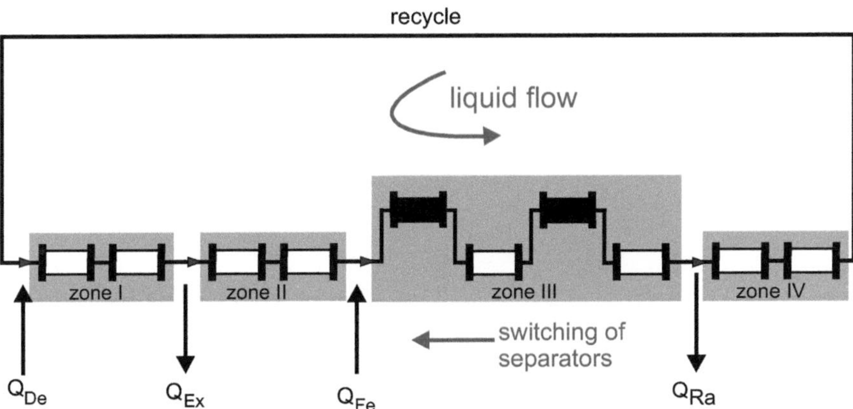

Fig. 7.6: Four-zone Hashimoto SMB process

The four-zone Hashimoto SMB process with an additional product stream that contains a large amount of the feed components can be modified by an internal recycle of the raffinate stream. The raffinate stream is fed back to the feed port and is then converted to the reactive equilibrium by the first reactor, illustrated by Figure 7.7 for a [1/1/1+1/1] column configuration. The notation [1/1/1+1/1] indicates one chromatographic column in each separation zone and one reactor in zone three. For a zone that contains reactors and separations, the units are placed in alternating order starting with a reactor.

Another extension is to consider a classic SMB process with a reactor positioned within a raffinate recycle line as depicted in Figure 7.8 for a [1/1/1/1]+1 configuration. The notation [1/1/1/1]+1 defines one reactor in the raffinate recycle for a SMB process with one chromatographic column in each separation zone. The feed is mixed with the raffinate stream and then converted to the reactive equilibrium in the reactor that is positioned in the additional raffinate recycle. The product and the feed are then separated in the SMB process. All variants of the Hashimoto SMB process can also be operated in the VARI*COL* mode with asynchronous port switching. The reactors must be moved together with the ports in front of them in order to maintain their positions in the zones, see Figure 7.9.

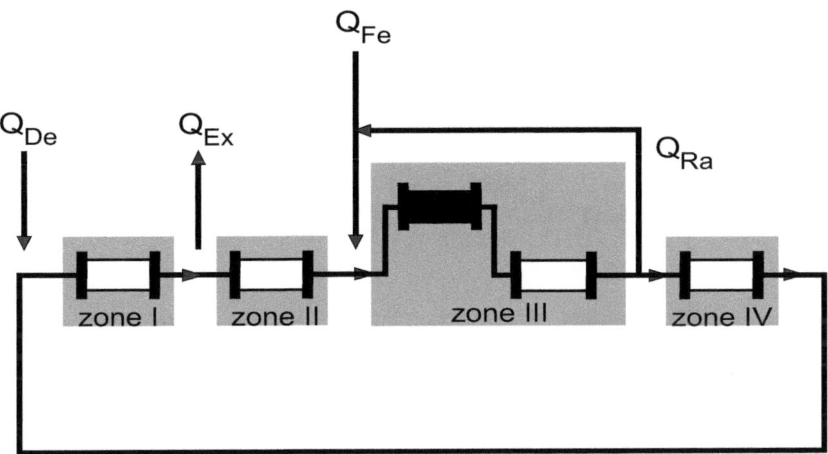

Fig. 7.7: Four-zone Hashimoto SMB process with recycle of the raffinate stream, [1/1/1+1/1] configuration

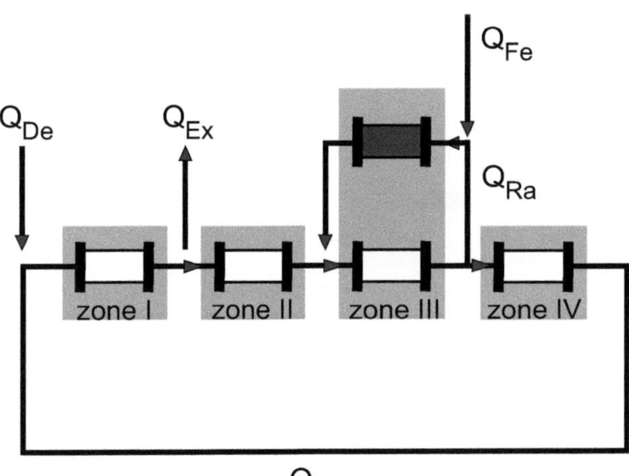

Fig. 7.8: Four-zone Hashimoto SMB process with a reactor placed in the recycle of the raffinate stream, [1/1/1/1]+1 configuration

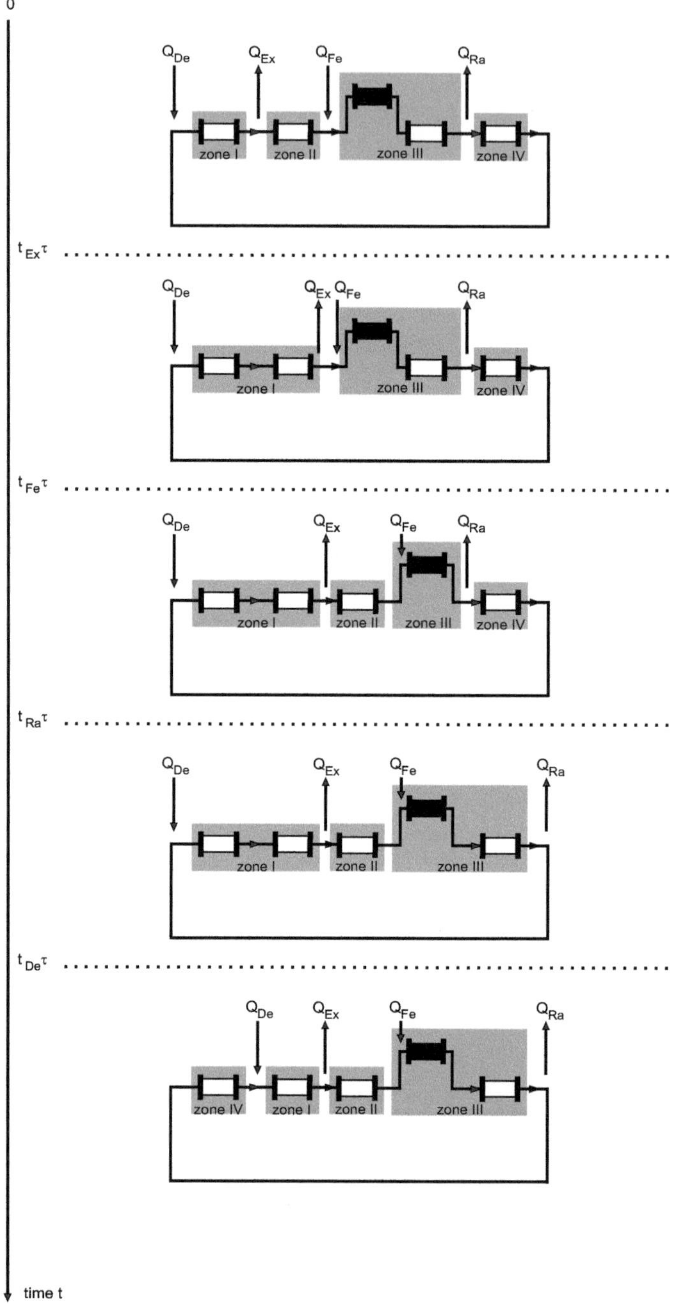

Fig. 7.9: Port positions and column distributions of a four-zone Hashimoto-VARI*COL* process during one period

7.4 Mathematical modelling

Accurate dynamic models of multi-column continuous chromatographic processes consist of dynamic process models of the single chromatographic columns and of the reactors, and node balances which describe the connections of the columns and the switching of the ports. A single chromatographic column with radially homogeneous flow distribution is described accurately by the *General Rate Model* (Guiochon et al. 1994, Dünnebier and Klatt 2000) which accounts for all important effects within the column: adsorption equilibrium, mass transfer between the liquid and the solid phase, pore diffusion, and axial dispersion. It is assumed that the particles of the solid phase are uniform, spherical, and porous with a constant void fraction ε_P, and that the mass transfer between the particle and the surrounding layer of the bulk is in a local equilibrium. The concentration of component i is denoted by c_i in the liquid phase and q_i in the solid phase. D_{ax} is the axial dispersion coefficient, u the interstitial velocity, ε_b the void fraction of the bulk phase, $k_{l,i}$ the film mass transfer resistance, and $D_{p,i}$ the diffusion coefficient within the particle pores. The liquid concentration of the bulk is denoted by $c_{b,i}$, within the pores by $c_{p,i}$. A reaction term r_{kin} can be considered both in the liquid phase and for the solid phase. Assuming that u and $c_{b,i}$ are uniformly distributed over the radius, the following set of partial differential equations can be obtained from a mass balance around an infinitely small cross-section of the column:

Model of a chromatographic column

$$\frac{(1-\varepsilon_b)3k_{l,i}}{\varepsilon_b}(c_{b,i}-c_{b,i|r=R_p})+Xr_{kin,i}^{liq}=D_{ax}\frac{\partial^2 c_{b,i}}{\partial z^2}-u\frac{\partial c_{b,i}}{\partial z} \tag{7.5}$$

$$(1-\varepsilon_p)\frac{\partial q_i}{\partial t}+\varepsilon_p\frac{\partial c_{p,i}}{\partial t}-\left[\frac{1}{r^2}\frac{\partial}{\partial r}\left(r^2\frac{\partial c_{p,i}}{\partial r}\right)\right]-r_{kin,i}^{sol}=0, \tag{7.6}$$

with appropriate initial and boundary conditions

$$c_{b,i|t=0} = c_{b,i}(t=0,z), \quad c_{p,i|t=0} = c_{p,i}(t=0,z,r),$$

$$\left.\frac{\partial c_{b,i}}{\partial z}\right|_{z=0} = \frac{u}{D_{ax}}(c_{b,i} - c_i^{in}), \quad \left.\frac{\partial c_{b,i}}{\partial z}\right|_{z=L} = 0,$$

$$\left.\frac{\partial c_{p,i}}{\partial r}\right|_{r=0} = 0, \quad \left.\frac{\partial c_{p,i}}{\partial r}\right|_{r=R_p} = \frac{k_{l,i}}{\varepsilon_p D_{p,i}}(c_{b,i} - c_{p,i|r=R_p}),$$

$$(7.7)$$

where X denotes the catalyst fraction in the fixed bed of the column, and $1-X$ is the adsorbent fraction. In the case of the Hashimoto SMB process, the catalyst fraction of a reactor is given by $X_r = 1$ and the catalyst fraction X_s of a separator is zero.

The concentrations at the surface of the particles and in the liquid in the surrounding pores are in an equilibrium described by the adsorption isotherm:

$$q_i = f(c_{p,1},...,c_{p,n_{sp}}), \quad i = 1,...n_{sp} \quad . \tag{7.8}$$

The reactive terms considered are of equilibrium type

$$r_{kin} = \upsilon_i k_m \left(\frac{c_{b,i}}{k_{eq}} - c_{b,j}\right), \tag{7.9}$$

where k_m, k_{eq} and υ_i denote the reaction rate constant, the equilibrium constant and the stoichiometry, respectively.

From mass balances around the inlet and the outlet nodes the following expressions for the internal flow rates Q_I, Q_{II}, Q_{III}, Q_{IV} and the inlet concentrations c^{in} after the mixing nodes can be derived:

Desorbent node:
$$Q_{IV} + Q_{De} = Q_I$$
$$c_{i,IV}^{out} Q_{IV} = c_{i,I}^{in} Q_I$$
$$i = A, B$$

Extract node:
$$Q_I - Q_{Ex} = Q_{II}$$

$$(7.10)$$

Feed node:
$$Q_{II} + Q_{Fe} = Q_{III}$$
$$c_{i,II}^{out} Q_{II} + c_{i,Fe} Q_{Fe} = c_{i,III}^{in} Q_{III}$$
$$i = A, B$$

Raffinate node:
$$Q_{Ra} + Q_{IV} = Q_{III} \quad .$$

$Q_{De}, Q_{Ex}, Q_{Fe}, Q_{Ra}$ denote the external flow rates while c^{out} denotes the concentrations of the streams leaving the respective zone.

The approach proposed in (Gu 1995) that combines a finite element discretization of the bulk phase with orthogonal collocation of the solid phase is employed here to discretize the partial differential equations, giving rise to about 100 state variables per column for an appropriate discretization grid. The resulting ode system is solved by the solver DVODE (Hindemarsh 1983).

7.5 Steady state optimization of SMB processes

This section deals with the optimization of the degrees of freedom of SMB processes at the periodic steady state. The formulation and the solution of the optimization problem are described and examples of the application to SMB and Hashimoto SMB processes are given.

7.5.1 General approach

There exist several short-cut methods for the design of SMB processes. The most established method of this type is the triangle theory derived by Mazzotti, Storti and Morbidelli (Storti et al. 1993). It has been extended to systems with nonlinear isotherms in (Migliorini et al. 1998). However, since the method is based on the True Moving Bed (TMB) process, it is only suitable for systems with a relatively large number of columns and it cannot take the additional effects introduced by the $VARICOL$ operation into account.

The optimization approach presented here is based on a rigorous model of the SMB or $VARICOL$ processes and fully exploits the economic potential of the process under consideration. It can be applied to the Hashimoto SMB process as well.

The optimization problem is stated as follows:

$$\min_{Q_i(t),\overline{N}_i,\tau,x^{css}_{smb}} J$$

$$s.t. \quad \left\| \Gamma(c_{ax,k}) - c_{ax,k} \right\| \leq \delta$$
$$Pur_{Ex} \geq Pur_{Ex,min}$$
$$Pur_{Ra} \geq Pur_{Ra,min} \tag{7.11}$$
$$0 \leq x_{smb}$$
$$0 \leq Q_j \leq Q_{max}, \qquad j = 1,...,N_{col},$$

where J is the (economic) objective that is minimized and x_{smb} denotes the concentrations in the liquid phase and at the solid phase in all zones of the SMB process. The goal of the optimization is to determine the optimal cyclic steady-state operation point with minimal separation costs while satisfying product purity requirements and restrictions on the flow rates. The process is required to be at the periodic steady state which is denoted by x_{smb}^{css}. The operator Γ summarizes the mixed discrete continuous dynamics of the respective SMB process over the period k:

$$x_{smb}^{k+1} = \Gamma(x_{smb}^k) : \begin{cases} \widetilde{x}_{smb}^k = x_{smb}^k + \int_{t=0}^{\tau} f_{smb}(x_{smb}(t,u_{smb}(t),p))dt \\ x_{smb}^{k+1} = P\widetilde{x}_{smb}^k, \end{cases} \qquad (7.12)$$

x_{smb}^k denotes the axial concentration profile at the end of period k. The matrix P, see the appendix, represents the shift of the concentration profile backwards by one column. The periodic steady state satisfies $x_{smb}^{css} = \Gamma(x_{smb}^{css})$. The product purity requirements for the extract and the raffinate streams

$$Pur_{Ex,k} = \frac{\int_{t_k}^{t_{k+1}} c_{A,Ex}(t)dt}{\int_{t_k}^{t_{k+1}} \left(c_{A,Ex}(t) + c_{B,Ex}(t)\right)dt} \qquad (7.13)$$

and

$$Pur_{Ra,k} = \frac{\int_{t_k}^{t_{k+1}} c_{B,Ra}(t)dt}{\int_{t_k}^{t_{k+1}} \left(c_{A,Ra}(t) + c_{B,Ra}(t)\right)dt} \qquad (7.14)$$

as well as flow-rate limitations due to the maximum pressure drop over a column are enforced by additional constraints. The main difficulty of the solution of the optimization problem is the computation of the periodic steady state for a model derived from first principles. A robust and simple method is the sequential approach where the process is simulated from initial values until the periodic steady state condition is fulfilled.

The numerical tractability is improved by translating the natural degrees of freedom of the process $Q_{De}(t), Q_{Fe}(t), Q_{Ra}(t), Q_{Re}(t), \tau$ into the so-called β-factors via a nonlinear transformation (Hashimoto et al. 1993). These factors represent the ratios between the flow rates Q_i in each zone i and the effective solid flow rate Q_s:

$$Q_{solid} = \frac{(1-\varepsilon)V_{col}}{\tau} \qquad (7.15)$$

$$\beta_1 = \frac{1}{H_A}\left(\frac{Q_1}{Q_s} - \frac{1-\varepsilon}{\varepsilon}\right) \qquad (7.16)$$

$$\beta_2 = \frac{1}{H_B}\left(\frac{Q_2}{Q_s} - \frac{1-\varepsilon}{\varepsilon}\right) \qquad (7.17)$$

$$\frac{1}{\beta_3} = \frac{1}{H_A}\left(\frac{Q_3}{Q_s} - \frac{1-\varepsilon}{\varepsilon}\right) \qquad (7.18)$$

$$\frac{1}{\beta_4} = \frac{1}{H_A}\left(\frac{Q_4}{Q_s} - \frac{1-\varepsilon}{\varepsilon}\right) \; . \qquad (7.19)$$

The optimization problems discussed in the following sections are solved by the solver FFSQP (Zhou et al. 1997).

7.5.2 Examples

In the following subsections, we will present the results of two optimization studies, a pure separation problem and a reactive process which is investigated in all variants discussed above. The optimization is performed with respect to the full set of operating parameters while the plant setup (number and dimensions of the columns, choice of the adsorbent and thus the adsorption isotherm) is assumed to be given and fixed.

7.5.2.1 Optimization of the separation of a mixture of propanolol isomers in a VARICOL process

The propanolol isomers mixture considered here is a precursor of a beta blocker known as Inderal. The investigation was conducted in collaboration with the company *Novasep* (Toumi et al. 2003). The adsorption equilibrium of the propanolol isomers mixture can be described by a modified competitive Langmuir adsorption isotherm (the involved components are referred to as *A* and *B*):

$$q_i = H_i^1 c_{p,i} + \frac{H_i^2 c_{p,i}}{1 + \sum_k k_k c_{p,k}}, \quad i = A, B \quad . \tag{7.20}$$

The model of the separation system was parametrized and experimentally validated during the collaboration. The parameters of the separation system are given in Table 7.1.

Table 7.1: Model parameters for the separation of propanolol isomers

Parameter		Value	Units
Column distribution		[2/2/2/2]	
Column length	L_c	10	cm
Column diameter	d_c	1	cm
Feed concentration	c_{feed}	[7.5 7.5]	g l^{-1}
Column void fraction	ε_b	0.4	-
Particle void fraction	ε_p	0.5	-
Viscosity	η	6.85 10^{-4}	g cm^{-1} s^{-1}
Density	ρ	1.0	g cm^{-3}
Particle diameter	d_p	0.2 10^{-6}	cm
Particle diffusion coefficient	D_p	10^{-5}	cm^2 s^{-1}
Axial diffusion coefficient	D_{ax}	10^{-6}	cm^2 s^{-1}
Adsorption coefficients	H_A^1	2.68	-
	H_B^1	2.2	-
	H_A^2	0.9412	-
	H_B^2	0.4153	-
	k_A^2	340	cm^3 g^{-1}
	k_B^2	262	cm^3 g^{-1}
Film mass transfer resistances	$k_{1,A}$	0.56 10^{-2}	cm s^{-1}
	$k_{1,B}$	0.33 10^{-2}	cm s^{-1}

The throughput Q_F is optimized for desired extract and raffinate purities of 98.0%. The numerical optimization results are shown in Table 7.2 for different numbers of chromatographic columns. A considerably higher feed throughput can be achieved by the VARICOL operation compared to the SMB operation. The difference in the performances of the VARICOL and the SMB process becomes larger for plants with a smaller number of columns, since the optimal VARICOL column length distribution then differs more from integer values. It can be seen that the productivity per column is higher for larger numbers of columns, but the VARICOL operation may give the same throughput utilizing one column less than the SMB process..Since the column distribution of the VARI*COL* process depends on the initial distribution of the columns at the beginning of a period and the optimization problem is non-convex, the optimization has to be performed with carefully chosen starting values.

Table 7.2: Comparison of the optimized SMB and VARI*COL* processes for different total numbers of columns

number of columns	6		5		4		3
Process	SMB	VARI-COL	SMB	VARI-COL	SMB	VARI-COL	VARI-COL
Q_{De} [ml/min]	7.87	10.75	3.28	6.23	1.85	2.60	4.38
Q_{Ex} [ml/min]	4.77	7.84	2.95	4.31	1.61	1.51	3.22
Q_{Fe} [ml/min]	0.79	0.93	0.35	0.67	0.15	0.39	0.25
Q_{Ra} [ml/min]	3.89	3.83	0.68	2.59	0.39	1.48	1.41
τ [min]	1.18	1.19	2.16	1.52	4.23	2.55	3.14
N_1	1	0.49	1	0.55	1	0.83	0.29
N_2	2	2.68	2	2.10	1	1.45	1.21
N_3	2	2.23	1	1.86	1	1.39	1.15
N_4	1	0.60	1	0.49	1	0.34	0.35

7.5.2.2 Optimization of the Hashimoto process for glucose isomerization

The isomerization of glucose to fructose is a very important process in the sugar industry with a world market of over 5 million tons. Glucose has only about 70% of the sweetness of sucrose and is less soluble. Fructose is 30% sweeter than sucrose and twice as soluble as glucose. Using enzyme technology, the conversion of glucose to a syrup that contains 55% fructose overcomes both problems, giving a stable high fructose corn syrup (HFCS55) that is as sweet as sucrose. HFCS55 is produced by hydrolysis of glucose followed by enzyme-catalyzed isomerization of glucose. The isomerisation of glucose is an equilibrium-limited reaction yielding a syrup that contains 42% of fructose. The production of HFCS55 from HFCS42 requires an integrated process containing reaction and chromatographic re-action. The reactive Hashimoto SMB process is an interesting option to re-duce the production cost in comparison to a classic sequential process.

The natural objectives in the optimization of the Hashimoto process are to maximize the feed throughput, thus increasing the profit by selling more product or to minimize the desorbent consumption that dilutes the product and has to be evaporated causing high energy costs. Both aims are in com-petition, since for a higher feed flow a higher desorbent flow is required to reach the specified product purities, and a smaller desorbent flow enables less feed flow. For this study, a mixed objective was formulated that in-corporates both goals, weighted according to economic considerations to maximize profit. The revenue is generated by selling the product HFCS55 that is converted from HFCS42 while energy for the evaporation of the de-sorbent (water) has to be paid for. The two terms are proportional to the feed flow rate and the desorbent flow, respectively. US prices of natural gas for evaporation and for HFCS55 and HFCS42, and the heat of vapori-zation r_v and the specific heat capacity c_p, as well as the boiling tempera-ture T_b of water and the operating temperature T_p of the Hashimoto process are given in Table 7.3.

Table 7.3: Parameters and US prices for the profit function of the glucose isom-erization

Parameters	prices
$r_v{=}540\,\dfrac{\text{cal}}{\text{g}}$	$p_{HFCS42}{=}0.0288\,\dfrac{\text{cent}}{\text{g}}$
$c_p{=}1\,\dfrac{\text{cal}}{\text{g}^{0}\text{C}}$	$p_{HFCS55}{=}0.0367\,\dfrac{\text{cent}}{\text{g}}$
$T_p{=}60\,^{0}\text{C}$	$p_{ernergy}{=}0.0013777\,\dfrac{\text{cent}}{\text{kcal}}$
$T_b{=}100\,^{0}\text{C}$	

Assuming an efficiency factor λ of 0.8 of the heat applied for the evaporation, the economic profit [$/h] results as:

$$profit = (p_{HFCS55} - p_{HFCS42})\sum c_{HFCS42}Q_{Fe} - ((T_b - T_p)c_p + r_e)\rho_w \frac{1}{\lambda}p_{energy}Q_{De} \quad (7.21)$$

From eq. (7.21), the objective function J that was used in the optimization had the form

$$J = \alpha Q_{De} - Q_{Fe}, \quad (7.22)$$

where the objective J was minimized. Using the values in Table 7.3, the weighting factor α results as 0.5626.

The adsorption isotherm of the glucose(G)/fructose(F) system is of parabolic type (Jupke 2003):

$$q_i = H_i c_{p,i} + k_i c_{p,i}^2 + k_{i,j} c_{p,i} c_{p,j}, \quad i,j = G,F \quad i \neq j. \quad (7.23)$$

The isomerization is treated as a homogenous, equilibrium limited reaction that takes place in the liquid phase only. The reaction kinetics are described by

$$r_{kin}^{liq} = \upsilon_i k_m \left(\frac{c_{b,i}}{k_{eq}} - c_{b,j} \right)$$
$$r_{kin}^{sol} = 0. \quad (7.24)$$

The parameters of the reactive separation system are listed in Table 7.4.

Table 7.4: Parameters for the glucose isomerization

Parameter		Value	Units
Column length	L_c	57.5	cm
Column diameter	d_c	2.6	cm
Feed concentration	c_{feed}	[126 174]	g l^{-1}
Column void fraction	ε_b	0.4	-
Particle void fraction	ε_p	0.01	-
Viscosity	η	5.8 10^{-3}	g cm^{-1} s^{-1}
Density	ρ	1.0	g cm^{-3}
Particle diameter	d_p	0.163 10^{-6}	cm
Particle diffusion coefficient	D_p	10^{-3}	cm^2 s^{-1}
Axial diffusion coefficient	acc. to (Chung and Wen 1968)		
Adsorption coefficients	H_F	0.2545	-
	H_G	0.1958	-
	k_F	1.46 10^{-1}	-
	k_G	1.33 10^{-1}	cm^3 g^{-1}
	$k_{F,G}$	2.9 10^{-1}	cm^3 g^{-1}
	$k_{G,F}$	9.3 10^{-2}	cm^3 g^{-1}
Film transfer resistances	$k_{1,A}$	2.05 10^{-5}	cm s^{-1}
	$k_{1,B}$	3.8 10^{-5}	cm s^{-1}

Stoichiometry	ν	$[1\,-1]$	-
Reaction rate constant	k_m	$4.70\ 10^{-3}$	s^{-1}
Equilibrium constant	k_{eq}	1.0798	-

The isomerization of glucose was investigated for the different variants of the Hashimoto SMB process presented in section 4.3 in standard SMB and in VARICOL mode of operation. The optimization of each VARICOL-Hashimoto configuration was performed with 15 different initializations to account for the non-convexity of the optimization problem. In Figure 7.17, the performance figures of the various Hashimoto SMB processes with four separative columns and one reactor are shown for the standard Hashimoto SMB process with three and with four zones, the extension by a raffinate recycle for the four-zone configuration and the fully integrated SMBR (Schmidt 2005). For the glucose isomerization, the four-zone process with raffinate recycle in VARICOL mode gives the best performance, closely followed by the three-zone Hashimoto process in VARICOL mode and the three-zone Hashimoto process in standard mode. For the glucose isomerization with an objective that combines solvent consumption and feed throughput, the VARICOL mode improves the performance only by 2.5%. By the VARICOL operation the solvent consumption is reduced by almost 16% compared to the standard three-zone Hashimoto process, but this effect is reflected in the economic objective function only by a moderate increase. The economic objective is influenced mostly by the amount of feed. The capacity of the plant to convert the feed into the desired product in zone III is mainly limited by the interaction of reaction and separation. The asynchronous switching of the ports has only a moderate influence on the reactor performance. Compared to the SMBR, the four-zone Hashimoto process with recycle in VARICOL mode improves the performance by 138%. The Hashimoto variants with a reactor in the raffinate stream showed a 11% lower performance than the optimal configuration.

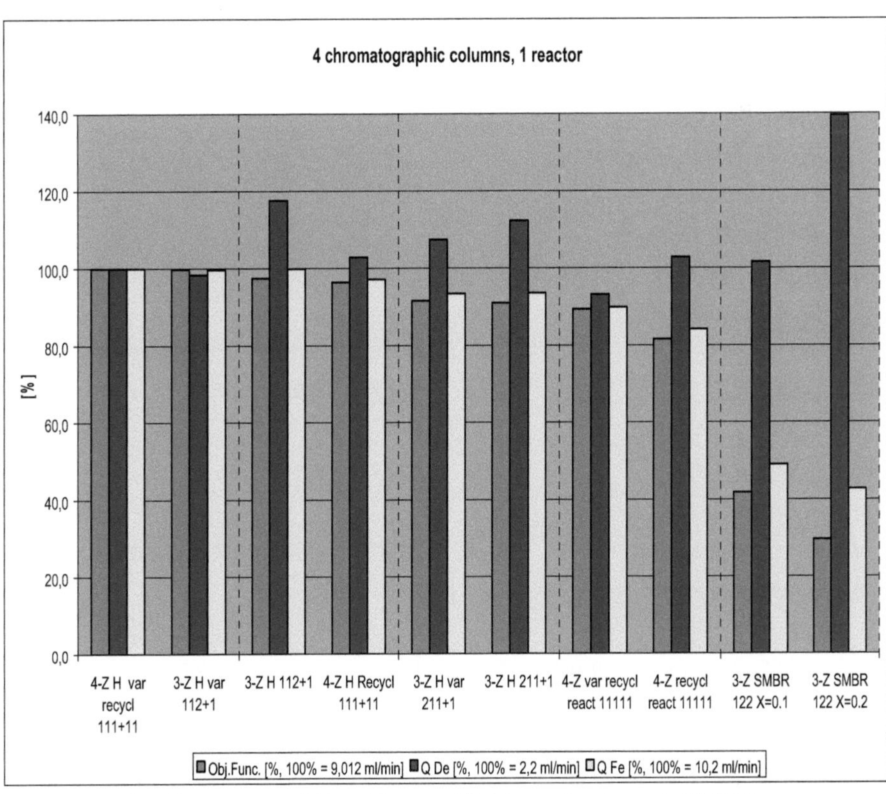

Fig. 7.10: Performance of Hashimoto SMB variants and SMBR

Table 7.5: Economic performance and operating parameters of the investigated SMB configurations (the notation is explained in section 3)

Configuration	Obj.	Q_{De} [ml/ min]	Q_{Ex} [ml/ min]	Q_{Fe} [ml/ min]	Q_{Re} [ml/ min]	Q_{Ra} [ml/ min]	τ [min]
3-zone-Hashimoto [1/1/2+1]	8.79	2.57	12.80	10.23	17.43	-	8.69
4-zone-Hashimoto Raffinate recycle [1/1/1+1/1]	8.70	2.24	12.20	9.95	17.76	0.01	8.61
3-zone-Hashimoto [2/1/1+1]	8.22	2.45	12.04	9.59	17.54	-	8.47
4-zone-SMB raffinate recycle with reactor [1/1/1/1]+1	7.36	2.24	10.86	8.62	15.14	4.82	9.96
3-zone-SMBR [1/2/2] X=0.1	3.78	2.14	7.12	4.98	17.86	-	9.23
3-zone-SMBR [1/2/2] X=0.2	2.68	3.04	7.43	4.39	16.96	-	9.54
4-zone-Hashimoto varicol raffinate recycle [1/1/1+1/1]	9.01	2.18	12.42	10.24	17.82	0.02	8.52
3-zone-Hashimoto varicol [1/1/2+1]	9.00	2.15	12.35	10.20	17.85	-	8.51
3-zone-Hashimoto varicol [2/1/1+1]	8.26	2.34	11.92	9.57	17.67	-	8.43
4-zone-SMB varicol raffinate recycle with reactor [1/1/1/1]+1	8.07	2.03	11.24	9.21	15.24	4.65	9.80

7.6 Optimization of the design of a Hashimoto SMB process

In this section, we will discuss the optimization of the design of the three zone Hashimoto SMB process for the isomerization of glucose focusing on the optimal positions and the optimal lengths of the reactors.

The Hashimoto process considered here is a three-zone process with a distribution of separating columns over the zones of [2,2,4] and possible reactor positions as indicated in Figure 7.11, i.e. potential reactor positions everywhere in zones II and III. The separative columns are uniform with column lengths L while the length of each reactor is regarded as a degree of freedom for the optimization. A reactor with a length of zero (or close to zero) is redundant and can be omitted.

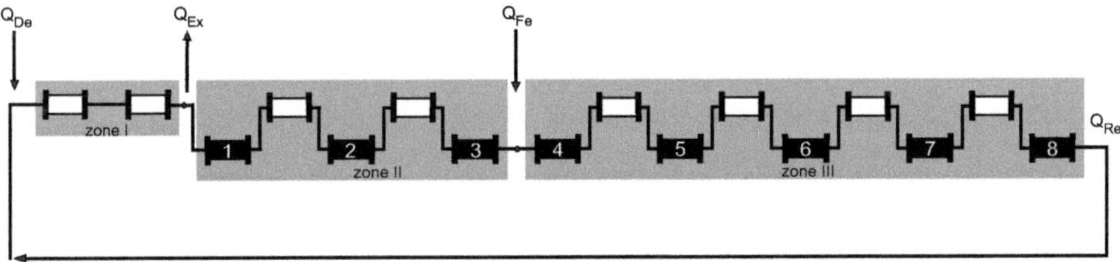

Fig. 7.11: Three-zone Hashimoto process with 8 potential reactors

The influence of the lengths of the reactors on the dynamic optimization of the process is rather complex. If a reactor is sufficiently long, the reaction equilibrium is reached at the outlet of the reactor. Since the reactors are stationary within the respective separation zone and are not shifted together with the separating columns, they provide a fixed hold-up and delay element with a non-negligible influence on the dynamics of the process. There exist many local optima in the solution of the design problem. The objective function was extended by an additional term that penalizes the lengths of the reactors in order to provide a driving force that reduces unused reactor volumes. The design problem is stated as:

$$\min_{Q_I,\beta_{I,II,III},l_{r,i}} J = \alpha Q_{De} - Q_{Fe} + \gamma \sum_{i=1}^{N_r} l_{r,i}$$

$$s.t. \quad \left\| \Gamma(c_{ax,k}) - c_{ax,k} \right\| \leq \delta$$

$$Pur_{Ex} \geq Pur_{Ex,min}$$

$$Q_I \leq Q_{max} \tag{7.25}$$

$$Q_{De}, Q_{Ex}, Q_{Fe}, Q_{Re} \geq 0$$

$$l_{r,min} \leq l_{r,i} \leq l_{r,max},$$

where $l_{r,i}$ denote the lengths of the corresponding reactors. The weight γ relates the two different objectives: economic yield and catalyst volume. The weight γ could be fixed such that it relates the two counteracting objectives in an economic manner. The intention here, however, is to study the trade-off between the overall reactor volume and the process performance and the identification of favorable reactor positions. The design task is transformed into the problem to determine the Pareto optimum and the optimization is performed with different weights γ in order to compute the Pareto curve.

The values of the two objectives (economic yield and total reactor length) are shown for all solutions that were obtained by varying γ in Figure 7.9. It can be seen that there is a large number of suboptimal solutions due to the non-convex nature of the optimization problem, but the shape of the Pareto curve is visible quite well. A total reactor length of 2-3 times the length of a separating column gives approximately the best economic performance, and the reactors should be placed in positions 2, 3, and 4. At the other positions, an additional reactor is not efficient. Table 7.7 displays the best values of the reactor lengths and the corresponding economic performance that were found for each Pareto weight γ. The solution displayed in Figure 7.13 is a sensible solution regarding both objectives with an economic performance of 11.7 ml/min. If less catalyst is used (total reactor length less than 1.5 times the length of a separator), the productivity deteriorates sharply whereas for higher values it remains almost constant. As can be seen from Table 7.7, reactor 4 has a dominant influence on the economic performance. The lengths of the remaining reactors 1, 5, 6, 7, and 8 were driven to zero by the optimization and are redundant. The reactors 2 and 3 play an important role for the economic performance. Operation without reactor 3 leads to a significant reduction in process performance as can be seen from Table 7.7. Similarly, elimination of reactor 2 gives a considerably worse performance as well, as further optimization studies have shown. The operating parameters of the best solutions are very similar, the result for the solution with reactors in positions 2, 3, and 4 and a total reactor length of 1.88 is given in Table 7.6.

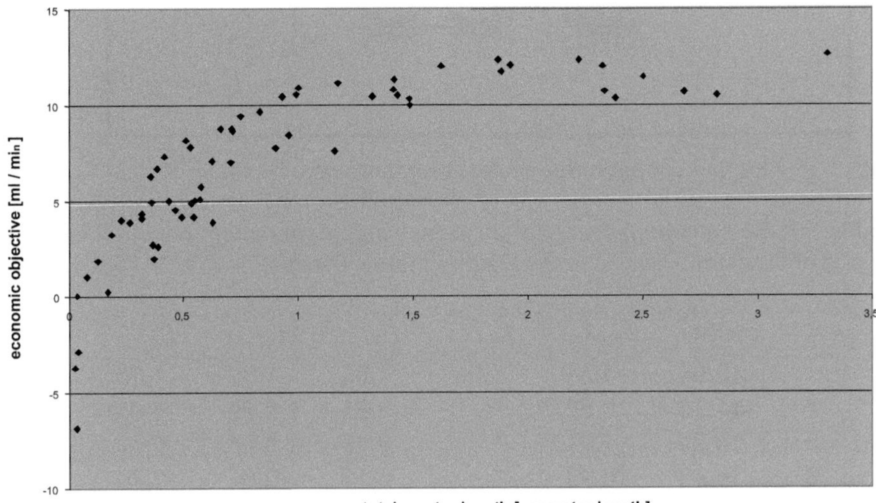

Fig. 7.12: Economic objective versus total length of reactors

Table 7.6: Optimal operating parameters for weight $\gamma=4$

Operating parameter		Value	Units
Period length	T	8.46	min
Desorbent stream	Q_{De}	2.26	ml min^{-1}
Extract stream	Q_{Ex}	15.24	ml min^{-1}
Feed stream	Q_{Fe}	12.98	ml min^{-1}
Recycle stream	Q_{Re}	17.73	ml min^{-1}

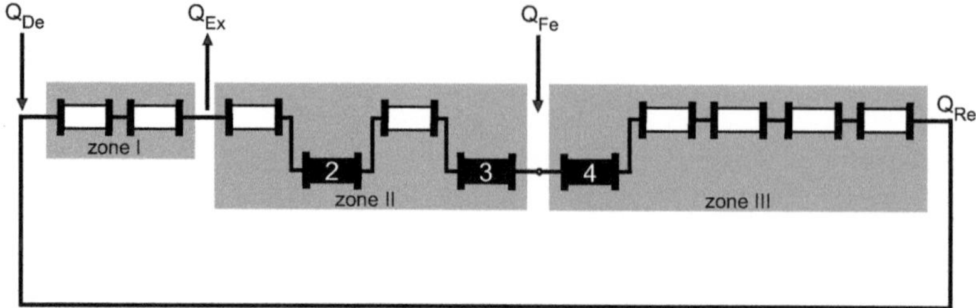

Fig. 7.13: Optimal configuration of the Hashimoto process for γ=4

Table 7.7: Best solutions found for different weights gamma (*: not considered as degree of freedom; + scaled to the length of one separator)

γ	profit	L_{total}^{+}	L_1^{+}	L_2^{+}	L_3^{+}	L_4^{+}
1	12,64	3,31	0,00	1,04	1.00	1,18
4	11,70	1,88	*	0,25	0,59	1,03
5	10,90	1,01	*	0,15	0,04	0,82
8	10,43	0,93	*	0,13	0,07	0,73
10	8,16	0,51	0,00	0,19	0,00	0,30
11	7,33	0,42	*	0,23	0,00	0,19
12	6,29	0,36	*	0,22	0,00	0,14
15	3,99	0,23	*	0,13	0,00	0,10
16	1,84	0,13	*	0,11	0,00	0,00
20	1,03	0,08	*	0,05	0,02	0,00
200	-3,70	0,02	0,00	0,00	0,00	0,02
2000	6,87	0,03	*	0,02	0,00	0,01

L_5^{+}	L_6^{+}	L_7^{+}	L_8^{+}
0,01	0,03	0,03	0,02
*	*	*	*
*	*	*	*
*	*	*	*
0,01	0,01	0,00	0,00
*	*	*	*
*	*	*	*
*	*	*	*
*	*	*	*
*	*	*	*
0,00	0,00	0,00	0,00
*	*	*	*

7.7 Control of reactive SMB processes

Classic feedback control strategies are not directly applicable to SMB processes due to their unconventional characteristics with mixed discrete and continuous dynamics, spatially distributed state variables with steep slopes, and slow and strongly nonlinear responses of the concentration profiles to changes of the operating parameters.

In (Klatt et al., 2000), a two-layer control strategy for SMB processes was proposed. On the upper layer, the steady-state operating parameters are optimized by using a rigorous plant model. The low-level control task is to keep the process on the optimal periodic trajectory despite disturbances and plant/model mismatch. This is realized by controlling the positions of the concentration fronts (Wang et al. 2003). On the upper layer, the model parameters are periodically re-estimated based on the available online measurements, and the model used on the lower level is adapted accordingly. However, the stabilization of the front positions does not guarantee the product purities in the presence of a structural plant/model mismatch, therefore an additional purity control layer must be introduced (Hanisch 2002). The resulting control structure is relatively complex with three interacting layers.

(Schramm et al. 2003) presented a model-based approach to control the product purities. From wave theory, they derived relationships between the inputs and the movements of the concentration fronts of an equivalent True Moving Bed model. A simple control concept employing PI controllers was developed based upon this approximate model. However, similar relationships are difficult to determine for nonlinear chromatographic SMB process with integrated chemical reaction.

Optimization-based approaches to the control of SMB processes were reported in recent years by groups at ETH Zürich (Erdem et al. 2004a, Erdem et al. 2004b, Abel et al., 2005) and at Universität Dortmund (Toumi and Engell 2004a, Toumi and Engell 2004b, Toumi 2004), (Engell and Toumi 2005). In the work by Erdem et al. (2004a, 2004b) and Abel et al. (2005), a moving horizon online optimization is performed based on a linear reduced-order model that is obtained from linearizing a rigorous model around the periodic steady state. The state variables of the model are estimated by a Kalman Filter that uses product concentration measurements as inputs. As a consequence of the chosen control strategy, the switching period that can be translated into the simulated flow of the solid bed and has a considerable influence on the process performance is assumed to be fixed and not taken into account as a degree of freedom for the online optimization in this approach.

In (Toumi and Engell 2004a, Toumi and Engell 2004b, Toumi 2004), a nonlinear moving-horizon optimizing control scheme that is based on a rigorous model and takes all degrees of freedom of the process into account was proposed and successfully applied to a three-zone reactive SMB process for glucose isomerization. In each switching period, the operating parameters are optimized to minimize an economically motivated cost function while the product purities appear as constraints in the optimization problem. In (Toumi et al. 2005), this control concept was extended to the more complex processes VARI*COL* and PowerFeed (Kearney and Hieb 1992), where the flow rates are varied in the subintervals. These process variants offer an even larger number of degrees of freedom that can be used for the optimization of the process economics while satisfying the required product purities. In the optimizing control scheme of Toumi et al., the states of the process model are determined by forward simulation starting from measurements in the recycle stream and in the product streams. In (Küpper and Engell 2006), a combined parameter and state estimation scheme with error feedback is reported for a nonlinear SMB process with individual column parameters based on measurements of the concentrations in both product streams and one internal measurement in the recycle stream.

In this section, we present the direct online optimizing control scheme for reactive and non-reactive SMB processes and its application to the Hashimoto SMB process. The online optimizing control scheme computes control variables (flow rates and the switching time) that optimize an economic objective over a moving horizon of several future switching periods while the purity requirements of the product streams are considered as constraints. In the optimization, a rigorous model of the general rate type is used. Plant/model mismatch is taken into account by error feedback of the predicted and the measured purities. In addition, the model parameters are updated by a parameter estimation scheme.

7.7.1 Online optimizing control

The basic idea of the control algorithm is to perform an optimization of the operational degrees of freedom at future switching periods based upon a rigorous model of the plant with respect to an economic cost function (rather than e. g. a cost function involving a tracking error) in which the specifications of the SMB process (purity requirements, limitations of the pumps) as well as the process dynamics are handled as constraints. The inputs located within the control horizon are considered as degrees of freedom of the optimization while the remaining inputs within the larger prediction horizon are set equal to the values in the final control interval. The computed inputs in the first sampling interval are applied to the plant, and the optimization is then repeated for the next time interval with the control and prediction horizon shifted forward by one time interval, using new measurement data and eventually new estimated model parameters. Due to the slow dynamic response of the concentration profiles of SMB processes to changes in the operating parameters, a modern PC is sufficient to solve the online optimization problems within a process cycle (length of a switching period times the number of chromatographic columns). In the application of optimizing control to SMB processes, the sampling time is chosen equal to the length of a cycle and hence varies during the operation of the process.

We here consider a four-zone Hashimoto SMB process with raffinate flow that is described by the nonlinear discrete dynamics (7.27). The objective of the optimizing controller is to minimize the eluent consumption Q_D for a constant feed flow and a given purity requirement of 99% in the presence of a plant/model mismatch. The inevitable mismatch between the model and the behavior of the real plant is taken into account by feedback of the difference of the predicted and the measured product purities. A regularization term is added to the objective function (7.26) to obtain smooth trajectories of the input variables. The controller has to respect the purity requirement for the extract flow (7.28) which is averaged over the prediction horizon, the dynamics of the Hashimoto SMB model (7.27) and the maximal flow rate in zone I due to limited pump capacities (7.30). In order to guarantee that at least 70% of the mass of the components fed to the plant averaged over the prediction horizon leaves the plant in the extract product stream, an additional productivity requirement (7.29) is added. The deviation between the prediction of the model and the plant behavior is considered by the error feedback term Δm_{ex} (7.34). The resulting mathematical formulation of the optimization problem is:

$$\min_{\beta_I,\beta_{II},\beta_{III},\beta_{IV}} \sum_{i=1}^{H_P} Q_i + \Delta\beta\, R\, \Delta\beta \tag{7.26}$$

$$s.t. \quad \begin{cases} \widetilde{x}_{smb}^{k} = x_{smb}^{k} + \int_{t=0}^{\tau} f_{smb}(x_{smb}(t,u_{smb}(t),p))dt \\ x_{smb}^{k+1} = P\widetilde{x}_{smb}^{k}, \end{cases} \tag{7.27}$$

$$\frac{\sum_{i}^{H_{P1}} Pur_{Ex,i}}{H_{P1}} \geq (Pur_{Ex,\min} - \Delta Pur_{Ex}) \tag{7.28}$$

$$\frac{\sum_{i}^{H_{P1}} m_{Ex,i}}{H_{P1}} \geq 0.7 m_{Fe,\min} - \Delta m_{Ex} \tag{7.29}$$

$$Q_I \leq Q_{\max} \tag{7.30}$$

$$Q_{De}, Q_{Ex}, Q_{Fe}, Q_{Re} \geq 0, \tag{7.31}$$

where the purity error, the mass inflow and outflow, and the mass error are calculated according to:

$$\Delta Pur_{Ex} = Pur_{Ex,plant,i-1} - Pur_{Ex,\mathrm{mod}el,i-1}, \tag{7.32}$$

$$m_i = \frac{\int_0^{\tau} (c_A + c_B)\cdot Q_i dt}{\tau}, \tag{7.33}$$

$$\Delta m_{Ex} = m_{Ex,plant,i-1} - m_{Ex,\mathrm{mod}el,i-1}. \tag{7.34}$$

For the solution of the optimization problem, the feasible path solver FFSQP (Zhou et al. 1997) is applied. It first searches for a feasible operating point and then minimizes the objective function. In the simulation of the controller, a problem was encountered. The gradient-based solver calculates the gradients of the objective and of the constraints by perturbation of the last solution point. Since the process is operated at a high product purity near 100% and the component that pollutes the extract stream can vanish as it is illustrated by Figure 7.14, the FFSQP solver can reach solution points with 100% purity. At such a point, it is not possible to calculate the gradient for the constraint (7.28) and the solver becomes "blind" regarding this constraint. Therefore, the constraint (7.35) is added that enforces that the purity set point for the optimizer is below 99.9% averaged over a cycle.

$$\frac{\sum_{i}^{8} Pur_{Ex,i}}{8} \leq 99.9\% \tag{7.35}$$

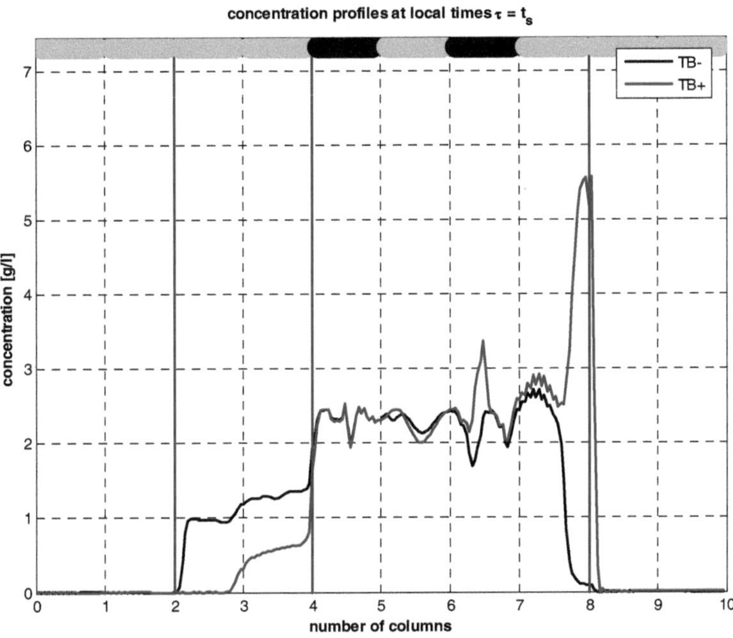

Fig. 7.14: Steady-state concentration profile of a four-zone Hashimoto process (column distribution [2/2/2+2/2]) at the end of a switching period

It is important to allow a maximum number of iterations of the solver that is large enough for convergence, but short enough to be solvable within the sampling interval of the controller.

7.7.2 Parameter estimation

The parameter estimation scheme is performed using the measurements from two sensors that are fixed behind the physical separation column in front of the recycle line. These measurements are collected over one full process cycle and then used to estimate the reaction rate constant k_m and the Henry coefficients H_A and H_B via a least-squares minimization according to:

$$\min_{p}(J(p) = \sum_{i=1}^{nsp} (\int_{0}^{N} (c_{i,meas}(t) - c_{i,re}(t))^2 dt), \tag{7.36}$$

where $c_{re}(t)$ is the predicted concentration profile.

7.7.3 Application study – racemisation of Troeger's Base

The racemization of Troeger's Base (TB) in combination with chromatographic separation is used for the production of TB- that is used for the treatment of cardiovascular diseases. The chromatographic columns are packed with Chiralcel OJ as adsorbent. The eluent is an equimolar mixture of acetic acid that acts as the catalyst for the reaction and 2-propanole that increases the solubility of the eluent mixture. The Troeger's Base system is described by an adsorption isotherm that is of the multi-component Langmuir type:

$$q_i = \frac{H_i c_{p,i}}{1 + \sum_{k} b_{i,k} c_{p,k}} . \tag{7.37}$$

The reaction takes place in single-phase plug flow reactors described by the differential equation

$$\frac{\partial c_{b,i}}{\partial t} + r_{kin,i}^{liq} = D_{ax} \frac{\partial^2 c_{b,i}}{\partial z^2} + u \frac{\partial c_{b,i}}{\partial z} . \tag{7.38}$$

The plug flow reactors are operated at 80°C where the catalyst is thermally activated. In the chromatographic columns that have a temperature of 25°C the catalyst is virtually deactivated. Hence, a alternating order of reactors and separators to increase the conversion of the feed can be achieved. In the simulation run shown below, a four-zone Hashimoto process with 8 chromatographic columns, 2 reactors, and a column distribution [2/2/2+2/2] as shown in Figure 7.6 is operated. The model parameters for the Troeger's Base system are given in Table 5.2 of chapter 5. The number of iterations of the SQP solver was limited to 8. The inputs to the virtual process are obtained from the optimizer within the sampling rate - the control scheme is applicable to real time.

In the control scenario, an exponential decrease of the catalyst activity is assumed that occurs in the case of a malfunction of the reactor heatings. A model/plant mismatch is introduced by disturbing the initial Henry coefficients H_A and H_B of the model by +10% and -15%. The parameters of the controller are displayed in Table 7.8.

Table 7.8: Parameters of the controller and of the estimator

Sampling time		8 periods = 1 cycle
Prediction horizon	H_P	6 intervals
Control horizon	H_C	1 interval
Regularization	R	[0.3 0.3 0.3 0.3]
Controller start		72nd period
Estimator start		72nd period

Figure 7.15 demonstrates that the parameter adaptation scheme estimates the parameters of the plant well.

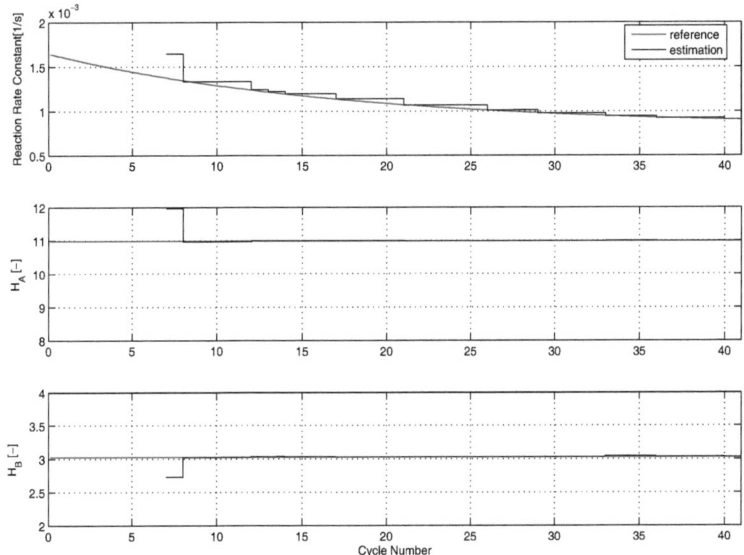

Fig. 7.15: Estimation of the reaction rate constant and of the Henry coefficients

The performance of the controller is illustrated by Figure 7.16. The controller manages to keep the purity and the productivity above their lower limits, while it improves the economic operation of the plant by reducing the desorbent consumption. The process converges to a stationary operating point. The optimizer converges to the optimum within the sampling time and hence can be applied in real-time.

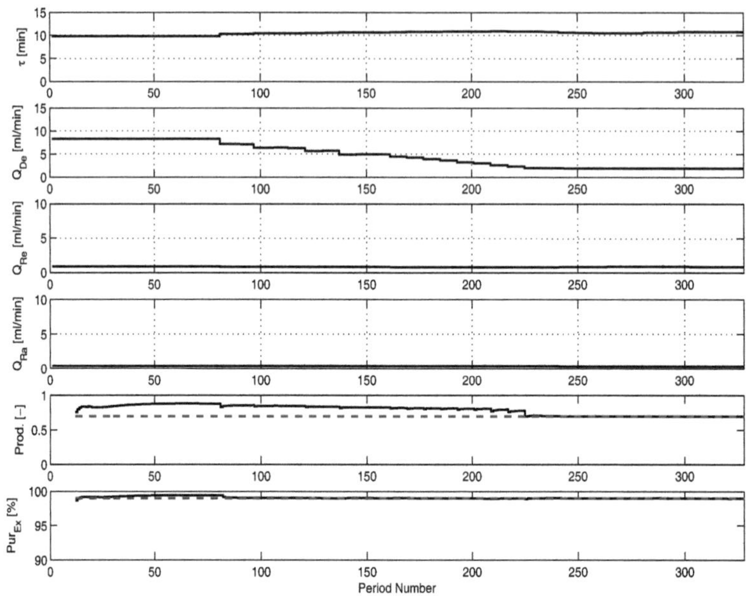

Fig. 7.16: Simulation of the optimizing controller

7.8 Conclusions

In this chapter, optimization and optimizing control approaches for advanced chromatographic separation processes were discussed. The processes considered were the simulated moving bed process that features a counter-current flow between the desorbent and the solid bed, the VARI*COL* operation that shifts the ports of the SMB process asynchronously, and reactive SMB processes that integrate reaction and separation. The focus was on the Hashimoto process that performs reaction and separation in dedicated but strongly coupled units.

Based on rigorous models, dynamic optimizations were performed and optimal operating points for different separation systems were identified. The optimization is based on an economic cost function and a rigorous model of the process dynamics. Purity requirements and plant limitations are included as constraints. For the separation of the propanolol isomers it was demonstrated that the VARI*COL* operation can lead to a considerably higher feed throughput compared to the SMB mode. The example of glucose isomerization was used to evaluate the performance of different variants of the Hashimoto process using an objective function that incorporates the desorbent consumption and the feed throughput, weighted according to economic considerations. The Hashimoto process (the four-zone process as well as the three-zone process) leads to large improvements in process performance compared to a fully integrated reactive SMB process, where the columns are uniformly packed with catalyst and with adsorbent. In a reactor design study, the reactors that have the greatest influence on the economic performance of the process were identified and optimal reactor lengths for different trade-offs between productivity and cost of catalyst were computed. .

A model-based optimizing control approach was proposed for the Hashimoto process. The optimizer determines the optimal flow rates and optimal switching times over a finite horizon and the first values are applied to the plant. Then the optimization is repeated. The goal is to minimize a cost function (here the desorbent consumption) so that the plant is operated as economically as possible while the product purities as well as the desired conversion are achieved. Concentration measurements from the plant are used to perform parameter estimation and to provide a model error feedback to the optimizer. A simulation study of the proposed control concept for the racemisation of Troeger's Base was presented. Despite the large dynamic nonlinear plant model used, the optimizer converged within each switching period and the system converges to the economic optimum while keeping the purity of the product and the conversion above the specified values.

7.9 Notation

Symbols

Symbol	Description	Units
$b_{i,j}$	Langmuir coefficients	$cm^3\,g^{-1}$
$c_{b,i}$	Concentration in the liquid phase	$g\,ml^{-1}$
$c_{p,i}$	Concentration inside the particle pores	$g\,ml^{-1}$
c_p	Heat capacity	$cal\,g^{-1}K$
D_{ax}	Axial Dispersion coefficient	$cm^2\,s^{-1}$
d_p	Particle diameter	cm
D_p	Particle diffusion coefficient	$cm^2\,s^{-1}$

H_i	Henry coefficient	-
H_c	Control Horizon	Interval length
H_p	Prediction Horizon	Interval length
J	Objective	
$k_{l,i}$	Film transfer resistance	$cm^2\,s^{-1}$
k_{eq}	Equilibrium constant	-
l_i	Length of reactor	Length of separator
L_{total}	Total length of reactors	Length of separator
L_c	Column length	cm
m_i	Mass flow	g
N	Number of measurements	
N_i	Number of columns in zone i	-
N_{col}	Total number of columns	-
nsp	Number of species	-
q_i	Solid concentration	$g\,ml^{-1}$
Q_i	Internal stream in zone i	$cm^3\,s^{-1}$
Q_{Ex}	Extract flow	$cm^3\,s^{-1}$
Q_{De}	Desorbent flow	$cm^3\,s^{-1}$
Q_{Fe}	Feed flow	$cm^3\,s^{-1}$
Q_{Ra}	Raffinate flow	$cm^3\,s^{-1}$
Q_S	Solid flow	$cm^3\,s^{-1}$
p	Parameter	
P	Permutation matrix	
p	Prices	$cent\,g^{-1}$
Pur_i	Purity	
r	Radial coordinate in the particle	cm
r^{sol}_{kin}	Reaction (solid)	$g\,cm^3\,s^{-1}$
r^{liq}_{kin}	Reaction (liquid)	$g\,cm^3\,s^{-1}$
r_v	Heat of vaporization	$cal\,g^{-1}$
t	Time	S
T_b	Boiling temperature	K
T_p	Operating temperature	K
u	Velocity	$cm\,s^{-1}$
	Input	
V_{col}	Column volume	cm^3
X	Catalyst fraction	-
z	Axial coordinate	cm

Greek Symbols

Symbol	Description	Units
α	Economic weight	-
β	β – factor	-
γ	Weight (reactor design)	
δ	Small number	
ε	Total porosity	-
ε_p	Particle void fraction	-
ε_b	Column void fraction	-
η	Viscosity	$g\,cm^{-1}\,s^{-1}$
λ	Heating efficiency factor	-
ν_i	Stoichiometry	-
ρ	Density	$g\,cm^{-3}$
τ	Switching period	min
Γ	Discrete dynamics	
ΔPur	Purity error feedback	-
Δm	Mass flow error feedback	g

$$P = \begin{pmatrix} 0 & I & 0 & \cdots & 0 \\ \vdots & \ddots & I & \ddots & \vdots \\ \vdots & 0 & \ddots & I & 0 \\ 0 & \cdots & \ddots & \ddots & I \\ I & 0 & \cdots & 0 & 0 \end{pmatrix}$$

7.10 Literature

Abel S, Erdem G, Amanullah M, Morari M, Mazzotti M, Morbidelli M (2005) Optimizing control of simulated moving beds-experimental implementation. Journal of Chromatography A, 1092(1) 2-16

Broughton D, Gerhold C (1961) Continuous sorption process employing fixed bed of sorbent and moving inlets and outlets. US Patent 2985586

Chung S, Wen C (1968) Longitudinal diffusion of liquid flowing through fixed and fluidized beds. AIChE Journal. 14, 875-866

Dünnebier G, Klatt K-U (2000) Modelling and simulation of nonlinear chromatographic separation processes: a comparison of different modelling approaches. Chem. Eng. Science, 55:373-380

Engell S, Toumi A (2005) Optimization and control of chromatography. Computers and Chemical Engineering. 29, 1243-1252

Erdem G, Abel S, Morari M, Mazzotti M, Morbidelli M (2004) Automatic control of simulated beds. Ind. and Eng. Chem. Res., 43:405-421

Erdem G, Abel S, Morari M, Mazzotti M, Morbidelli M (2004) Automatic control of simulated moving beds-II: nonlinear isotherms. Ind. and Eng. Chem. Res., 43:3895-3907

Gu T (1995) Mathematical Modelling and Scale Up of Liquid Chromatography. Springer Verlag, New York

Guiochon G, Golshan-Shirazi S, Katti A (1994) Fundamentals of preparative and non-linear chromatography. Academic Press, Boston

Hanisch, F (2002) Prozessführung präparativer Chromatographieverfahren. Dr.-Ing. Dissertation, Department of Biochemical and Chemical Engineering, Universität Dortmund, and Shaker-Verlag, Aachen

Hashimoto K, Adachi S, Noujima H, Ueda Y (1983) A New Process Combining Adsorption and Enzyme Reaction for Producing Higher-Fructose Syrup. Biotechnology and Bioengineering 25:2371-2393

Hashimoto K, Adachi S, Shirai Y (1993) Development of new bioreactors of simulated moving-bed type. *in* G. Ganetsos & P. Barker (eds.), Preparative and Production Scale Chromatography, Marcel Dekker, New York, 395-417

Hindemarsh A (1983) ODEPACK, a systematized collection of ODE solvers. in R. Stepleman (ed.), Scientific Computing, North-Holland, Amsterdam, 55-64

Jupke A (2003) Experimentelle Modellvalidierung und modellbasierte Auslegung von Simulated Moving Bed (SMB) Chromatographieverfahren. Dr.-Ing. Dissertation, Department of Biochemical and Chemical Engineering, Universität Dortmund, and VDI Verlag, Düsseldorf

Kearney M, Hieb K (1992) Time variable simulated moving bed process. US Patent 5.102.553

Klatt K-U, Hanisch F, Dünnebier G, Engell S (2000) Model-based optimization and control of chromatographic processes. Journal of Process Control, 24, 1119-1126

Küpper A, Engell S (2006) Parameter and state estimation in chromatographic SMB processes with individual columns and nonlinear adsorption isotherms. Accepted to ADCHEM 2006

Küpper A, Engell S (2005) Non-linear model predictive control of the Hashimoto simulated moving bed process. Contributed paper, International Workshop on Assessment and Future Directions of Nonlinear Model Predictive Control. Freudenstadt, August 2005

Ludemann-Hombourger O, Nicoud R-M (2000) The VARICOL process: A new multicolumn continuous separation process. Separation Science and Technology 35, 1829-1862

Ludemann-Hombourger O, Nicoud R-M, Bailly M (2003) New techniques to optimize the eluent consumption of the Varicol process, 16th International Symposium on Preparative Chromatography. San Francisco, USA, p. 45

Migliorini C, Mazzotti M, Morbidelli M (1998) Continuous chromatographic separation through simulated moving beds under linear and nonlinear conditions. Journal of Chromatography A, 827, 161-173

Schmidt C (2005) Bewertung von Fahrweisen- und Schaltvarianten des Simulated Moving Bed Prozesses zur Glukoseisomerisierung. Diplomarbeit, Lehrstuhl für Anlagensteuerungstechnik, Universität Dortmund

Storti G, Mazzotti M, Morbidelli M, Carra S (1993) Robust Design of Binary Countercurrent Adsorption Separation Processes. AIChE Journal 39:471-492

Toumi A, Engell S, Ludemann-Hombourger O, Nicoud R M, Bailly M (2003) Optimization of simulated moving bed and Varicol processes. Journal of Chromatography, 1006(1-2), 15-31

Toumi A (2004) Optimaler Betrieb und Regelung von Simulated-Moving-Bed-Prozessen. Dr.-Ing. Dissertation, Department of Biochemical and Chemical Engineering, Universität Dortmund, and Shaker Verlag, Aachen

Toumi A, Diehl M, Engell S, Bock H G, Schlöder J (2005) Finite horizon optimizing of control advanced SMB chromatographic processes. IFAC World Congress, Fr-M06-TO/2

Toumi A, Engell S (2004a) Optimal operation and control of a reactive simulated moving bed process. Prepints IFAC Symposium on Advanced Control of Chemical Processes (ADCHEM 2003), Hongkong, 2004, 243-248

Toumi A, Engell S (2004b) Optimization-based control of a reactive simulated moving bed process for glucose isomerization. Chemical Engineering Science, 59, 3777-3792

Schramm H, Grüner S, Kienle A (2003) Optimal operation of simulated moving bed chromatographic processes by means of simple feedback control. Journal of Chromatography, 1006(1-2) 3-13

Wang C, Klatt K-U, Dünnebier G, Engell S, Hanisch F (2003) Neural network-based identification of SMB chromatographic processes. Control Engineering Practice 11, 949-959

Zhou J, Tits A, Lawrence C (1997) User's Guide for *FFSQP* Version 3.7: A Fortran Code for solving constrained nonlinear (minmax) optimization problems, generating iterates satisfying all inequality and linear constraints. University of Maryland

8 Controlling reactive distillation

Marten Völker, Christian Sonntag, Sebastian Engell

8.1 Introduction

In recent years, integrated reactive separation processes (see e.g. (Agreda et al. 1990)) have attracted considerable attention in both academic research and industrial applications (see also chapter 3). In this contribution, the operation of a pilot-plant-scale reactive semi-batch distillation column at Dortmund University is studied. In previous work (Noeres et al. 2003), rigorous modelling of the process was pursued. In (Fernholz et al. 2000), a nominally optimal operating policy was determined by means of dynamic optimisation on the basis of an equilibrium stage model. Based on the rigorous model, potential approaches to control the plant at its nominally optimal operating point were investigated. In this context, in (Dadhe et al. 2002, Gesthuisen et al. 2002), control structures were identified by the procedure described in (Engell et al. 2004) that provide good performance with respect to the rejection of disturbances. It was shown by simulation of an equilibrium stage model as well as of a rate-based model that linear control is possible in the vicinity of the optimal operating point. However, when the controllers based on linearisations of the rigorous models were tested at the real plant, they either had to be adjusted manually, or their control performance degraded strongly from what was expected from simulation (Fernholz 2000). This is due to a considerable mismatch between model and reality. Hence, the question arose how to increase the control-relevant model accuracy, and, equally important, how to quantify the model accuracy in order to design controllers that are robust against inevitable mismatches.

In this contribution, rather than further refining the rigorous models and determining their parameters in expensive experiments, the accuracy problem is tackled more straightforwardly. Combining the previous results by (Fernholz et al. 2000) on steady-state process operability, and by (Dadhe et al. 2002) on dynamic controllability, a suitable control structure is determined. In this step, only the basic physico-chemical information provided by a relatively simple equilibrium stage model is employed. The nonlinear model is used to compute the incremental process gains for the assessment of steady-state operability, while linearisations of the rigorous model are used to assess the dynamic process controllability.

After a suitable control structure has been identified, linear system identification is used to refine the quantitative knowledge about the process dynamics in the vicinity of the nominally optimal operating conditions. The identification data is also used to compute bounds of the uncertainties which account for the time-varying and nonlinear process behaviour. This uncertainty description is then used to set up a control performance optimisation problem in which robustness to the model uncertainty from the previous step as well as process constraints such as actuator saturation are imposed. The specifications are met by using a two-step procedure: In the first step, the method described in (Völker and Engell 2004, 2005) is employed to design a high-order multi-objective controller which satisfies the specifications, while in a second step, a low-order controller is synthesised using an order reduction scheme that ensures that the specifications are also met for the reduced controller (Völker et al. 2005). The reduction step is based on frequency response approximation (Engell 1988b; Engell and Müller 1991; Engell and Pegel 2000). It is shown that the reduced-order controller meets the original multi-objective specifications and, due to its low order and its transparency for the plant operators, the control scheme is applicable also in industrial plants. Finally, the efficiency of the design procedure is substantiated by experiments performed at the pilot plant.

The remainder of this chapter is organised as follows: First, the reactive distillation process is introduced and the problem of finding a suitable control structure is discussed in depth. The next section is devoted to the model refinement by means of linear system identification. In the sequel, the model error bound computation is explained and the controller synthesis step is detailed. A major part is concerned with a discussion of the experimental validation at the plant. Finally, the overall methodology for the development of the process control scheme for the reactive distillation column is summarised and some conclusions are drawn.

8.2 The reactive distillation column

8.2.1 Chemical preliminaries

In this contribution, the heterogeneously catalysed esterification of acetic acid (HAc) and methanol (MeOH) to methyl acetate (MeAc) and water (H_2O) in a medium-scale semi-batch pilot plant is studied. This plant is operated by the Chair of Fluid Separation Processes at the Department of Biochemical and Chemical Engineering at Dortmund University in cooperation with our group.

Table 8.1: Boiling points of the components in the distillation column.

compound	boiling point [°C]
azeotrope MeAc/MeOH	53.8
azeotrope MeAc/H_2O	56.7
MeAc (CH_3COOCH_3)	56.9
MeOH (CH_3OH)	64.6
H_2O	100
HAc (CH_3COOH)	118

Values taken from "http://www.mpi-magdeburg.mpg.de"

MeAc is produced in the equilibrium reaction:

$$MeOH + HAc \underset{k}{\longleftrightarrow} MeAc + H_2O. \tag{8.1}$$

Thermodynamically, the system exhibits two dominating azeotropes. The first azeotrope occurs between MeAc and H_2O, and the second between MeAc and MeOH (see Table 8.1).

8.2.2 The reactive distillation column

The pilot column is 9 meters high and has an inner diameter of 100 millimeters. A photograph of the column is shown in Figure 8.1.

Fig. 8.1: The pilot plant.

A process scheme of the plant is depicted in Figure 8.2. In this chapter, a short description of the pilot plant is given. More details can be found in Chapter 3 of this book. The pilot plant consists of three parts, the reboiler, the condenser and reflux, and the column itself. Within the column, three sections of structured packings are installed, two catalytic ones at the bottom and a separating section at the top of the column. Each packing has a length of 1 meter.

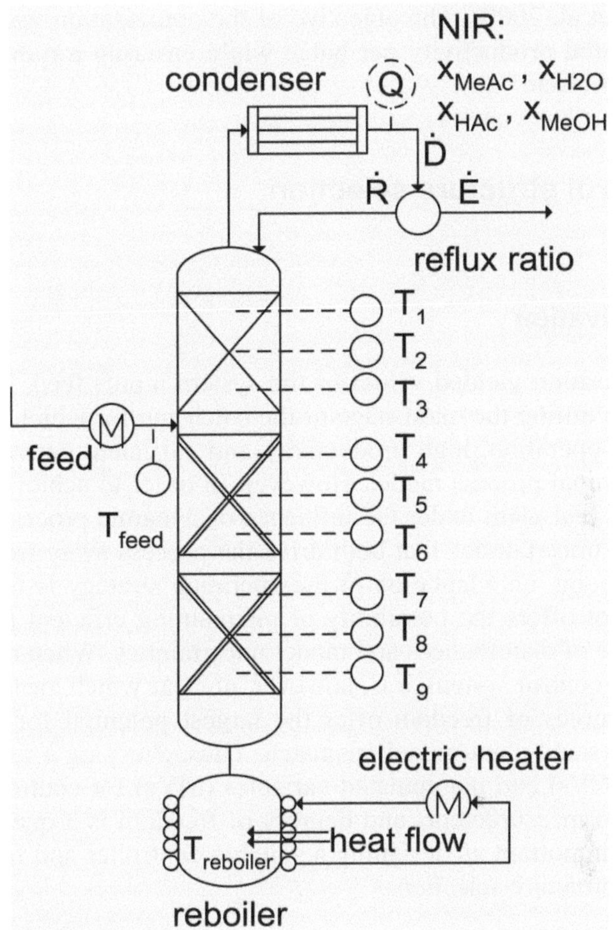

Fig. 8.2: Scheme of the plant.

The plant is operated in semi-batch mode. First, the reboiler is filled with methanol which is then evaporated. After gaseous methanol has ascended within the column body, the acetic acid feed is switched on, and MeOH is consumed by the reaction. The batch run is terminated when the concentration of methanol in the reboiler is too low to achieve the desired product concentration. The essential degrees of freedom are the reflux ratio (here defined as the ratio $\dot{R}/\dot{D}$ in the interval $[0,1]$, see Figure 8.2), the acetic acid feed (feed), and the heat flow supplied to the reboiler via an electrically heated water pipe system. These degrees of freedom were used to obtain an efficient nominal operating point by means of nonlinear optimisation of the productivity using a rigorous nonlinear equilibrium model

(Fernholz et al. 2000). The objective of the optimisation was to maximise the cumulated productivity per batch while ensuring a minimum conversion of acetic acid.

8.3 Control structure selection

8.3.1 Motivation

The optimisation yielded values of the system inputs feed, heat flow, and reflux ratio during the main stage of the batch run for which the objectives of process operation, high productivity and sufficient conversion are met for the nominal process model. However, in order to achieve the same results at the real plant under the influence of dynamic process disturbances and model uncertainties that both drive the process away from its optimal operating point, an adaptation of the operating strategy is required. Feedback control offers the possibility of maintaining efficient process operation in spite of disturbances and model uncertainties. When designing such a feedback control system, it is, however, unclear which measurements and process degrees of freedom offer the largest potential for counteracting disturbances and plant-model mismatch. Thus, choosing a set of controlled variables (CVs) and manipulated variables (MVs) for control from the set of available measurements and degrees of freedom is a question which is at least as important as designing a suitable controller and is here denoted as control structure selection.

8.3.2 Degrees of freedom and measurement equipment

As mentioned earlier, the process offers three major degrees of freedom: the heat flow, the acetic acid feed flow rate, and the reflux ratio. In the vicinity of the nominally optimal operating point, the ratio of heat flow to feed determines the reaction conditions and can therefore be lumped as one manipulated variable. Due to the long pipes in the indirect heating system which are responsible for a considerable time delay, the heat flow is kept constant by a simple auxiliary control loop, and the acetic acid feed represents the first manipulated variable (MV) for quality control. To adjust the separation efficiency, the reflux ratio is chosen as the second MV. The measurement equipment offers as potential controlled variables three temperature measurements per packing in the vapour phase (T_1-T_9 in Figure 8.2) and one temperature measurement at the liquid distributors in each packing. Additionally, the liquid phase compositions in the reflux at the top of the column (x_{MeAc}, x_{H2O}, x_{MeOH} and x_{HAc}) which are measured online by near-infrared (NIR) spectroscopy can be used. From this set of measurements, two controlled variables have to be chosen to match the number of degrees of freedom for quality control. One obvious choice is x_{MeAc}, since it determines the product quality. The second choice is less trivial, because no exact relations of the measured variables to process efficiency or operability are obvious. Hence, a quantitative investigation of the operability was carried out in two steps consisting of the analysis of the steady-state operability and the assessment of the dynamic controllability using the method described in (Engell et al. 2004) which is based on linearised models.

8.3.3 Steady-state process operability

Due to the semi-batch mode of operation, the process does not reach a steady state. However, extensive simulations have shown that the qualitative behaviour of the plant changes only moderately during the complete batch, with the exception of start-up and shut-down. To illustrate this, in Figure 8.3 the profiles of the vapour phase compositions in the column body are shown which were recorded by simulation of the nonlinear process model used in (Fernholz et al. 2000). The model was started up, and at time t:=0, the nominal operating conditions (reflux ratio = 0.6, feed = 0.04 mole/s, heat flow = 3.667 kW) were used as the simulation inputs. The diagram shows the profiles at different times of the batch. It can be seen that, qualitatively, the system behaviour is very similar at all times enabling an approximate quasi-steady-state analysis at a fixed time during the batch. Throughout the quasi-steady-state analysis, the model representing the plant at 30000 s was used.

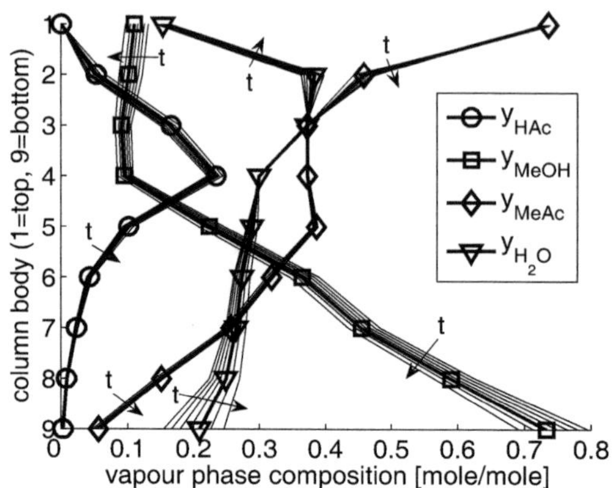

Fig. 8.3: Profile of the vapour phase composition over the column as computed by the rigorous model: 3 linearly interpolated measurement locations per packing, t=[10,15,20,25,30,35,40]*1000 s (indicated by arrows), the bold line refers to 30000 s.

A further analysis was performed by simulation at values of the different process degrees of freedom. The resulting quasi-steady-state characteristic diagrams in Figure 8.4 and Figure 8.5 depict static mappings $\mathbf{f(u)} \rightarrow \mathbf{y}$ at t = 30000 s, where, depending on the illustration purpose, the inputs $\mathbf{u}$ and the outputs $\mathbf{y}$ are either scalars or vectors. These diagrams give insight into the process gains, given by the respective partial derivatives $\partial \mathbf{y} / \partial \mathbf{u}$.

Due to the thermodynamic properties and the reaction kinetics, the quasi-steady-state diagrams show that the process gains in the individual channels exhibit multiple changes of sign with respect to the manipulated variables reflux ratio, acetic acid feed, and heat flow to the reboiler. Figure 8.4 and Figure 8.5 exemplify these gain changes for fixed feed and reflux ratio, respectively, and fixed heat flow at the nominally optimal operating point after a simulation time of 30000 s. In addition to the MeAc and water compositions (black line), the upper temperature in the separation section of the column T_1 is shown (grey line) in order to facilitate an approximate allocation of the temperatures to the azeotropes in Table 8.1. The circles indicate the nominal operating conditions. In Figure 8.4, the feed is fixed at 0.04 moles/s, and the reflux ratio is gradually increased from a reflux ratio of 0.4 to a reflux ratio of 1 where no product is withdrawn from the column. Evidently, for lower values of the reflux ratio, with increasing reflux ratio and therefore increasing separation efficiency, the composition of x_{MeAc} rises and T_1 decreases. At first, the azeotrope MeAc/H_2O can be withdrawn at the top of the column. For larger values of the reflux ratio, the azeotrope MeAc/MeOH is approached which leads to a change of the sign of the process gain in x_{MeAc} and a degradation of product purity. Fortunately, x_{H2O} does not exhibit this sign change, and by sufficiently tight control of x_{H2O}, deviations to the azeotrope MeAc/MeOH with lower purities and reduced productivity can be avoided. In Figure 8.5, the quasi-steady-state diagram for variations in the feed is shown, where the feed is gradually increased from 0.01 moles/s to 0.09 moles/s. To maintain the conversion constraint for acetic acid, the process must be operated on a slight acetic acid shortage. Since at this point small changes in the feed have a considerable effect on the product composition, accurate control of the feed is needed.

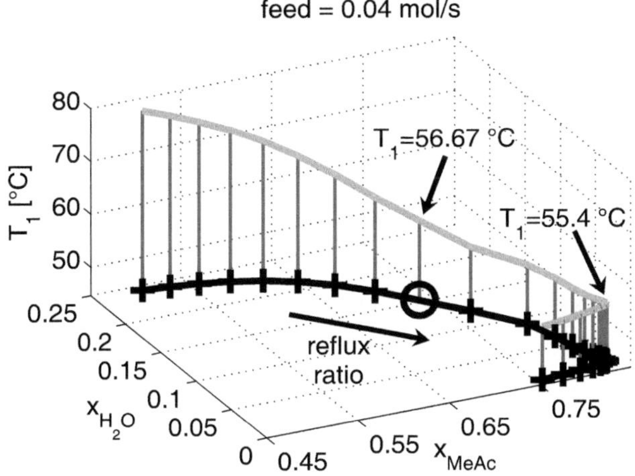

Fig. 8.4: Quasi-steady-state characteristic diagram in the MeAc-H$_2$O-plane after 30000 s for constant inputs (heat flow = 3.667 kW, feed = 0.04 moles/s) Additionally, the corresponding values of the upper temperature in the separation section are shown.

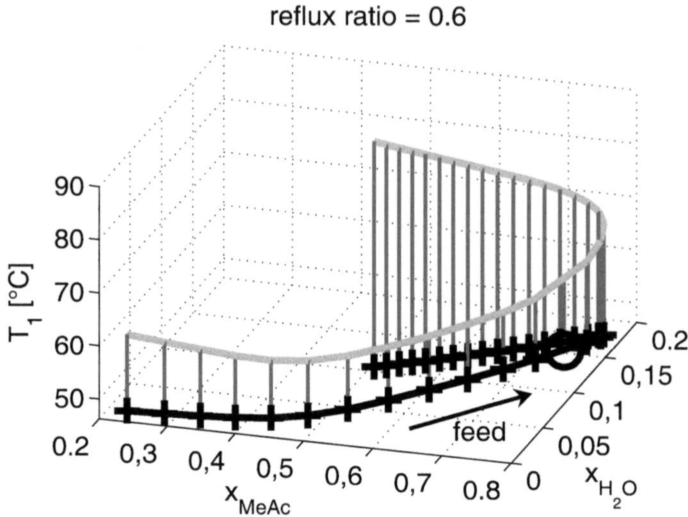

Fig. 8.5: Quasi-steady-state characteristic diagram in the MeAc-H_2O-plane after 30000 s for constant inputs (heat flow = 3.667 kW, reflux ratio = 0.6). Additionally, the corresponding values of the upper temperature in the separation section are shown.

The quasi-steady-state process gains show that the process at hand exhibits several steady-state nonlinearities. They also show that gain sign changes do not occur in close vicinity of the optimal operating point, and hence, the necessary condition for the application of linear control is fulfilled. However, the diagrams do not indicate the dynamic implications for linear control which can be limiting as well, cf. (Engell 1988a). To this end, in a complementary fashion, the dynamic controllability was investigated on the basis of linearised dynamic models.

8.3.4 Dynamic process operability

The dynamic process controllability is investigated, employing linearised models which are represented as transfer matrices $\mathbf{G}(s)$ in the Laplace domain, where bold font indicates the multivariate case. Linear systems admit a rigorous generic analysis of the limitations imposed by a control structure (Engell 1988a). As mentioned in section 8.3.2, one further controlled variable (CV) had to be chosen in addition to the x_{MeAc} composition measurement. In total, 13 candidate measurements were analysed, the 9 temperature measurements in the vapour phase of the packings, the temperature in the reboiler, and the three composition measurements of the

NIR except x_{MeAc}, yielding 13 potential control structures as shown in Table 8.2.

Table 8.2: Potential control structures.

No	CVs	MVs
1	x_{MeAc} and T_1	reflux ratio and feed
2	x_{MeAc} and T_2	
3	x_{MeAc} and T_3	
4	x_{MeAc} and T_4	
5	x_{MeAc} and T_5	
6	x_{MeAc} and T_6	
7	x_{MeAc} and T_7	
8	x_{MeAc} and T_8	
9	x_{MeAc} and T_9	
10	x_{MeAc} and $T_{reboiler}$	
11	x_{MeAc} and x_{H2O}	
12	x_{MeAc} and x_{HAc}	
13	x_{MeAc} and x_{MeOH}	

The structures were analysed, using qualitative performance indicators and a quantitative control performance calculation on the basis of linearisations of the rigorous nonlinear process model at the nominal operating point (reflux ratio = 0.6, feed = 0.04 moles/s, heat flow = 3.667 kW).

The procedure (Engell et al. 2004) tackles the problem of control structure selection in two steps. In the first step, candidate structures are discarded by computationally inexpensive qualitative indicators such as transfer matrix zeros in the right Laplace half-plane which limit the achievable bandwidth or the condition number which determines the sensitivity of a control structure to the direction of e. g. a disturbance. In the second step, the computation of the optimal control performance enables a quantitative comparison of the structures which pass the first step. Applied to the example at hand, the control structure selection procedure indicated that, besides controlling x_{MeAc}, the control of either the middle temperature in the separation packing (T_2) or x_{H2O} in the reflux enables high performance linear control. As was shown in several rigorous simulation studies in which the robustness against typical uncertainty scenarios was checked (Völker 2002; Sonntag 2004), both control structures lead to efficient linear control of the rigorous model. Unfortunately, due to plant-model mismatch and non-ideal operating conditions, these results could not be transferred to the control of the real process with immediate success. Particularly, controlling x_{MeAc} and the middle temperature in the separation packing (T_2) led to disappointing results, since the temperature showed much stronger deviations from the model than expected. To illustrate this, in Figure 8.6 the experimental data recorded in an experiment on September 9th, 2004, which was later used to obtain a more accurate linear model, is shown in comparison to the simulation of the rigorous model with the same set of inputs. The agreement with respect to the first controlled variable x_{MeAc} is reasonably good.

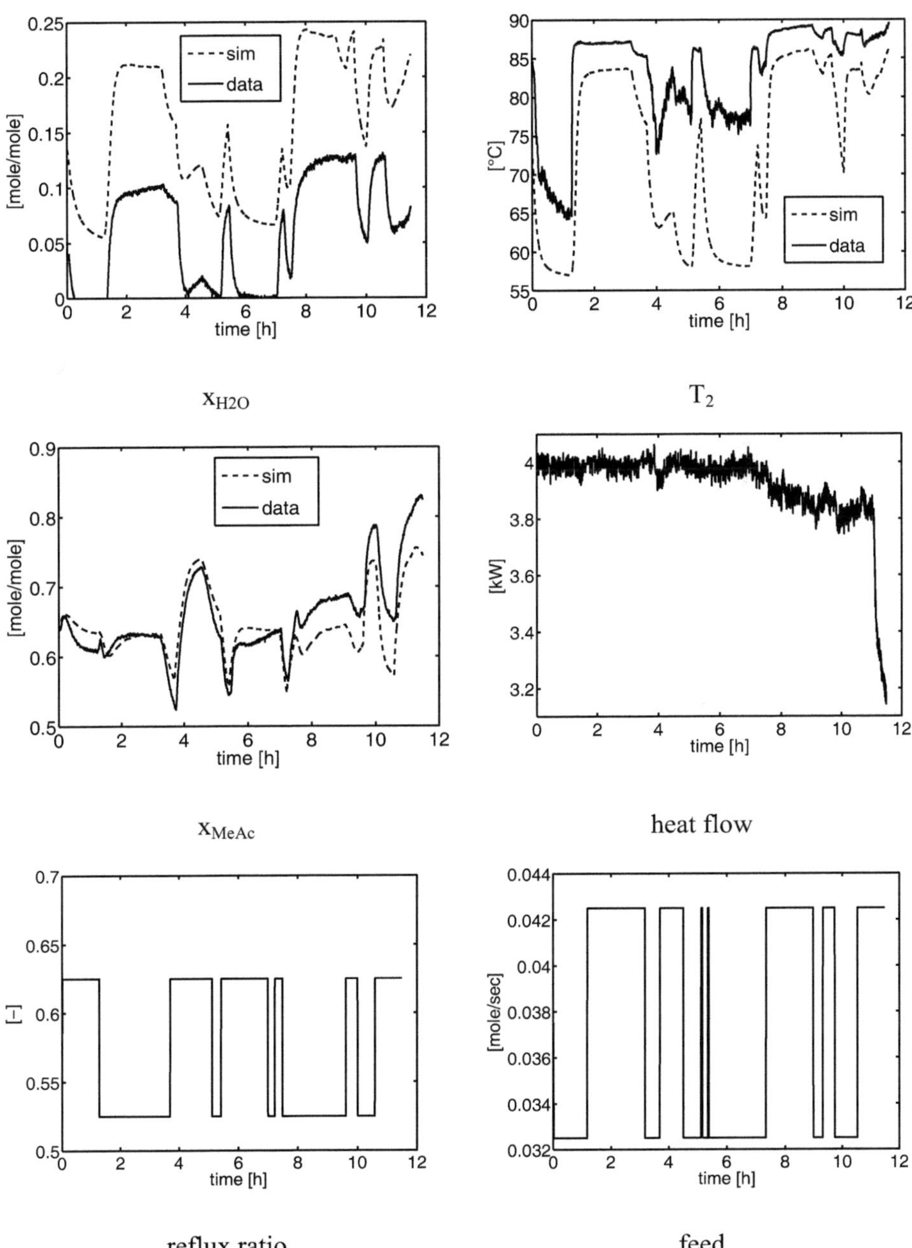

Fig. 8.6: Comparison between the simulation of the rigorous model (sim) and the measured experimental curves (data).

In contrast, the other potential CVs, x_{H2O} and T_2, exhibit much stronger deviations. While x_{H2O} shows a large and approximately constant offset between data and simulation, possibly caused by errors in the calibration of the NIR measurement, but reasonably accurate dynamic process gains, the simulated T_2 trajectory indicates too high dynamic process gains.

Hence, it was decided to use the control structure with the CVs x_{MeAc} and x_{H2O}. In order to obtain a quantitatively more accurate prediction of the dynamic behaviour of x_{H2O}, linear system identification was used rather than to further increase the accuracy of the rigorous model by expensive independent experiments. Using the identification data, a computation of dynamic and static model uncertainties can be performed in order to systematically derive model error bounds for robust control.

8.4 Model refinement by linear system identification

Linear system identification consists of three major steps. The first step deals with the determination of a suitable signal to excite the process such that the identifiability of all model parameters is guaranteed. In this step, the process nonlinearities as well as the control goal have to be considered. In the second step, the experiment is conducted, and the recorded data is used to obtain a linear model. In this context, a regression scheme is set up, and issues like the optimal model order and numerically efficient regression have to be addressed. Finally, a reduction step is performed in which the preservation of the control-relevant model accuracy is ascertained.

8.4.1 Choice of the identification signal

The first step in choosing the identification signal is to determine its range by means of a quasi-steady-state analysis. Figure 8.7 shows the quasi-steady-state behaviour of the controlled variables as a function of the reflux ratio and the feed as obtained by the simulation of the nonlinear model as in section 8.3.3.

To obtain accurate gains of the identified model, the excitation of nonlinearities, especially of gain sign changes must be avoided during the identification taking actuator uncertainties into account, since the actuators of the real plant cannot be assumed to be perfectly accurate. For example, the calculation of the heat flow transferred to the reboiler which is controlled by a simple auxiliary control loop involves the measurement of two temperatures and a flow rate. The overall accuracy depends on the accuracies of the flow and temperature measurements. Likewise, the feed and the reflux ratio cannot be actuated precisely. This issue was addressed by choosing a slightly suboptimal operating point (reflux ratio = 0.575, feed = 0.0375 mole/s, heat flow = 4 kW), as indicated by the crosses in Figure 8.7, where the grey-shaded areas mark the ranges of the identification signals around the suboptimal operating point (Δ reflux ratio = ± 0.05, Δ feed = ± 0.005). As can be seen in the figures, the grey identification areas are located within a region of nearly constant process gains, even for deviations in the reflux ratio and the feed, and include the nominally optimal operating point.

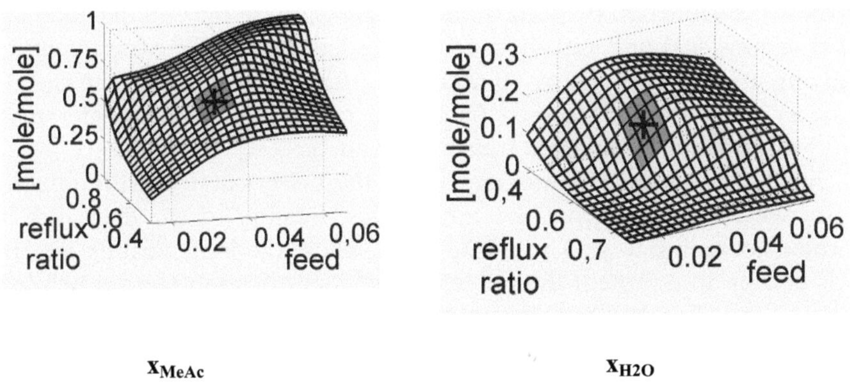

Fig. 8.7: Quasi-steady-state characteristic diagrams of the CVs (heat flow = 3.667 kW).

In addition to the operating point and the deflection of the signal, the dynamic signal properties have to be determined. It is a well-known fact from linear system identification (Hjalmarrson 2003; Gevers 2003) that the power spectrum of the identification signal is strongly correlated with the accuracy of the identified model so that at frequencies of high signal energy, the identified model is in general more accurate and vice versa. This fact is here exploited by choosing a signal with most of its energy concentrated in the estimated gain crossover frequency range of the open control loop - the frequency range around which the model accuracy has the strongest impact on control performance (Engell 1988a). Using the control performance calculation and rigorous simulation results from (Sonntag 2004), the maximal achievable crossover frequency is approximately 10^{-3} rad/s. The considerations led to the choice of a generalised binary noise signal (see Tulleken 1990) the frequency spectrum of which can be designed such that most of its energy is concentrated in the desired frequency range. On September 9th, 2004, an identification experiment was conducted at the real plant employing the designed signals. The recorded data is shown in Figure 8.6.

8.4.2 Linear model identification: Data pretreatment and regression

The goal of the identification was to compute a multivariable linear transfer function model $\mathbf{G}(s)$ mapping the vector valued signal $\mathbf{u}(t)$, where $u_1(t)$ is the reflux ratio, and $u_2(t)$ is the feed, to the vector of controlled variables $\mathbf{y}(t)$, where $y_1(t)$ is x_{MeAc}, and $y_2(t)$ is x_{H2O}, First, a linear discrete-time model was estimated by performing a regression on the equidistantly sampled signals $\mathbf{u}(t)$ and $\mathbf{y}(t)$ that were recorded during the experiment, employing a sampling time of 10 seconds. After the regression, the discrete-time system was transformed back to a continuous-time transfer matrix $\mathbf{G}(s)$ using a zero-order-hold transformation.

Since a linear system was identified, the offsets defining the operating point were removed from the time series. Scaling of the data was performed according to $\mathbf{y}_s = \mathbf{L}\mathbf{y}$ and $\mathbf{u}_s = \mathbf{R}^{-1}\mathbf{u}$, where the input scaling matrix $\mathbf{R}$ was chosen to represent the ranges of the identification signals, while the output scaling matrix $\mathbf{L}$ represents the control goals. The rationale behind this procedure is that the scaling influences the ratios of the accuracies of the transfer functions that represent the channels of the multivariate system. Therefore, the scaling was chosen such that the accuracy of the multivariate model corresponds to the control objectives for the two output channels. The scaling matrices are given by:

$$\mathbf{L} = \begin{pmatrix} \dfrac{1}{0.06} & 0 \\ 0 & \dfrac{1}{0.03} \end{pmatrix}, \mathbf{R} = \begin{pmatrix} 0.05 & 0 \\ 0 & 0.005 \end{pmatrix}, \tag{8.2}$$

where the second diagonal entry of $\mathbf{L}$ represents the goal of more accurate control of x_{H2O} than of x_{MeAc}. Tight control of x_{H2O} aims at avoiding the transition to the azeotrope MeAc/MeOH and the corresponding sign changes in the process gains. In order to reduce the effect of slow drifts on the estimated model due to the batch characteristics, the inputs and outputs of the data set were high-pass-filtered using a fourth-order lead element with the continuous- time transfer function:

$$G_F(s) = \left(\frac{s + 3.162 \cdot 10^{-5}}{s + 3.162 \cdot 10^{-4}} \right)^4 .$$

The filter provides a stopband attenuation of 80 dB so that the negative effect of drifts with dynamics slower than 10^{-4} rad/s on the estimation result is decreased. Furthermore, these dynamics are not important for the control scheme, since their effect only becomes visible beyond the time range of an entire batch run, and it is therefore natural to remove them. Model estimation was performed employing an estimation algorithm based on orthonormal basis functions (van den Hof et al. 1995). The method leads to a simple convex quadratic optimisation problem that can be solved numerically efficiently. The order of the estimated model was computed using the final output error criterion from (Zhu 2001, p. 136):

$$\frac{N+d}{N-d} \frac{1}{N} \sum_{t=T_s}^{N \cdot T_s} \left(\mathbf{y}(t) - \hat{\mathbf{y}}(t) \right)^T \left(\mathbf{y}(t) - \hat{\mathbf{y}}(t) \right),$$

where N is the number of sampling points, T_s is the sampling time of 10 seconds, $\mathbf{y}(t)$ is the measured data, $\hat{\mathbf{y}}(t)$ is the model prediction, and d is the model order. The criterion aims at avoiding overfitting of the data. Applying the criterion led to a linear discrete-time model of order 44.

8.4.3 Model order reduction

To circumvent numerical drawbacks due to the high order, model reduction by means of frequency response approximation was employed. In this scheme, a weighted approximation aims at minimising the difference of the frequency responses of the high- and low-order plants:

$$\Delta(\omega_k) := \mathbf{G}_w(j\omega_k) \left[\mathbf{G}_{ho}(j\omega_k) - \mathbf{G}_{lo}(j\omega_k) \right],$$

where ω_k is a frequency point, $\mathbf{G}_{ho}$, $\mathbf{G}_{lo}$, $\mathbf{G}_w$ are the frequency responses of the high-order model, of the low-order model and of a weighting function that is chosen to emphasise the frequency range in the proximity of the desired gain crossover frequency. The difference is expressed in terms of the absolute squared deviation summed up over all channels and over all considered frequencies ω_k

$$\sum_i \sum_j \sum_k \left| \Delta_{ij}(\omega_k) \right|^2, \tag{8.3}$$

where the indices ij refer to the channel from the j-th input to the i-th output of Δ. Similar to the choice of the identification signal, this reduction technique takes the control-relevant accuracy into account. $\mathbf{G_w}$ was chosen as

$$\mathbf{G_w}(s) = \begin{pmatrix} 1 & 0 \\ 0 & 1 \end{pmatrix} \frac{s + 1.581 \cdot 10^{-4}}{(s + 0.005)(s + 1.581 \cdot 10^{-3})},$$

having its maximum magnitude around 10^{-3} rad/s, as shown in the frequency response plot in Figure 8.8. For the approximation, 50 discrete frequencies ω_k in a logarithmically spaced interval between 10^{-4} to 10^{-1} rad/s were chosen.

Similar to the choice of the weight $\mathbf{G_w}$, the approximation on a finite frequency interval was utilised to cut off unimportant and/or undesirable frequencies.

The choice of the lower interval bound aimed at removing slow elements from the high-order model, the effect of which only becomes visible beyond the end of a batch run. The choice of the upper bound was motivated by the targeted gain crossover frequency of 10^{-3} rad/s and by the observation that the system attenuates larger frequencies such that above this upper frequency bound the identification signals are dominated by noise.

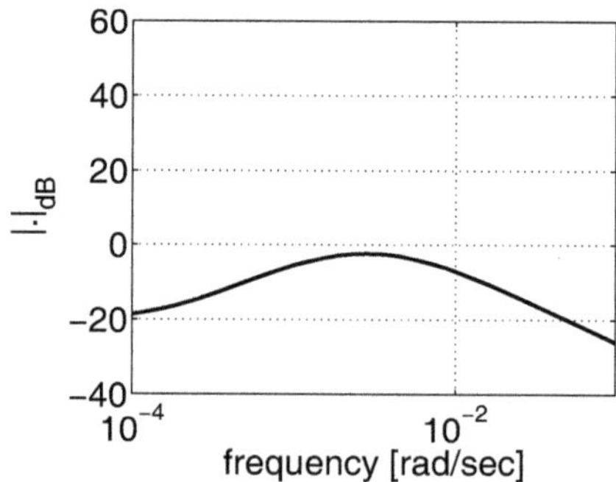

Fig. 8.8: Frequency response of $\mathbf{G_w}$ (11-element).

The frequency response approximation was performed in discrete time. After the reduction, the estimated pure time delay was considered by multiplying the reduced discrete-time transfer function with the discrete-time version of a first-order Padé approximation given by

$$\text{zoh}\left\{\mathbf{G}_{\text{Pade}}(s)\right\} = \text{zoh}\left\{\dfrac{\dfrac{2}{T_D} - s}{\dfrac{2}{T_D} + s}\right\},$$

where zoh denotes zero-order-hold transformation, and T_D is the time delay of 170 s so that the discrete-time model including the delay can be directly transformed back to continuous time (recall that ideal delays can be represented by real rational functions in discrete time but not in continuous time). This procedure might seem unintuitive but aims at maintaining the consistency between the discrete-time model which is later on used for the computation of error bounds and the continuous-time model which is used during the controller synthesis step.

The continuous-time unscaled version of the corresponding transfer matrix of the reduced system including the delay (time unit is given in seconds), mapping changes in the reflux ratio and the acetic acid feed to changes in x_{MeAc} and x_{H2O} around the operating point (reflux ratio = 0.575, feed = 0.0375 mole/s, heat flow = 4 kW), is given by:

$$G(s) =$$

$$\begin{bmatrix} 0.0026\dfrac{(s-0.0105)(s-6.73\cdot10^{-6})}{(s+0.0118)(s+0.0025)(s+4.46\cdot10^{-5})} & 0.0070\dfrac{(s-0.0105)(s+0.0012)}{(s+0.0118)(s^2+0.0010s+8.28\cdot10^{-7})} \\ 0.0011\dfrac{(s+0.0483)(s-0.0105)}{(s+0.0188)(s+0.0118)(s+0.0030)} & -0.0011\dfrac{(s-0.0105\ )(s+0.0012\ )}{(s+0.0118\ )(s^2+0.0011s+5.39\cdot10^{-7})} \end{bmatrix}. \qquad (8.4)$$

That the weighted model reduction indeed preserves the accuracy of the high-order model in the desired frequency range can be seen in Figure 8.9, which depicts the frequency responses of both systems, including the delay.

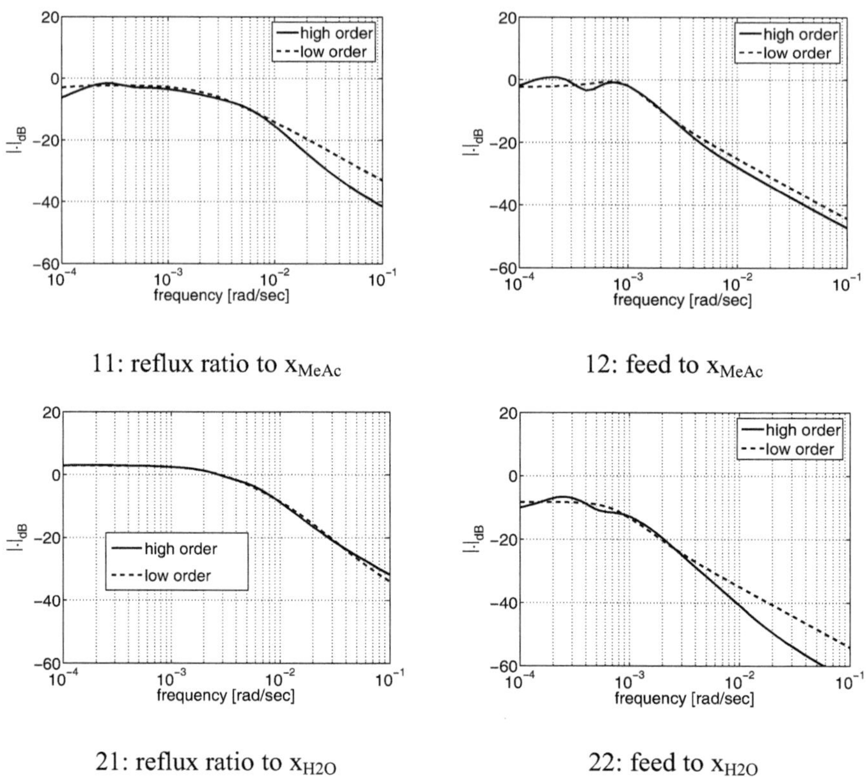

Fig. 8.9: Comparison of the frequency response of the high- and of the low-order system (scaled according to eq. (8.2)).

Figure 8.10 shows a comparison of the identification data and the outputs of the reduced-order unscaled model. It can be seen that the reduced-order linear system describes x_{MeAc} with similar quality as the rigorous model, and that x_{H2O} is modelled more accurately. Figure 8.10 also shows that the time-varying and nonlinear process is not represented completely by the linear model. This remaining plant-model mismatch must be accounted for when designing the controller.

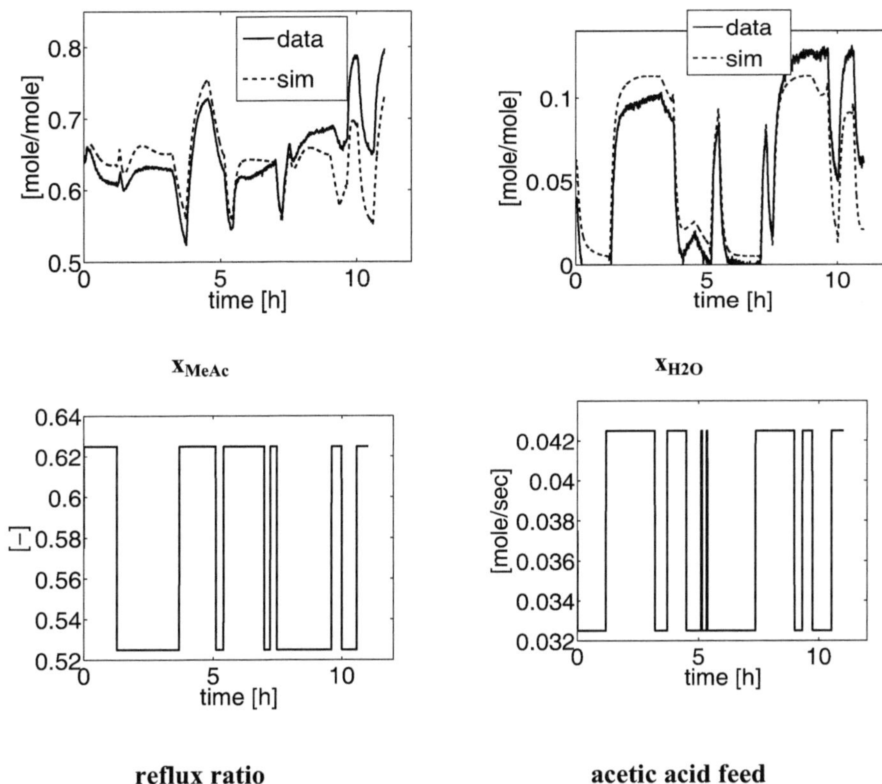

Fig. 8.10: Identification data (data) and the corresponding simulated outputs (sim) of the reduced-order linear model.

8.5 Model uncertainty assessment

When a model is used to design a controller, its quality has a large impact on the control performance at the real process. This section deals with the quantification of the uncertainty of the model derived in the previous section such that, during the later controller design step, plant model mismatch can be taken into account systematically. The first part is concerned with the description of a model set that is compatible with the controller design framework. The second part describes a way to compute the required model set from measured data.

8.5.1 Model error model

One way to express the control-relevant plant model mismatch is to introduce a model error model, as shown in Figure 8.11.

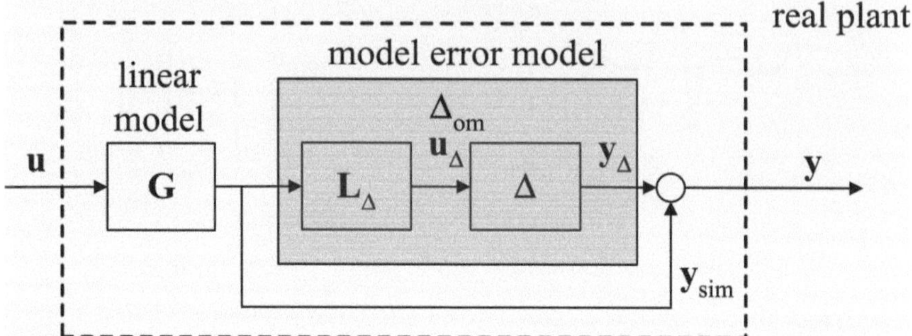

Fig. 8.11: Output-multiplicative uncertainty setup.

Therein, $\mathbf{u}(t)$ and $\mathbf{y}(t)$ are the multivariate input resp. output signals generated by the true process, $\mathbf{y}_{sim}(t)$ is the signal obtained by simulation of the linear model $\mathbf{G}$, and its uncertainty is expressed by the model error model Δ_{om}, where the index om refers to its output-multiplicative form. To discriminate uncertainties with respect to their frequency contents, the model error model is decomposed into a linear filter $\mathbf{L}_\Delta$ and a potentially nonlinear unknown perturbation Δ. The degree of model uncertainty is quantified in terms of bounds on the gain of Δ. Its gain is expressed by (see e. g. (Vidyasagar 1993)):

$$\|\Delta\|_{i2} := \sup_{u(t)\neq 0} \frac{\|\mathbf{y}_\Delta(t)\|_2}{\|\mathbf{u}_\Delta(t)\|_2} := \gamma_2(\Delta) \tag{8.5}$$

which is the worst-case ratio of the energy contained in the signal $\mathbf{y}_\Delta$ to the energy contained in the signal $\mathbf{u}_\Delta$. It is also called the induced 2-norm of a system. The 2-norm or rms (root mean square) value of a vector signal is given by

$$\left\| \mathbf{u}(t) \right\|_2 := \sqrt{\int_0^\infty \mathbf{u}(t)^\mathrm{T} \cdot \mathbf{u}(t)\,dt}. \tag{8.6}$$

The parameters γ_2, $\mathbf{G}$, and $\mathbf{L}_\Delta$ describe the model uncertainty as the set of all systems, described by the block diagram in Figure 8.11 for which the gain of the unknown perturbation Δ is less than or equal to γ_2. This description of a model set is then used in the computation of a controller that stabilises not only the nominal model $\mathbf{G}$ but all models contained in the set. This systematic guarantee on the control performance at the real plant is denoted as robust stability, see e. g. (Skogestad and Postlethwaite 1996) or (Zhou and Doyle 1998).

8.5.2 Data-driven computation of uncertainty bounds

Gain estimation from time series

In our case, the gain γ_2 in eq. (8.5) is unknown and must be estimated from the data at hand. To this end, first the times series $\mathbf{u}_\Delta$ and $\mathbf{y}_\Delta$ are obtained by filtering the measured data $\mathbf{u}$ and $\mathbf{y}$, scaled and with removed offset as described in section 0. It can be readily verified from Figure 8.11 that the input of the model error model Δ_{om} is $\mathbf{y}_{\mathrm{sim}}$. Clearly, $\mathbf{u}_\Delta$ can be obtained by filtering $\mathbf{y}_{\mathrm{sim}}$ according to $\mathbf{u}_\Delta = \mathbf{L}_\Delta \cdot \mathbf{y}_{\mathrm{sim}}$, and $\mathbf{y}_\Delta = \mathbf{y} - \mathbf{y}_{\mathrm{sim}}$. For a perfect match of model output and measured data it follows that $\mathbf{y}_\Delta = 0$. Of course, no model is perfectly accurate, therefore $\mathbf{y}_\Delta$ is not equal to zero and, by employing $\mathbf{u}_\Delta$ and $\mathbf{y}_\Delta$, an estimate of γ_2 can be obtained by means of the gain validation method described in (Poolla et al. 1994). In this approach, the consistency of the bound γ_2 on the unknown perturbation Δ and the observed data is verified. The method yields a minimum gain γ_2 that is consistent with the data so that a model error model of minimal conservatism can be constructed from $\mathbf{L}_\Delta$ and γ_2.

Computation of L_Δ

The linear weighting matrix $\mathbf{L}_\Delta$ plays a key role when using the proposed uncertainty description for robust control, because the distribution of the uncertainty in the frequency domain crucially influences the feasibility of the control design. So far, no systematic methods to choose the linear filter $\mathbf{L}_\Delta$ are available. In order to solve this problem, we propose to calculate the filter by means of the asymptotic theory employing the algorithm from (Zhu 1989). The asymptotic theory can be used to compute the stochastic error of the identified model under the assumption that the true process can be completely described by a linear system. Clearly, these assumptions can only be fulfilled approximately, and nonlinearities or time-varying process behaviour are not allowed in this setting. The gain estimation method of the previous section is used to compensate for the errors in these assumptions.

For the computation of the uncertainty of the linear model we proceed as follows. The asymptotic theory yields a frequency domain description of the magnitude of $\mathbf{L}_\Delta$ at a gridded set of frequency points. Phases corresponding to minimum phase behaviour were generated by the algorithm described in (Oppenheim and Schaffer 1975) so that a linear transfer matrix of $\mathbf{L}_\Delta$ of order 40 is obtained by frequency response approximation, see section 8.4.3. Using this transfer matrix, the gain estimation technique of the previous section is applied yielding a gain γ_2 that is not contradicted by the experimental data. Hence, the wrong assumptions of the asymptotic theory do not matter, since the final validation step accounts for the nonlinearities and the time-varying behaviour. As usual for the robust control framework, scaling of $\mathbf{L}_\Delta$ and Δ according to $\mathbf{L}_\Delta \rightarrow \mathbf{L}_\Delta \cdot \gamma_2$, $\Delta \rightarrow \Delta / \gamma_2$ is performed to end up with the unified representation in which the unknown perturbation is bounded by 1, $\|\Delta\|_{i2} < 1$. In the last step, the frequency response of $\mathbf{L}_\Delta$ is upper bounded such that it is monotonically increasing and smooth which has a positive effect on the numerical tractability of the controller design step. Note that the smoothed upper bound includes the consistent model set. The smoothed frequency response of the scaled output-multiplicative uncertainty weight $\mathbf{L}_\Delta$ is shown in Figure 8.12.

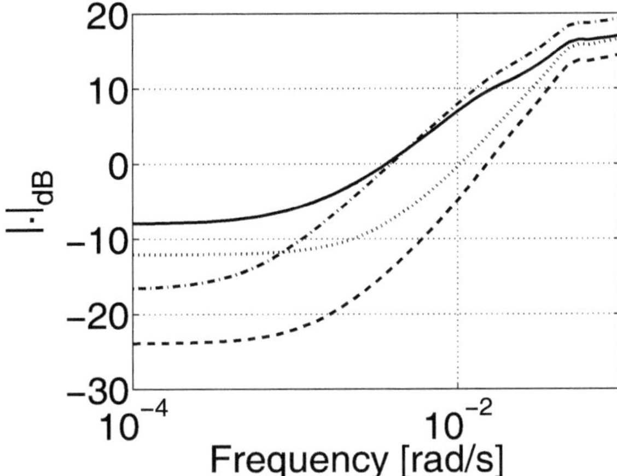

Fig. 8.12: Frequency response of the scaled ($\gamma_2 = 1$) output-multiplicative uncertainty weighting function $\mathbf{L}_\Delta$ (channel 11: solid, channel 12 dashed, channel 21: dotted, channel 22: dash-dotted).

8.6 Controller design

A controller that achieves good control performance has to meet a set of constraints such as avoidance of actuator saturation, robustness to model uncertainties, and efficient disturbance rejection. In practice, such constraints are usually met approximately by following simple rules of thumb, e. g. Ziegler Nichols tuning rules, (Ziegler and Nichols 1942), often followed by an online retuning step. However, such procedures do not guarantee optimality and require a certain degree of expertise, and it is difficult, if at all possible, to straightforwardly extend them to the multivariable case that needs to be considered here. In contrast, optimal controller synthesis techniques enable a systematic control design by optimising rigorous mathematical criteria. Optimal linear control with respect to one single mathematical criterion has become very mature and well-founded. However, one mathematical criterion often does not suffice to capture all relevant specifications. In order to facilitate the formulation of practically relevant specifications, multiple mathematical criteria have to be combined which is generally more demanding, (Boyd and Barrat 1991; Scherer 1995; Scherer et al. 1997; Hindi et al. 1998; Sznaier 2000; Webers and Engell 1996). In the approach presented here, the controller synthesis includes three different mathematical criteria and is based on a finite-dimensional Youla parametrisation, (Youla et al. 1976), leading to a convex optimisation problem. The finite-dimensional Youla parametrisation causes a high controller order so that, in a second step, low-order controllers are synthe-sised using an order reduction scheme such that the specifications are also met for the reduced controller and the control scheme becomes practically applicable. During the design, the standard feedback loop depicted in Figure 8.13 is used.

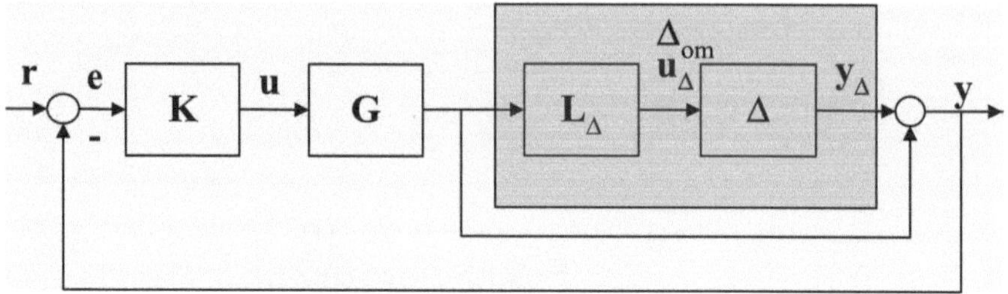

Fig. 8.13: Feedback control loop.

8.6.1 Control performance specification

For the design, three different objectives are considered. The objective function aims at minimising an integral-squared error criterion. In addition, hard amplitude constraints in the time domain and robustness constraints in the frequency domain are formulated.

Integral-squared error objective

In the objective function, the minimisation of the integral-squared control error $e(t)$ to unit steps in the setpoints $r(t)$ for the nominal control loop in Figure 8.13 ($\Delta = 0$)

$$\min \int_0^T e^T(t) \cdot e(t)\, dt, \tag{8.7}$$

is considered which implies the rejection of disturbances at the output of the plant. The minimisation of such integral-squared objectives can be formulated as a quadratic optimisation problem (Pegel and Engell 2000) of the form:

$$\min_{\mathbf{x}} \frac{1}{2} \cdot \mathbf{x}^{\mathrm{T}} \cdot \mathbf{H} \cdot \mathbf{x} + \mathbf{c}^{\mathrm{T}} \cdot \mathbf{x}, \tag{8.8}$$

where the matrix $\mathbf{H}$ is positive definite so that the problem is convex and therefore amenable to numerically efficient solution techniques. Since the matrices $\mathbf{H}$ and $\mathbf{c}$ are obtained for each reference-control-error pair and can be superimposed in the optimisation, it is possible to weight the coupling, i. e. the reaction of e_i to r_j for $i \neq j$, less than the setpoint tracking ($i=j$). In the context of the application to the reactive distillation column, the weighting of the coupling was chosen to be smaller than that of the setpoint tracking by a factor of 2, since disturbance rejection rather than decoupled setpoint tracking is desired for process operation.

Hard amplitude constraints in the time domain

To account for the physical limitations of the actuators, hard amplitude constraints on the manipulated variables $\mathbf{u}$ in the nominal case (see Figure 8.13) are imposed by

$$\max_{t} |u_i(t)| < 2 \quad \forall i, \tag{8.9}$$

where all entries of the manipulated variables $\mathbf{u}(t)$ were restricted to a modulus of 2 for unit setpoint steps $\mathbf{r}(t)$ in all possible directions for a time horizon $t < 6$ h. These limitations are formulated as linear constraints $\mathbf{Ax} < \mathbf{b}$ during the optimisation of eq. (8.8). Physically, in unscaled coordinates, this means that the reflux ratio is restricted to the interval 0.575 ± 0.1, and the feed is restricted to the interval 0.0375 ± 0.01 mole/s for all possible setpoint step-changes of ± 0.06 in x_{MeAc} and ± 0.03 in x_{H2O}.

Robustness constraints in the frequency domain

As shown in section 8.5, the model uncertainty is described by all models for which the gain eq. (8.5) of Δ is bounded by 1. By design, the system shown in Figure 8.13 is stable for the nominal unperturbed case ($\Delta=0$). Instability might then only result if the loop is closed between $\mathbf{u}_\Delta$ and $\mathbf{y}_\Delta$ via some unknown perturbation Δ with $\|\Delta\|_{i2} < 1$. A sufficient condition for stability under all such perturbations can be obtained using the small gain theorem, see e. g. (Engell 1988a; Skogestad and Postlethwaite 1996). The theorem says that the uncertain feedback loop via $\mathbf{u}_\Delta$ and $\mathbf{y}_\Delta$ can only cause instability if its loop gain is larger than 1. By following the path from $\mathbf{y}_\Delta$ to $\mathbf{u}_\Delta$ in Figure 8.13, it is readily verified that the sufficient condition for stability under all possible perturbations is given by

$$\left\| \mathbf{L}_\Delta(s) \cdot \mathbf{G}(s)\mathbf{K}(s) \cdot (\mathbf{I} + \mathbf{G}(s)\mathbf{K}(s))^{-1} \right\|_{i2} \cdot \underbrace{\left\| \Delta \right\|_{i2}}_{<1} < 1, \tag{8.10}$$

where we have dropped a negative sign that does not matter for the induced 2-norm. Hence, stability for all possibly nonlinear plant perturbations $\|\Delta\|_{i2} < 1$ can be guaranteed by imposing an induced 2-norm (see eq. (8.5)) constraint on the linear system described by $\mathbf{M}(s) := \mathbf{L}_\Delta(s) \cdot \mathbf{G}(s)\mathbf{K}(s) \cdot (\mathbf{I} + \mathbf{G}(s)\mathbf{K}(s))^{-1}$. A helpful result of linear systems theory is that the worst case gain eq. (8.5) of a linear system equals to its largest amplification of a sinusoidal input and can be expressed in terms of its frequency response matrix. Therefore, the constraint eq. (8.10) can be evaluated as:

$$\left\| \mathbf{M}(s) \right\|_{i2} = \sup_\omega \overline{\sigma}\left(\mathbf{M}(j\omega) \right) < 1, \tag{8.11}$$

where $\overline{\sigma}$ is the largest singular value of the frequency response matrix $\mathbf{M}(j\omega)$. In the solution of eq. (8.7) under the constraint eq. (8.9), the constraint eq. (8.11) is considered by a recently developed method of solving a sequence of quadratic optimisation problems, see (Völker and Engell 2005).

That the computed uncertainty bounds indeed limit the achievable bandwidth of the optimal closed-loop system can be seen in Figure 8.14 which depicts the largest singular value of $\mathbf{L}_\Delta(j\omega)$. From eq. (8.10) it can be inferred that

$$\mathbf{M}(s) = \mathbf{L}_\Delta(s) \cdot \mathbf{T}(s),, \tag{8.12}$$

where $\mathbf{T}(s) = \mathbf{G}(s)\mathbf{K}(s)\cdot(\mathbf{I}+\mathbf{G}(s)\mathbf{K}(s))^{-1}$ is the transfer matrix from $\mathbf{r}$ to $\mathbf{y}$ in Figure 8.13 if $\mathbf{\Delta}=0$, describing the setpoint tracking behaviour of the nominal control loop.

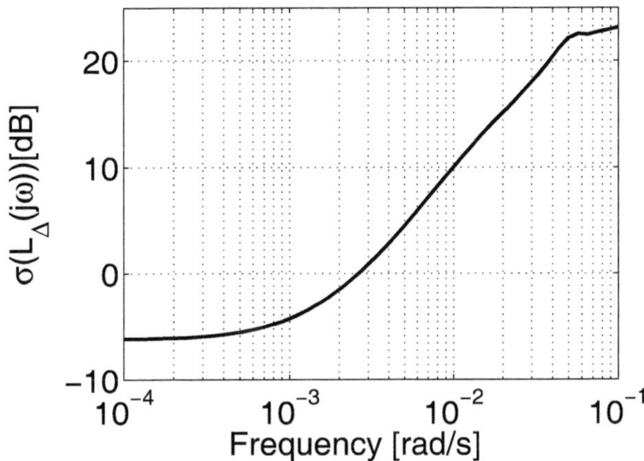

Fig. 8.14: Largest singular value of the unstructured output-multiplicative uncertainty weight $\mathbf{L}_\Delta$.

For good setpoint tracking in the frequency region below a limiting frequency ω_{BT}, it is required that

$$\bar{\sigma}\left(\mathbf{T}(j\omega)\right)\overset{!}{\approx}\underline{\sigma}\left(\mathbf{T}(j\omega)\right)\approx 1 \; \forall \; \omega < \omega_{BT}, \tag{8.13}$$

where $\underline{\sigma}$ denotes the smallest singular value of a matrix. Condition eq. (8.13) ensures that the setpoint vector $\mathbf{r}$ is tracked irrespective of its direction. From the robustness condition eq. (8.11) and from eq. (8.12) it can be inferred that eq. (8.13) can only be satisfied for those frequencies at which

$$\bar{\sigma}\left(\mathbf{L}_\Delta(j\omega)\right)<1.$$

In Figure 8.14 it can be seen that the largest singular value of $\mathbf{L}_\Delta$ crosses the 0 dB line at roughly $2\cdot10^{-3}$ rad/s which implies that ω_{BT} must be less than this frequency and thus the achievable bandwidth is limited.

The optimisation of eq. (8.8) under consideration of the constraints eq. (8.9) and eq. (8.11) yielded an optimal controller $\mathbf{K}_0(s)$ of 72nd order. This controller is subsequently simplified using again a frequency response approximation scheme.

8.6.2 Controller reduction

The main idea of the frequency response approximation technique used here is that the closed control loop containing the low-order controller should behave like the closed loop containing the high-order controller. It was shown in (Engell 1988b; Engell and Müller 1991) that this goal can be adequately addressed if the difference between the closed loops is described by the quadratic deviation of its frequency response as mentioned in section 8.4.3. For simple controllers, this criterion is convex in the controller parameters and stability of the approximated loop can be formulated as an optimisation constraint. The objective is thus to minimise the cumulated absolute square error eq. (8.3) of the difference between the control loops given by

$$\Delta_s\left(j\omega_k\right) := \mathbf{S}_0\left(j\omega_k\right) - \mathbf{S}_{red}\left(j\omega_k\right), \tag{8.14}$$

where $\mathbf{S}_0$, $\mathbf{S}_{red}$ are the frequency response matrices from the sepoint $\mathbf{r}$ to the control error $\mathbf{e}$ of the nominal control loop in Figure 8.13 with the optimal controller and the reduced-order controller, respectively. For simple controllers, this leads to a quadratic optimisation problem of the form eq. (8.8), (Völker et al. 2005). Constraints such as stability and robust stability to model uncertainties of the reduced closed loop can be included, again using the technique described in (Völker and Engell 2005).

From this procedure, the transfer matrix of the unscaled reduced-order PI-controller results as

$$\mathbf{K}(s) = \begin{bmatrix} 0.2168\dfrac{(s+0.0015)}{s} & -0.9493\dfrac{(s+0.0016)}{s} \\ 0.1223\dfrac{(s+0.0010)}{s} & -0.0058\dfrac{(s+0.0013)}{s} \end{bmatrix}.$$

The comparison of the high-order and the low-order controllers on a setpoint step scenario for the identified reduced linear plant model eq. (8.4) is depicted in Figure 8.15. It can be seen from Figure 8.15 and Figure 8.16 that the reduced-order controller satisfies the robustness constraint and the actuator saturation constraint, and that its nominal performance is similar to that of the high-order controller. It can also be seen that the manipulated variables exhibit some drift and are not tending to a constant finite value which can be attributed to the batch characteristics of the process. Despite the model reduction step and the high-pass filtering these drifts are still present in the linear model. Therefore, the actuator saturation constraint was formulated on a finite horizon, as for the infinite horizon case the deviations of the manipulated variables from their nominal values may become large, which would introduce unnecessary conservatism at times exceeding the entire batch run.

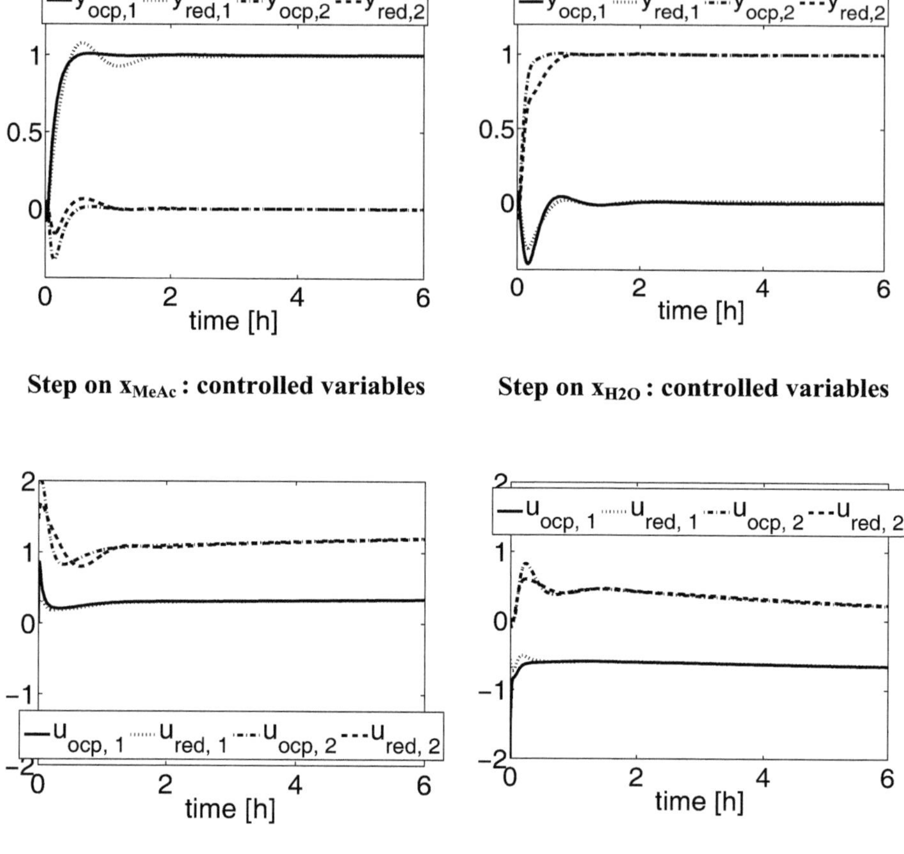

Step on x_{MeAc} : controlled variables   **Step on x_{H2O} : controlled variables**

Step on x_{MeAc} : manipulated variables   **Step on x_{H2O}: manipulated variables**

Fig. 8.15: Comparison of the high-order optimal controller (ocp) and the reduced-order controller (red) on the identified linear model, scaled according to eq. (8.2), for independent reference steps in both channels.

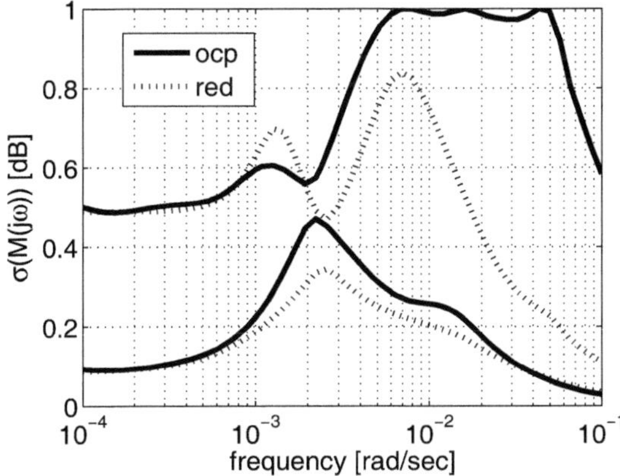

Fig. 8.16: Comparison of the high-order optimal controller (ocp) and the reduced-order controller (red) on the identified linear model, scaled according to eq. (8.2), with respect to the robustness constraint $\bar{\sigma}\big(\mathbf{M}(j\omega)\big) < 1$.

Controller validation at the pilot plant

The reduced-order controller was implemented at the experimental pilot plant and tested in a setpoint change and disturbance scenario. Figure 8.17 shows the main variables of the experiment recorded on September 28th, 2004. When the controller was activated after 2.15 hours, it could settle until, after 3.2 hours, a setpoint change was applied to drive the process from the slightly suboptimal operating point used for system identification, see section 8.4.1, towards the optimal operating point which is characterised by a higher purity of MeAc. After 4.4 hours, the heat flow to the reboiler was increased to about 4.8 kW. This increase in the heat flow represents a large disturbance to the nominal process operation. Finally, after 6.8 hours, the feed temperature was decreased by switching off the corresponding heating facility. The reduced-order controller is capable of coping with the setpoint change and the large increase in the heat flow while it can not completely compensate the disturbance in the feed temperature. This can be attributed to the lack of methanol in the reboiler towards the end of a batch run which cannot be completely counteracted by the reduction of the acetic acid feed flow. After 10.5 hours, the controller had to be switched off due to the depletion of the methanol holdup in the reboiler.

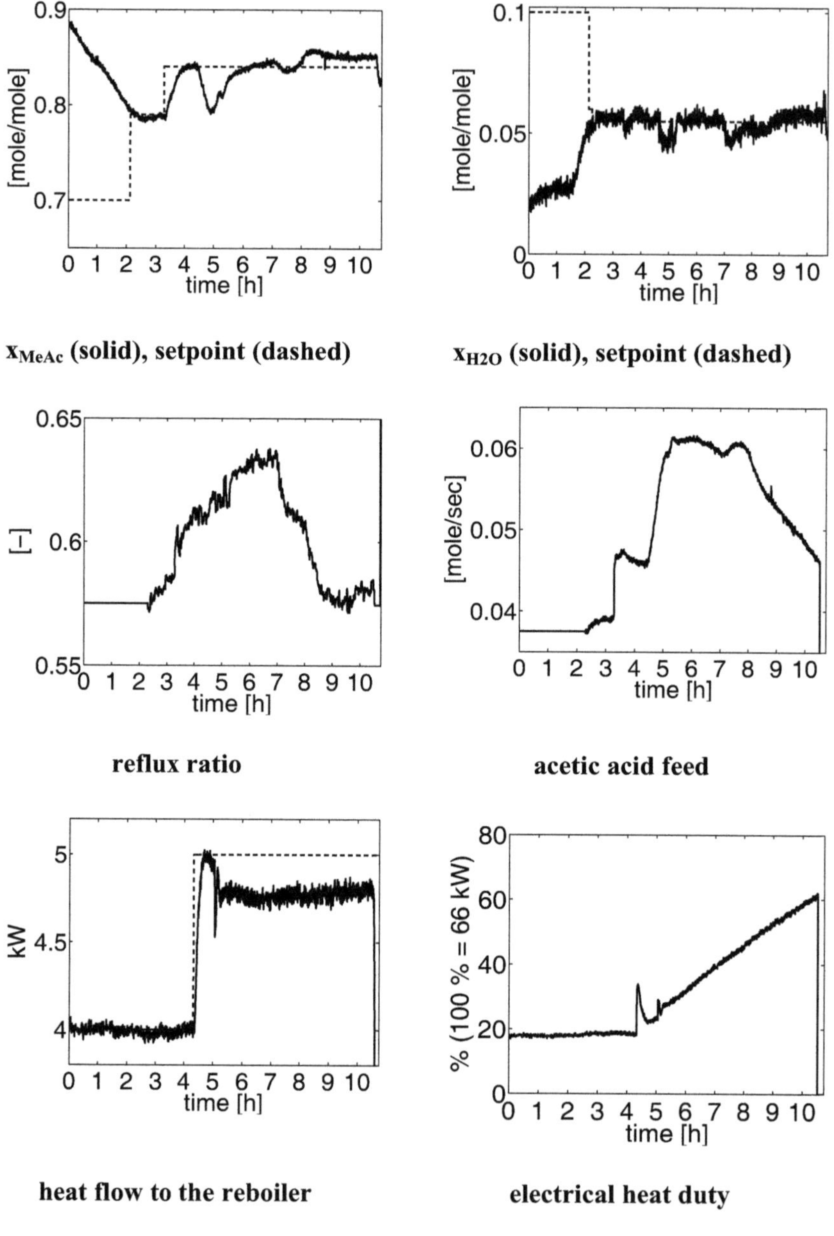

x_{MeAc} (solid), setpoint (dashed)

x_{H2O} (solid), setpoint (dashed)

reflux ratio

acetic acid feed

heat flow to the reboiler

electrical heat duty

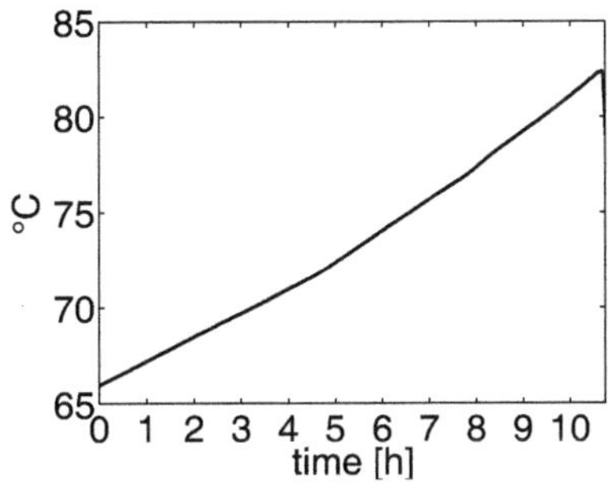
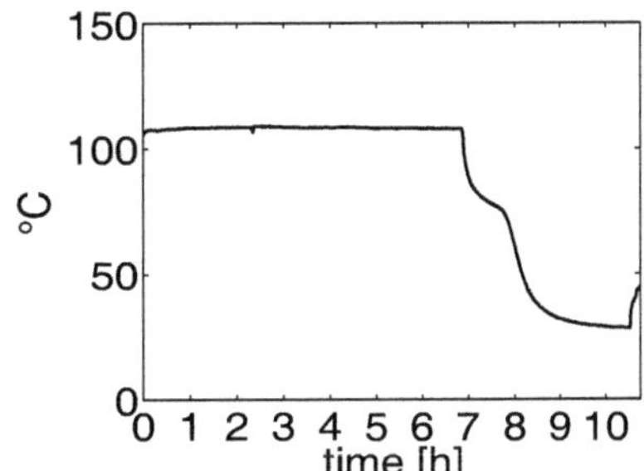

temperature in the reboiler   **temperature of acetic acid feed**

Fig. 8.17: Disturbance rejection scenario.

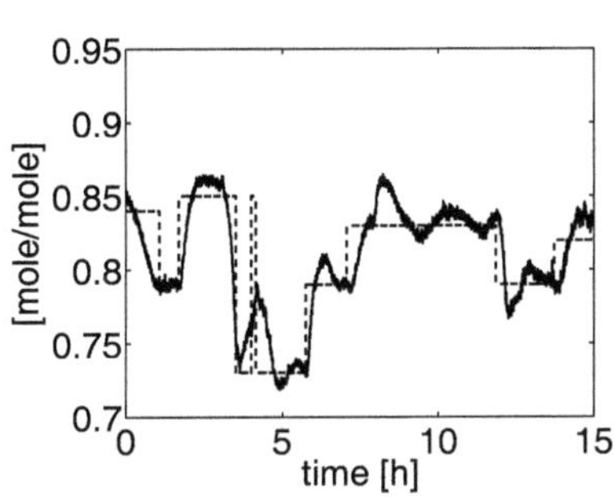
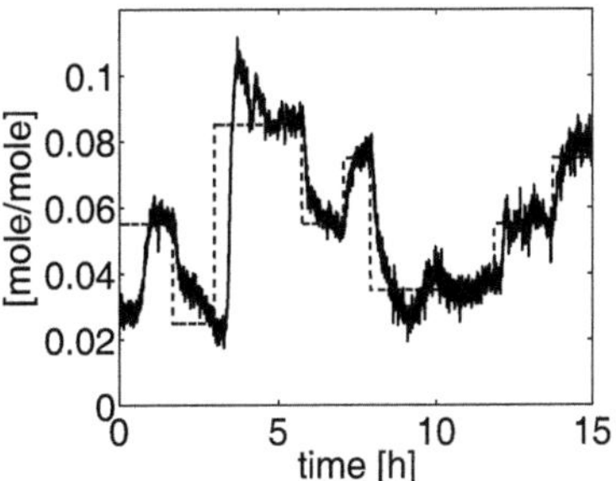

x_{MeAc} **(solid), setpoint (dashed)** x_{H2O} **(solid), setpoint (dashed)**

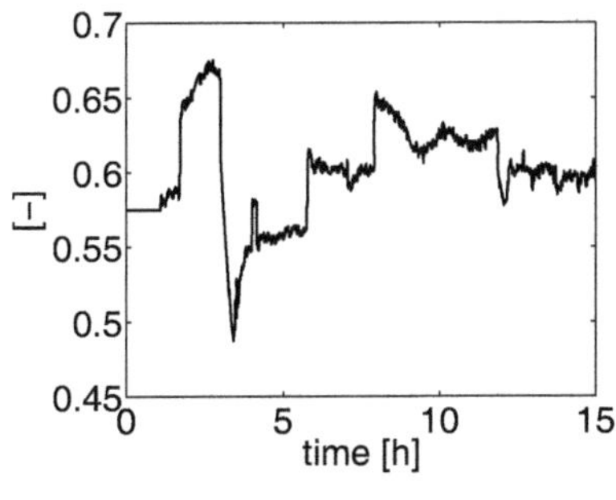
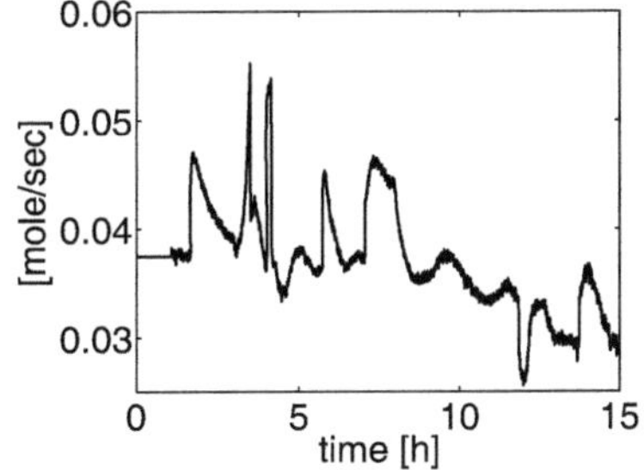

reflux ratio   **acetic acid feed**

Fig. 8.18: Setpoint change scenario.

On October 6th, 2004, the same controller was tested in a setpoint change scenario. As can be seen from Figure 8.18, the controller is able to track the setpoints despite the pronounced nonlinearity of the plant. The controller performance shown in Figure 8.17 and Figure 8.18 is achieved without any retuning of the controller.

8.7 Conclusions

We have described a method to develop a control strategy for a strongly nonlinear complex integrated reactive distillation process. For illustration, the overall methodology is depicted in Figure 8.19. Starting from a relatively simple equilibrium model that comprises the basic physico-chemical process properties, a control structure is chosen taking into account both steady-state operability and dynamic controllability, as represented by the box "Control Structure Selection" in Figure 8.19. The steady-state operability analysis yields a qualitative understanding of possible approaches. Specifically, it showed that the process exhibits multiple steady-state nonlinearities (input multiplicities) that have to be considered. However, the steady-state operability is a qualitative tool and due to its disregard of the process dynamics only yields a necessary but not a sufficient condition for the dynamic controllability of potential control structures. Hence, the dynamic limitations were analysed following the procedure described in (Engell et al. 2004). It corroborated that using the measurements of x_{MeAc} and x_{H2O} or x_{MeAc} and T_2 for control is recommended. However, control structures that were validated by simulations of the equilibrium model did not perform well in the experiments, showing that the model accuracy is not sufficient for control and a refinement of the model, especially concerning the dynamics, was required.

As it is costly to extend the rigorous model to account for neglected actuator dynamics, unknown or time-varying parameters, or to employ a more complex structure with even more unknown parameters, a suitable model for controller design was obtained in a different way. To refine the quantitative knowledge about the process dynamics for the chosen control structure, a linear model was identified. The identification experiment was designed under consideration of the steady-state diagrams obtained from the rigorous model and in the frequency range important for control so that the model accuracy in the vicinity of the optimal operating conditions could be increased. Additionally, the experimental results were used to derive an uncertainty description which accounts for nonlinearities and time-varying process behaviour and is compatible with the robust controller synthesis design framework.

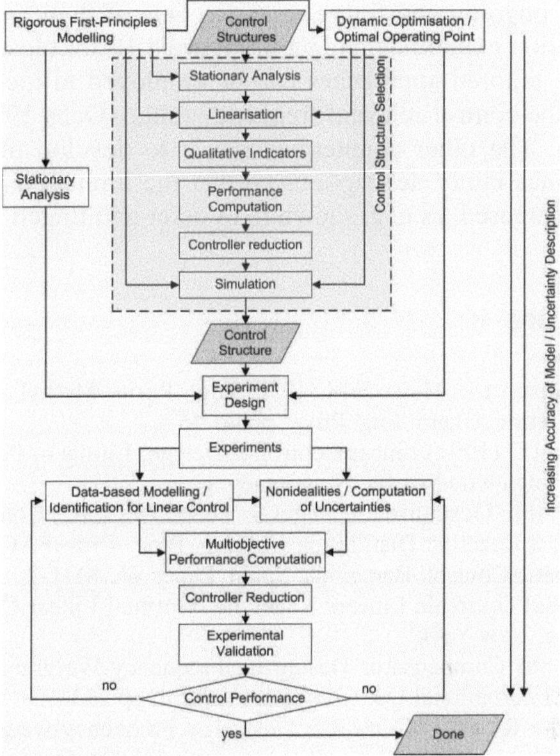

Fig. 8.19: Overview of the methodology.

In the controller synthesis step, a linear high-order controller was computed by means of a recently extended optimisation technique that can handle constraints in the time domain and in the frequency domain simultaneously. To obtain a reduced-order controller which is transparent and can be implemented in the PC-based digital control system at the plant, the optimal high-order controller was used as a specification for designing a low-order controller which fulfils the robust stability and actuator saturation specifications and only leads to a slight increase of coupling in the controlled variables. Finally, the controller was validated at the experimental plant. Both experiments showed that the controller performs well for large setpoint changes and in the face of process disturbances which in open-loop operation would have driven the process far away from its specified operating regime.

In this case study, the diamond-shaped decision node "Control Performance" in Figure 8.19 could in fact be left without any iteration, since a reasonable control performance was achieved. If the control performance is not sufficient, the proposed scheme offers potential countermeasures that are issues of ongoing and future research. One of these is to use the controller validation experimental data in Figure 8.18 for closed-loop identification where tailored approaches can be employed to increase the model accuracy in the control-relevant frequency range (Ochs 1997; Forssell and Ljung 1999). The other countermeasure is to develop methodologies to tighten the uncertainty descriptions so that the limitations for robust control can be decreased, as e. g. shown in (Völker and Engell 2005a).

8.8 Literature

Agreda VH, Partin LR, Heise WH (1990) High Purity Methyl Acetate via Reactive Distillation. Chem. Eng. Prog., pp 40-46

Boyd SP, Barrat C (1991) Linear Controller Design: Limits of Performance. Prentice Hall, Englewood Cliffs, New Jersey

Dadhe K, Engell S, Gesthuisen R, Pegel S, Völker M (2002) Control Structure Selection for a Reactive Distillation Column. Proc. 15th IFAC World Congress on Automatic Control, Barcelona, Spain, Paper We-M11-2

Engell S (1988a) Optimale Lineare Regelung (Optimal Linear Control). Springer, Heidelberg, New York

Engell S (1988b) Compensator Design by Frequency-Weighted Approximation. Proc. IEE International Conference on Control, pp 253-258

Engell S, Müller R (1991) Controller Design by Frequency-weighted Approximation: the Multivariable Case. IEE International Conference CONTROL-91, Edinburgh, pp 581-586

Engell S, Pegel S (2000) Design of PID Controllers via Frequency Response Approximation. IFAC Workshop on Digital Control, Terrassa, Spain, pp 65-72

Engell S, Trierweiler JO, Völker M, Pegel S (2004) Tools and Indices for Dynamic Controllability Assessment and Control Structure Selection. In Seferlis, P; Georgiadis, MC (eds.) Integration of Design and Control. Elsevier, pp 430-464

Fernholz G, Engell S, Kreul LU, Górak A (2000) Optimal Operation of a Semibatch Reactive Distillation Column. Proc. PSE 2000, Keystone, Colorado, Comp. Chem. Eng. 24: 1569-1575

Fernholz G (2000) Prozeßführung einer halbkontinuierlich betriebenen Kolonne zur Reaktivrektifikation. Dr.-Ing. Dissertation, Universität Dortmund and Shaker Verlag

Forssell U, Ljung L (1999) Closed-loop Identification Revisited. Automatica, pp 1215-1241

Gesthuisen R, Dadhe K, Engell S, Völker M (2002) Systematischer Reglerentwurf für eine semikontinuierliche Reaktivrektifikation. Chemie Ingenieur Technik, pp 1433-1438

Gevers M (2003) A Personal View on the Development of System Identification. Proc. 13th IFAC Symposium on System Identification, Rotterdam, pp 773-784

Hindi HA, Hassibi B, Boyd SP (1998) Multiobjective H_2/H_∞-Optimal Control via Finite Dimensional Q-Parametrization and Linear Matrix Inequalities. Proc. 17th American Control Conference, pp 3244-3249

Hjalmarsson H (2003) From Experiments to Closed-loop Control. Proc. 13th IFAC Symposium on System Identification, Rotterdam, pp 1-14

Noeres C, Kenig EY, Górak A (2003) Modelling of Reactive Separation Processes: Reactive Absorption and Reactive Distillation. Chemical Engineering and Processing, pp 157-178

Ochs S (1997) Synthese robuster Regelungen durch Iteration von Identifikation und Reglerentwurf (Synthesis of Robust Control Systems by Iteration of Identification and Controller Synthesis). Dr.-Ing. Dissertation, Universität Dortmund and Shaker Verlag

Oppenheim AV, Schaffer RW (1975) Digital Signal Processing. Prentice Hall, Englewood Cliffs, New Jersey

Pegel S, Engell S (2000) Computation of Achievable Performance of a Heat Integrated Distillation Column. Proc. IFAC Conference on Control Systems Design (CSD2000) Bratislava, pp 33-38

Poolla K, Khargonekar P, Tikku A, Krause J, Nagpal K (1994) A Time-domain Approach to Model Validation. IEEE Transactions on Automatic Control, pp 951-959

Scherer CW (1995) Multiobjective H_2/H_∞ Control. IEEE Transactions on Automatic Control, pp 1054-1062

Scherer CW, Gahinet P, Chilali M (1997) Multiobjective Output-Feedback Control via LMI Optimization. IEEE Transactions on Automatic Control, pp 896-911

Skogestad S, Postlethwaite I (1996) Multivariable Feedback Control. John Wiley & Sons Ltd

Sonntag C (2004) Control-relevant Identification of a Reactive Distillation Column. Dipl.-Ing. Thesis, Lehrstuhl für Anlagensteuerungstechnik, Universität Dortmund

Sznaier M, Rotstein H, Bu J, Sideris A (2000) An Exact Solution to Continuous-time Mixed $\mathcal{H}_2/\mathcal{H}_\infty$ Control Problems. IEEE Transactions on Automatic Control, pp 2095-2101

Tulleken HJAF (1990) Generalized Binary Noise Test-signal Concept for Improved Identification-experiment Design. Automatica, pp 37-49

van den Hof PMJ, Heuberger PSC, Bokor J (1995) System Identification with Generalized Orthonormal Basis Functions. Automatica, pp 1821-1834

Vidyasagar M (1993) Nonlinear Systems Analysis. Prentice-Hall, Englewood Cliffs, New Jersey

Völker M (2002) Robuste, Lineare Regelung einer Reaktiven Rektifikationskolonne (Robust, Linear Control of a Reactive Distillation Column). Dipl.-Ing. Thesis, Lehrstuhl für Anlagensteuerungstechnik, Universität Dortmund

Völker M, Engell S (2004) Quantitative I/O-Controllability Assessment for Uncertain Plants. Proc. IEEE Conference on Computer-aided Control System Design Taipei, Taiwan, pp 13-18

Völker M, Engell S (2005) Multi-objective Controllability Assessment by Finite Dimensional Approximation. Proc. 24th American Control Conference Portland, Oregon, Paper WeA02.2

Völker M, Engell S (2005a) Computation of Tight Uncertainty Bounds from Time-domain Data with Application to a Reactive Distillation Column. 44th Conference on Decision and Control, European Control Conference (CDC-ECC) Sevilla, Paper TuB06.4, pp 2939-2944

Völker M, Sonntag C, Engell S (2005) Design of Robust Low-order Controllers for Complex Processes: a Case Study on Reactive Distillation in a Medium-scale Pilot Plant. Proc. 16th IFAC World Congress Prague, Paper Mo-A21-TO/1

Webers K, Engell S (1996) Controller Structure Evaluation by Convex Optimization. Computers and Chemical Engineering, pp 949-954

Youla D, Jabr H, Bongiorno J (1976) Modern Wiener-Hopf Design of Optimal Controllers-Part II: The Multivariable Case. IEEE Transactions on Automatic Control, pp 319-338

Zames G (1966) On the Input-Output Stability of Time Varying Nonlinear Feedback Systems, Part I. IEEE Transactions on Automatic Control, pp 228-238

Zhou K, Doyle JC (1998) Essentials of Robust Control. Prentice Hall, Englewood Cliffs, New Jersey

Zhu YC (1989) Black-box Identification of MIMO Transfer Functions: Asymptotic Properties of Prediction Error Models. International Journal of Adaptive Control and Signal Processing, pp 357-373

Zhu YC (2001) Multivariable System Identification for Process Control. Pergamon, Amsterdam

Ziegler JG, Nichols NB (1942) Optimum Settings for Automatic Controllers. Transactions of the A.S.M.E., pp 759-768

9 Multifunctionality at particle level – Studies for adsorptive catalysts

Praveen S Lawrence, Marcus Grünewald

9.1 Introduction

Following the advent of multifunctional reactors (Agar 1988), a plethora of multifunctional integration possibilities at different scales within a reactor has been developed and analysed. Dautzenberg (Dautzenberg and Mukherjee 2001) has classified multifunctional reactors under four subcategories based on where the integration occurs and the multifunctional catalysts are grouped under Type A and B. According to this definition, multifunctional methods which are based on the integration of functionalities on particle scale and only linked by mass and heat transport mechanisms over microscopic dimensions are termed multifunctional catalyst systems. This work focuses on the concept of multifunctional catalyst where the second functionality within a particle operates as a source or sink of mass or heat in order to favourably manipulate the concentration and temperature profile within the particle.

Here, it should be noted that the term multifunctional catalyst has already been in use in literature under a different context. Though the title „The winning catalysts are multifunctional" may ideally suit this work, (Schuit and Gates 1983b; Schuit and Gates 1983a) has already presented an article with this title where the term multifunctional is used to describe bi-metallic catalysts. In contrast to multifunctional catalysts, a bimetallic catalyst does not consist of two single independently integrated functionalities, but the second integrated metal component is rather used to enable and assist the primary active metal respectively (Farkas 2001). Foley and Davis (Davis 1994; Foley 1994) have shown that by using new chemical principles of catalyst design and using chemical engineering methods, the result could be effective catalyst systems. Reschetilowski presents in

(Reschetilowski 2003) such an innovative system together with a well-known principle under the term of multifunctional catalysts.

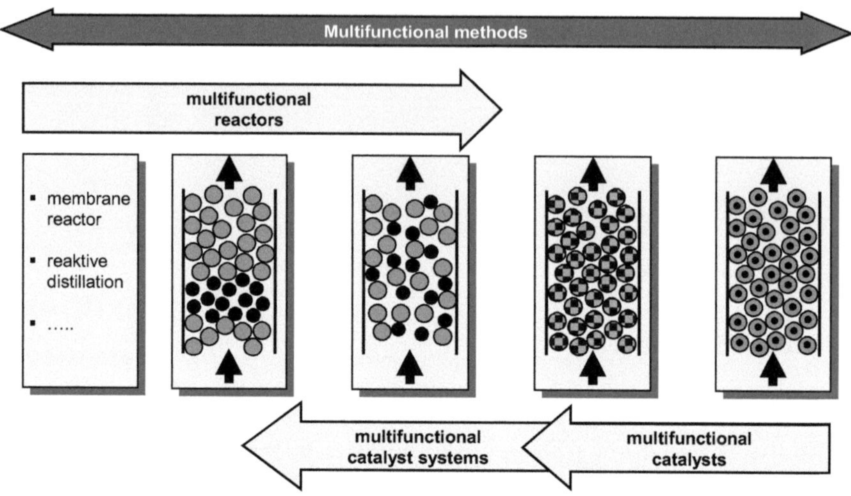

Fig. 9.1: Multifunctional reactors, catalysts and catalyst systems

On the basis of the definition given by Dautzenberg, multifunctional catalyst systems could be divided into subgroups characterised by the nature and impact of the second integrated functionality as follows:

– *Integrated mass storage*
  e.g. (Lawrence et al. 2003; Dietrich et al. 2005; Lawrence et al. 2005a; Lawrence et al. 2005b), - for the removal of limiting and inhibiting products,
    - for dosing reactants enhancing selectivity and/or conversion .
– *Integrated heat storage*
  e.g. (Grünewald and Agar 2004b; Richrath et al. 2005),
    - for heat removal from exothermic reactions,
    - for heat supply to endothermic reaction systems.
– *Integrated second catalytic activity in terms of unexhaustable mass and heat storage*
  (Grünewald and Agar 2004a; Hünnemann et al. 2005),
    - for the removal and dosing of reactants and products,
    - for tempering the main reaction.

- *Selective mass transport mechanisms*
  e.g. (Foley et al. 1994),
    - for a hindered access of components limiting catalysts' performance,
    - for the retention of components controlling selectivity and conversion,
    - for the spatial structuring of homogeneous catalysts.
- *switchable mass transport mechanism*
  e.g. (Wernke et al. 2004; Zirkwa et al. 2004),
    - for intensifying the transport of limiting component through pores,
    - for temporal distribution of reaction zones within a catalyst particle.

In our research group , the concept of multifunctional catalysts has been extended beyond the works of Foley and Davis. The citations listed above reveal the different concepts that have been studied (Davis 1994; Grünewald and Agar 1994). The previous chapters of this book offer a perspective on the different strategies one can adopt in the process design of multifunctional reactors. In this chapter, the concept of multifunctional catalysts is shown to provide an additional degree of freedom in multifunctional reactor design (Figure 9.1). Furthermore, we shall discuss adsorptive catalysts as an example of multifunctional concepts on particle scale.

9.2 Integration of adsorptive functionality on particle scale

It is known from textbooks that, in general, transport limitations are detrimental to the performance of heterogeneous catalysed reactions and adsorptive reactors are no exception to this phenomenon. In an adsorptive reactor, besides the possible lowering of catalyst effectiveness, transport limitations typically widen the adsorbate concentration front and thus influence the performance of adsorptive reactors (see also chapter 5). The concept of multifunctional catalysts can be used as an effective tool to circumvent transport limitations. By integrating functionalities within the same particle, the functionalities are brought in closer proximity and thus expected to reduce the influence of transport limitations.

In the case of conversion enhancement for an equilibrium-limited reaction via adsorptive product removal, the benefits provided by multifunctional catalysts can be easily illustrated as depicted in Figure 9.2. For a simple bimolecular equilibrium reaction the thermodynamic equilibrium is shifted towards the product side according to LE CHATELIER's principle if the reaction (by-) product C is adsorbed.

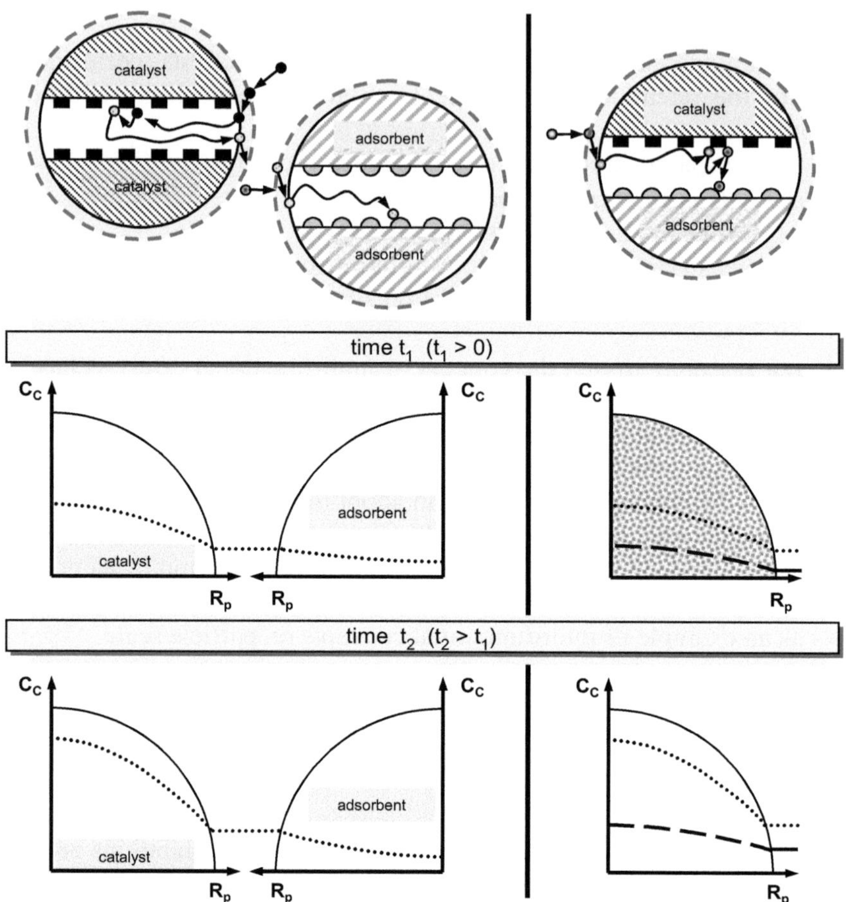

Fig. 9.2: Concentration profiles inside multifunctional catalysts compared with a conventional mixture of catalyst and adsorbent particles

In a multifunctional particle the component C can be adsorbed directly in the same particle in which it is formed, while in 'conventional' arrangement C has to be transported from the catalyst particle via the bulk fluid phase into a neighbouring adsorbent particle. As a result, a lower concentration of the product is achieved at the catalyst surface yielding a higher rate of reaction compared to a simple monofunctional catalyst particle. The more direct local product removal also results in an increased driving force for the adsorption process and thus better utilisation of the adsorbent capacity. In addition, the bulk phase concentration of C is diminished, so that the integrated functionalities at the particle scale lead to steeper concentration profiles in the reactor and, therefore, yield sharper breakthrough curves for both the adsorbed component and unconverted re-actants permitting longer cycle times in adsorptive reactor operation. For a given reactor volume and reactant flow rate an extended cycle time is equivalent to a higher space-time yield.

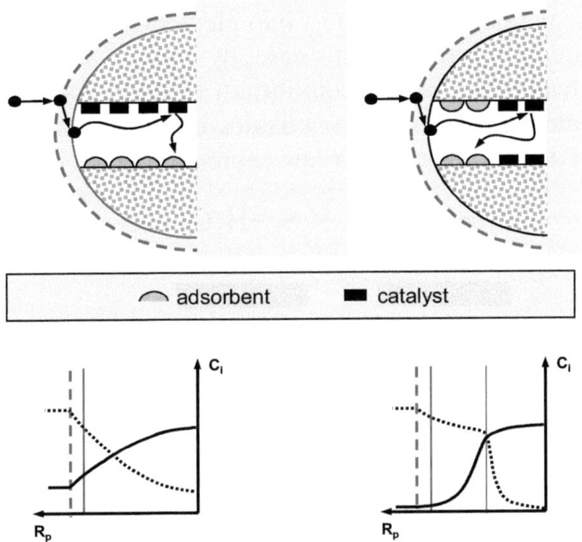

Fig. 9.3: Concentration profiles inside multifunctional catalysts compared with a conventional mixture of catalyst and adsorbent particles

Now, with the appreciation of the need for adsorptive catalysts, the natural question that arises is the approach to distribute the functionalities within the particle. For the distribution of a single catalytic activity within a particle, (Morbidelli et al. 1982) have shown that transport limitations are not always detrimental to process performance and that by prudent distri-bution of catalyst activity within a particle, the transport limitations can be exploited to enhance the process performance. This suggests that a more subtle and considered approach to the integration of functionalities at the particle level is necessary. With only limited knowledge of underlying mi-crostructured functionalities, a range of possibilities can be proposed for optimally distributing the functionalities within a particle, ranging from a uniform distribution to complete segregation of functionalities (Figure 9.3). Here we shall analyse the benefits of a non-uniform distribution of functionalities within the particle space and study the possible improve-ments achieved by this microstructuring of adsorptive reactors. For the sake of clarity, the concept of particle level distribution is subsequently re-ferred to as microstructuring.

9.3 Test reaction scheme

The single-stage Claus process, an adsorptive reactor concept proposed previously by our group (Elsner et al. 2002; Elsner 2004), has been chosen as a reference system for this study. The Claus process involves the conversion of sour gases (H_2S and SO_2) into elemental sulfur and water by a reversible chemical reaction. In this case, by-product adsorption results in conversion enhancement of the equilibrium reaction according to the Le Chatelier's principle. The reaction kinetics over an γ-Al_2O_3-catalyst has been described using simple power law expression given below:

$$2H_2S + SO_2 \; \leftrightarrows \; \frac{3}{n} S_n + H_2O \tag{9.1}$$

$$r_{V,i} = \nu_i \cdot k_1 \left(p_{H_2S}^{0.95} \cdot p_{SO_2}^{0,22} - \frac{p_{H_2O}^{0.99}}{K_{eq}} \right) \frac{mmol}{s\, kg_{cat}} \tag{9.2}$$

where $K_{eq} = {k_1}/{k_2}$ is the reaction equilibrium constant.

The water vapour formed is adsorbed on a 3A zeolite with the adsorption equilibrium being described using a Freundlich-type isotherm.

$$q_{H_2O} = K_{F,H_2O} \cdot c_{H_2O}^{0.75} \tag{9.3}$$

The model parameters required have been obtained from either literature data or calculated using empirical correlations (Wakao and Funazkri 1978) and are summarised in Table 9.1.

Table 9.1: Model parameters and correlations used in the case study

Parameter	Value	Parameter	Value
ε	0.4	T	250 °C
ε_p	0.5	p	1.013 bar
D_i	0.2 m²/h	p_{H2S}	0.1 bar
L	1.5 m	p_{SO2}	0.05 bar
u	0.15 - 0.35 m/s	p_{N2}	0.85 bar
d_p	2 – 8 mm		
D	0.06 m		
k_{film} & D_{ax}	(Wakao and Funazkri 1978)		

9.4 Modeling of adsorptive catalyst

An unsteady-state heterogeneous two-phase dispersion model with internal pore diffusion describing the dynamic behaviour of a fixed-bed adsorptive reactor with adsorptive catalyst particles has been developed. In addition, a model with separate catalyst and adsorbent phases has also been developed and used as a benchmark system. The models are based on the following simplifyied assumptions:
- the fluid in the particle pores and the internal adsorbent/ catalyst surface are in equilibrium,
- the adsorption process is assumed to exhibit only macropore and film transport limitations. The effect of micropore transport limitation is neglected. This assumption limits the applicability of the conclusions drawn from the study to adsorbents like alumina, silica, activated carbon and molecular sieve carbons, though not for zeolites (transport within micropores also needs to be considered for zeolites),
- the adsorption process is assumed to be highly selective. It should be noted that this assumption may not be realistic for physisorption processes, but is probably reasonable for chemisorption processes,
- the physical structure of the particles can be idealised as porous spheres with a uniform pore structure,
- particle size and structure were averaged for all particles in order to permit a direct comparison of different particle and reactor configurations,
- backmixing in bulk fluid phase flow is lumped into an axial dispersion coefficient term,
- the gas phase behaviour obeys the ideal gas law,
- heat effects are neglected and the reactor is operated isothermally,
- the volume change due to reaction is neglected. This is justified due to the high dilution of the reaction system with 85% inert,
- the bed is assumed to be free of any adsorbate at the start of the adsorption cycle.

Particle model

Inside the particles the mass transfer is described using an effective pore diffusion model. The following expression gives the mass balance for an adsorptive catalyst particle.

$$\varepsilon_p \frac{\partial c_p}{\partial t} = D_{eff}\left(\frac{\partial c_p}{\partial r}\frac{2}{r} + \frac{\partial^2 c_p}{\partial r^2}\right) - \left(1-\varepsilon_p\right)\left(f_{ads}(r,x)r_{V,ads} + f_{cat}(r,x)r_{V,cat}\right) \qquad (9.4)$$

For the benchmark system, the catalyst and adsorbent are present in separate particles. The mass balance for the particles are given below:

$$\varepsilon_p \frac{\partial c_{p,i}}{\partial t} = D_{eff,i} \left(\frac{\partial c_{p,i}}{\partial r} \cdot \frac{2}{r} + \frac{\partial^2 c_{p,i}}{\partial r^2} \right) - \left(1 - \varepsilon_p \right) \cdot r_{V,ads,i} \qquad \text{with}: H_2O, \ r_{v,ads,i} = 0 \tag{9.5}$$

$$\varepsilon_p \frac{\partial c_{p,i}}{\partial t} = D_{eff,i} \left(\frac{\partial c_{p,i}}{\partial r} \cdot \frac{2}{r} + \frac{\partial^2 c_{p,i}}{\partial r^2} \right) - \left(1 - \varepsilon_p \right) \cdot r_{V,cat,i} \tag{9.6}$$

The boundary conditions for the particle mass balance are derived by assuming linear concentration profiles in the particle boundary layer eq. (9.7) and symmetric profiles within the particles eq. (9.8).

$$k_{film,i} \left(c_i - c_{p,i} \big|_{r=R_p} \right) = D_{eff,i} \frac{\partial c_{p,i}}{\partial r} \bigg|_{r=R_p} \tag{9.7}$$

$$\frac{\partial c_i}{\partial r} \bigg|_{r=0} = 0 \tag{9.8}$$

The particle models are incorporated in a fixed-bed reactor and a one-dimensional axial dispersion model is used to describe the reactor fluid bulk phase eq. (9.9). The third term on the right hand side of eq. (9.9) accounts for the mass transfer from the bulk phase to the adsorptive catalyst particles.

$$\varepsilon \frac{\partial c_i}{\partial t} = -u \frac{\partial c_i}{\partial x} + D_{ax} \varepsilon \frac{\partial^2 c_i}{\partial x^2} - \left(1 - \varepsilon \right) \frac{3}{R_p} k_{film} \left(c_i - c_p \big|_{R_p} \right) \tag{9.9}$$

In case of the benchmark system, the reactor mass balance is given by the following expression. The third term on the right hand side of eq. (9.9) accounts for the mass transfer from the fluid bulk phase to the catalyst and adsorbent particle each weighted according to the contribution of the volume fractions of the particles in the fixed-bed.

$$\varepsilon \cdot \frac{\partial c_i}{\partial t} = -u \frac{\partial c_i}{\partial x} + D_{ax} \cdot \varepsilon \cdot \frac{\partial^2 c_i}{\partial x^2} \tag{9.10}$$

$$- \left(1 - \varepsilon \right) \cdot \frac{3}{R_p} \cdot \left(f_{ads}(r,x) \cdot k_{film} \left(c_i - c_{p,ads,i} \big|_{R_p} \right) + f_{cat}(r,x) \cdot k_{film} \left(c_i - c_{p,cat,i} \big|_{R_p} \right) \right)$$

At the outset of each reaction cycle the reactor contains just an inert component.

$$c_i(x, t=0) = 0, \ c_{p,i}(r, t=0) = 0 \tag{9.11}$$

Numerical solution

The model has been implemented in a commercial dynamic simulation tool with a modular model structure and solved using the method of lines. The spatial domain has been discretised using a backward finite difference method with 100 finite elements for the reactor length axis and on 28 elements over the particle dimension, with the central finite difference on the radial axis of the particle being allocated in such a manner as to obtain an equivalent volume increment grid. This procedure ensured a better numerical resolution of the particle mass balances by using incremental control volumes of equal size.

Optimisation procedure

As will be explained later, at any time, different sections of an adsorptive reactor tend to be dominated either by the reaction kinetics or sorption processes. This suggests that the optimal microstructure for a reactor is not unique, but may rather differ with axial position within the reactor. Thus to arrive at an optimal microstructure, the microstructure optimisation study cannot be considered in isolation but warrants a simultaneous optimisation at the reactor and particle level. For this reason, the optimisation space is divided into discrete elements and the optimal functionality distribution in each element is determined.

Though it is desirable to arrive at a continuous optimal functionality distribution profile along the reactor length and within a particle, the problem is simplified by a discrete approach which yields entirely adequate results. It should be noted that comparable studies on catalyst dilution for smoothing reactor temperature profiles have revealed only modest sensitivities for influence of the precise fixed-bed structure on the performance. In this study, the reactor length has been divided up into 13 discrete sections and, in each section the particle has been divided up into 5 layers. Within any discrete element, the optimiser is given the freedom to choose the fraction of catalyst/ adsorbent, subject to process constraints. The definition of the optimisation criteria are summarised in Table 9.2.

Dynamic simulation studies have been carried out using Aspen Custom Modeler® (ACM), while the feasible path successive quadratic programming optimiser, FEASOPT®, which is part of the ACM packet, has been used to carry out dynamic optimisation. FEASOPT® employs a reduced space optimisation method to arrive at the optimal solution. It evaluates the objective variable (cycle time) at the current point and moves the design variables (catalyst fraction values) to take the objective variable towards its optimum value. After solving with the new values of the design variables, FEASOPT® re-evaluates the objective variable. In this way, it steps towards the optimum solution. FEASOPT® solves the simulation at each step subject to simulation equations, variable bounds and any constraints applied to the optimisation

Table 9.2: Optimisation criteria

	Benchmark system	Adsorptive catalyst system
Objective function	Maximum process cycle time, $\square$	
Variable constraints	$0 \leq f_{cat}(x) \leq 0$	$0 \leq f_{cat}(r,x) \leq 0$
	$f_{cat}(x) + f_{ads}(x) = 1$	$f_{cat}(r,x) + f_{ads}(r,x) = 1$
Process constraints	Actual conversion, $X(t)$	
	$\geq$ Desired minimum conversion, X_{min}	

As one can see from the equations, this modelling study has been restricted to the adsorption step in an adsorptive reactor process. Any study on the optimal distribution of functionalities in an adsorptive reactor is deemed to be complete if and only if both the adsorption and desorption steps of an adsorptive reactor are considered. It is known from standard studies on adsorption processes (Yang 1997), that more than one method is available for regeneration and very often a combination of different methods is required for an efficient desorption. In addition to the well-known non-reactive desorption techniques, reactive desorption techniques(Xiu and Rodrigues 2002) have also been suggested to improve the performance of adsorptive reactor. This provides us with numerous options to consider when studying the optimal distribution of catalyst and adsorbent in an adsorptive reactor. Thus, the choice of a particular regeneration strategy for an overall optimisation study to develop the proposed structuring guidelines would have introduced an arbitrary element into the analysis. Furthermore, as any catalyst in an adsorptive reactor acts merely as an inert during the desorption step (unless the catalyst possesses strong adsorbing properties towards the adsorbate or under reactive desorption conditions), the inclusion of the desorption step is not expected to significantly alter the results presented in our article.

For any given reactor length and feed flow rate conditions, the cycle time is a direct measure of the extent of the adsorbent utilisation and thus deemed to be an appropriate optimisation criterion for the process. For the Claus process, the desired conversion is set by environmental legislation, which typically demands a minimum sulphur recovery of 99.5% (German environmental standards, (TA_Luft 1986)).

9.5 Results and discussion

9.5.1 Particle level integration vs. conventional particles

It is rather apparent that particle scale integration brings functionalities at closer proximity and this lowers the transport limitations in an adsorptive reactor. Figure 9.4 shows the effect of Thiele modulus on the process cycle time for the Claus process. Under low transport limitation conditions, the performance of a micro-integrated particle and functionalities in a separate particle are comparable. With increasing transport limitations, however, integration of functionalities at a particle level reduces the influence of transport limitations in an adsorptive reactor resulting in significantly higher cycle times over conventional adsorptive reactors. Thus, integration of functionalities at a particle level offers a useful tool to circumvent transport limitations in an adsorptive reactor without reducing the particle size.

$$\Phi = l_p \sqrt{\frac{r_V\left(c_0\right)}{D_{eff} \cdot c_0}} \tag{9.12}$$

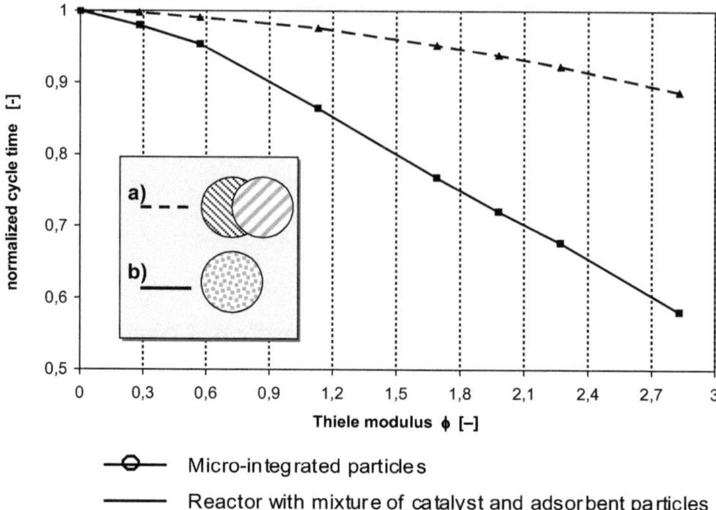

Fig. 9.4: Influence of transport resistance on process cycle time in Claus process: Comparison of performance by conventional adsorptive reactor to micro-integrated adsorptive reactor. Thiele modulus is defined under reactor entrance conditions , reproduced from (Lawrence 2005)):
   a: mixture of catalyst and adsorbent particles
   b: micro-integrated particles

9.5.2 Particle level integration vs. particle structuring

As adsorptive reactors are operated under unsteady state conditions, all catalyst and adsorbent particles are subjected to continuously varying reactant and product concentrations throughout the process cycle and this complicates the task of identifying the optimal microstructure within particle. To simplify this task, one needs to discern the different regimes that influence the performance of an adsorptive reactor. The approach to equilibrium Λ is a direct measure of the reaction driving force at any spatial position in the reactor and assists in the identification of the different regimes in an adsorptive reactor. It is defined as follows (Elsner 2004).

$$\Lambda = \frac{\prod_i c_i^{\upsilon_i}}{K_{eq}} = \frac{p_{H_2S}^{0.95}\, p_{SO_2}^{0.22}}{p_{H_2O}^{0.99}}\, \frac{k_1}{k_2} \tag{9.13}$$

For

$= 1$, the reaction is at thermodynamic equilibrium

> 1, forward reaction occurs

< 1, backward reaction occurs

Figure 9.5 shows the typical variation of Λ along the length of an adsorptive reactor at different cycle times before the process constraint is breached. The figure indicates that the value of Λ is a function of time and reactor space. This graph is typical for a reversible reaction taking place in an adsorptive reactor. Based on this figure, the reactor may be broadly classified into 3 segments. The following discussions focus on the nature of these segments and how they influence the choice of microstructure at any location in an adsorptive reactor. The results of only one of the studies ($u = 0.3$m/s, $R_p = 7$ mm) are discussed here for reasons of conciseness. The optimal process cycle time is calculated to be 0.29 h. For the following discussions normalised concentration profiles are plotted to facilitate comparison. The use of the H_2S concentration under inlet conditions as the reference value offers the possibility to compare the relative amounts of H_2S and H_2O present at any point within reactor. The normalised concentration is defined as follows.

$$c_{comp}(r,t) = \frac{c_{comp}(r,t)}{c_{H_2S,\,inlet}} \tag{9.14}$$

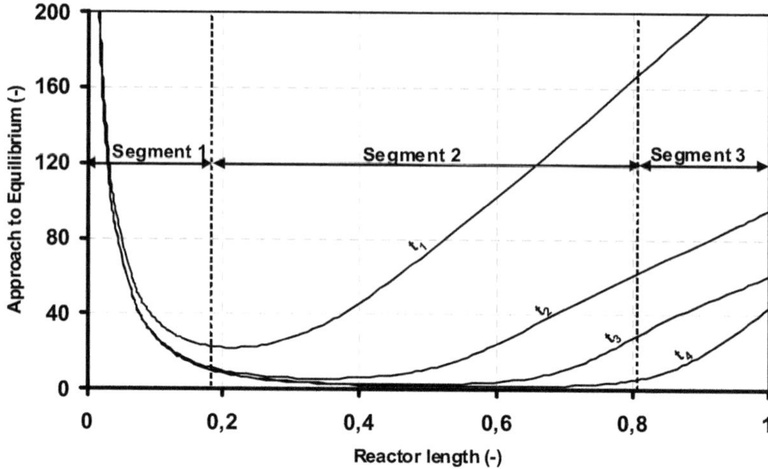

Fig. 9.5: Typical variation of approach to equilibrium Λ along the axial space at different process times for the Claus process (reprinted from (Lawrence 2005)) (R_p= 7mm, u = 0.25 m/s, $t_1 < t_2 < t_3 < t_4'$)

In segment 1, the value of Λ is always greater than one and is virtually constant with time. The positive value of Λ indicates that the forward reaction occurs at all times in this segment. It should be noted that of all the reaction occurring in an adsorptive reactor, roughly 80% takes place in the first 20% of the reactor.

Figure 9.6 shows the concentration profile of H_2S (reactant) and H_2O (adsorbate) within a microstructured particle located in this segment at different process times. The reactant concentration profile undergoes little evolution in time and profiles virtually overlap. By contrast, the adsorbate concentration varies in time, indicating the gradual saturation of the adsorbent. However, the product concentration is not significant enough to exert any influence on the reaction rate and the segment is kinetically controlled. Thus, any adsorbent present in this segment plays a secondary, if any, role in its performance.

Based on the work of Morbidelli (Morbidelli et al. 1982), for simple power law kinetics, it could be expected that a catalyst shell arrangement would offer the optimal performance. The shaded area in Figure 9.6 shows the optimal fraction of catalyst in microstructure recommended by the optimiser and the structure closely resembles a catalyst shell arrangement.

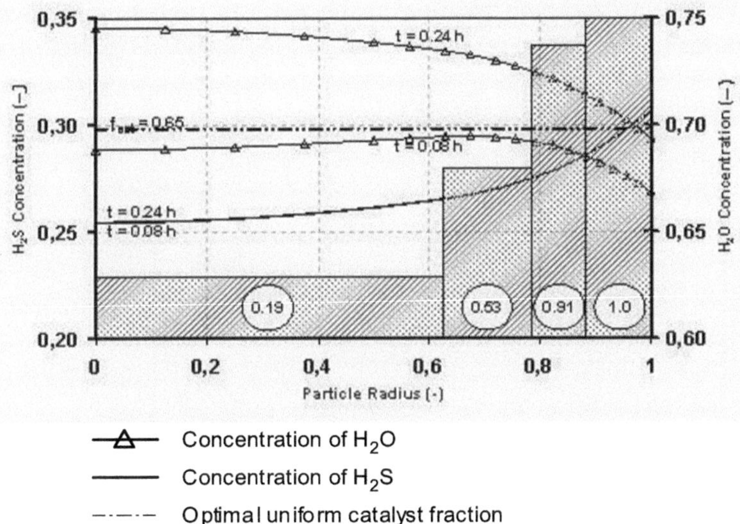

Fig. 9.6: H_2S and H_2O concentration profile within a microstructured particle located 135 mm from reactor entrance. The shaded areas and the circled numbers indicate the optimal catalyst fraction at any location within particle (segment 1). (reproduced from (Lawrence 2005))

Unlike the previous segment, segment 2 is characterised by a dynamic variation of Λ in space and time (Figure 9.5). At any point, while the adsorbent is unsaturated, Λ is greater than one. Upon adsorbent saturation, however, the value of Λ reaches one and this indicates the approach to reaction equilibrium. Thus, though the segment is initially kinetically controlled, the sorption regime gradually sets in. Figure 9.7 illustrates the concentration profile of H_2S (reactant) and H_2O (adsorptive) within a microstructured particle located in this segment at different process times.

As the predominant portion of all reaction taking place in an adsorptive reactor is limited to segment 1, the concentration of reactant is conspicuously low in the second segment. This, compounded with higher product concentrations, leads one to expect that the rate of reaction will be slower than in segment 1. Thus, the adsorption of the product is vital for the forward reaction to occur in this regime. The shaded area in Figure 9.7 shows the optimal microstructure of the particle recommended by the optimiser. The low reaction rates place lower demands on the need for a catalyst shell arrangement for the effective utilisation of the catalyst activity. For this reason, the functionality distribution within the particle tends to be rather uniform. The high adsorbent fraction in the particle shell is attributed to the importance of adsorptive removal to sustain the forward reaction.

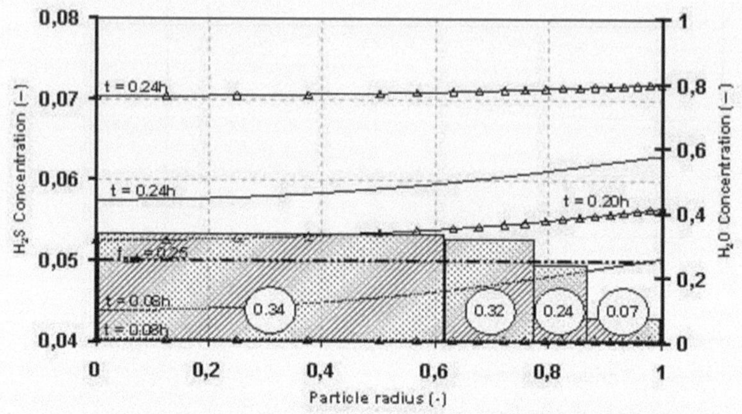

Fig. 9.7: H_2S and H_2O concentration profile and optimal microstructure within a microstructured particle located at 675mm from reactor entrance (segment 2). (refer Figure 9.6 for legend, reproduced from (Lawrence 2005))

The behaviour of segment 3 is basically similar to segment 2 and may be considered as its extension. What distinguishes these two segments is that, unlike segment 2, the adsorbent in section 3 remains unsaturated even up to the breach of the process constraint. The adsorptive concentration front does not really break through into this segment before the breach of process constraint. Thus, any adsorbate present in this segment primarily originates within it and hence the adsorbate concentration is not particularly significant here. This means that the segment is kinetically controlled throughout the process cycle. Figure 9.8 shows the concentration profile of H_2S (reactant) and H_2O (adsorbate) within a microstructured particle located within segment 3 at different process times and the corresponding optimal microstructure. As the segment is kinetically controlled, the optimal microstructure resembles a catalyst shell similar to that observed in segment 1.

It may be argued that the existence of segment 3 is a characteristic of the Claus reaction, by virtue of its very rigorous process termination conditions (Sulphur recovery efficiency > 99.5%). However, it should be noted that the existence of segment 3 is not uncommon in adsorptive reactors and a similar zone has been suggested for steam reforming reaction in adsorptive reactors (Xiu and Rodrigues 2003)

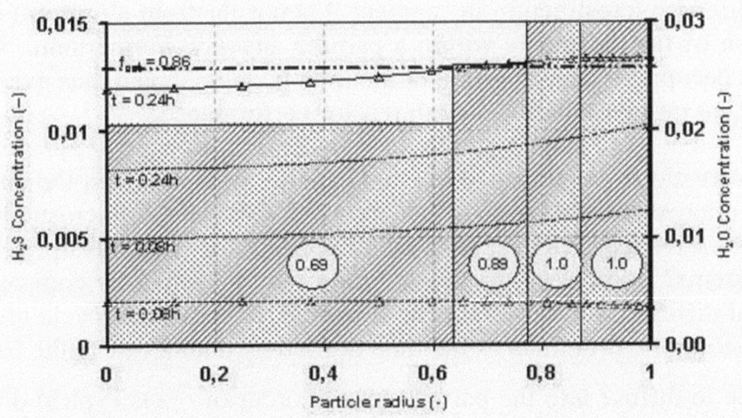

Fig. 9.8: H_2S and H_2O concentration profile and optimal microstructure within a microstructured particle located at 1335mm from reactor entrance (segment 3). (refer Figure 9.6 for legend, reproduced from (Lawrence 2005))

9.5.3 Relevance of macro- and microstructuring

From the previous discussion, it becomes obvious that the choice of particle microstructure is unique to the regime that governs particle behaviour. In order to study the importance of the non-uniform distribution of functionalities within the particle, an additional optimisation run for the above test case was carried out with only one particle layer (i.e. a uniform particle distribution case), while the same number of axial discrete sections was retained. The broken lines in Figure 9.6, 9.7 and 9.8 show the optimal uniform catalyst fraction recommended by the optimiser. It was found that though the functionality distribution profile is different, the performance of the uniform and non-uniform microstructures is comparable (less than 3% difference in cycle time). This suggests that a simple particle level integration of functionalities can perform as well as any optimal microstructure, and that the marginal difference in performance is attributable to the following causes:

– In general, the combined width of segment 1 and 3 is less than 20% of the total reactor length. Thus, even though a uniform distribution is far from a truly optimal microstructure, the relatively short span of these segments limits their influence on the overall performance of the reactor. In addition, though a kinetically controlled regime exists in segment 3, the reaction rate itself is too low to exert a significant influence on the overall performance of the reactor,

– The optimal microstructure in segment 2 is not far from a uniform distribution of functionalities within a particle anyway. Furthermore, segment 2 occupies a major portion of the total bed length and thus exerts a significant influence on the overall reactor performance.

All observations mentioned above indicate the limitations on the application of microstructuring to the Claus process. But can the microstructuring of adsorptive reactors ever be a useful concept for other adsorptive reaction systems? The answer to this question may be found by comparing the typical diffusion time constants in a particle to the process cycle times. The diffusion time is defined as the time needed by a molecule at the particle surface to diffuse into the particle (of the order of R_p^2 / D_{eff}). Typical diffusion times for gas phase systems are in the range of 0.01 to 2 s. In comparison to the relatively small diffusion times, the process cycle times are significantly higher (of the order of minutes) and thus provide ample time for the diffusion to take place, leaving the system lacking any significant diffusion limitations in the sorption regime. This is also the reason for satisfactory performance of Glueckauf's linear driving force approximation in predicting adsorption behaviour processes. If the diffusion time constants and the process cycle time were comparable, microstructuring would have a significant impact on the performance of the adsorptive reactor.

From the above discussions it is evident that the structuring of functionalities is beneficial only up to the point of integration of functionalities within the particle. In addition, one can also notice that the catalyst fraction across the reactor length is not the same, indicating that the optimal distribution need not be a homogenous distribution at reactor level. A detailed study on the non-uniform distribution of functionalities at a reactor level indicates that the optimal distribution is a function of the process constraint. Under lenient process constraints, a non-uniform distribution of functionalities offers a significant performance over uniform distribution at a reactor level. But, under stricter process constraints, the performance of a uniform distribution and an optimal non-uniform distribution are comparable. The above conclusions are drawn on the basis of studies on the Claus process and its generic kinetic variants. Figure 9.9 shows the influence of structuring on the performance of an adsorptive reactor for different values of reaction equilibrium constant. The continuous solid line shows the variation of equilibrium conversion against the reaction equilibrium constant. For any desired conversion below the equilibrium line, there is little justification for employing an adsorptive reactor and a conventional heterogeneously catalysed gas phase reactor is adequate. The data points correspond with the case studies that have been carried out, and the adjoining numbers indicate the increase in the performance of a macrostructured adsorptive reactor over homogeneously distributed adsorptive reactor. Lawrence et al. provide a detailed discussion on the macrostructuring of adsorptive reactors.

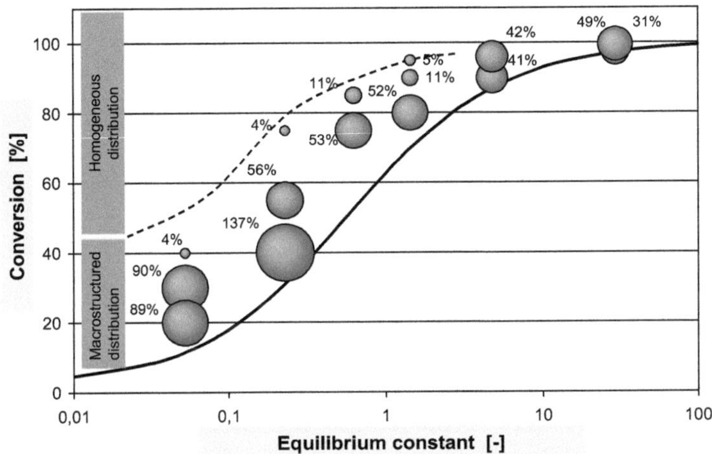

Fig. 9.9: Influence of structuring on the performance of an adsorptive reactor for different values of reaction equilibrium constant. The numbers indicate the increase in performance of macrostructured adsorptive reactor over homogeneously distributed adsorptive reactor.

9.6 Conclusions

In this work, an attempt has been made to demonstrate the relevance of multifunctional catalyst using an adsorptive catalyst as the model system. The Claus process was used as a reference reaction scheme to study the influence of distribution of catalyst and adsorbent functionalities at the particle level on the performance of an adsorptive reactor. The study reveals the existence of distinct regimes that govern the performance of an adsorptive reactor. The regimes determine the choice of the optimal microstructure at any location in the reactor. Though a non-uniform distribution of functionalities is found optimal, the performance of an optimal uniform distribution is nevertheless comparable. Thus the utility of functionality microstructuring may be limited to the micro-integration of functionalities at the particle level. Micro-integration of functionalities is found to be a useful tool to circumvent transport limitations in an adsorptive reactor. Though not explicitly described here, the optimal distribution of functionalities in an adsorptive reactor also involves a non-uniform distribution of functionalities at the reactor level.

9.7 Literature

Agar DW, Ruppel W (1988) Multifunktionale Reaktoren für die heterogene Katalyse. Chem.-Ing.-Tech. 60:731-741

Dautzenberg FM, Mukherjee M (2001) Process intensification using multifunctional reactors. Chem. Eng. Sci. 56:251-267

Davis ME (1994) Reaction chemistry and reaction engineering principles in catalyst design. Chem. Eng. Sci. 49:3971-3980

Dietrich W, Lawrence PS, Grünewald M, Agar DW (2005) Theoretical studies on multifunctional catalysts with integrated adsorption sites. Chem. Eng. J. 107:103-111

Elsner MP. (2004) Experimentelle und modellbasierte Studien zur Bewertung des adsorptiven Reaktorkonzeptes am Beispiel der Claus-Reaktion. Thesis, Logos-Verlag, Berlin

Elsner MP , Dittrich C , Agar DW (2002) Adsorptive reactors for enhancing equilibrium gas-phase reactions - Two case studies. Chem. Eng. Sci. 57: 1607-1616

Farkas A (2001) Production of Catalysis. Ullmann's Encyclopedia of Industrial Chem., Chap. 5

Foley HC (1994) Chapter 2, Catalyst preparation - design and synthesis. Catalysis Today 22:235-260

Foley HC, Lafyatis DS, Mariwala RK, Sonnichsen GD, Brake LD (1994) Shape selective methylamines synthesis: Reaction and diffusion in a CMs-SiO$_2$-Al$_2$O$_3$ composite catalyst. Chem. Eng. Sci. 49:4471-4786

Grünewald M, Agar DW (2004a) Enhanced catalyst performance using integrated structured functionalities. Chem. Eng. Sci. 59:5519-5526

Grünewald M, Agar DW (2004b) Intensification of regenerative heat exchange in chemical reactors using desorptive cooling. Ind. Eng. Chem. Res. 43:4773-4779

Hünnemann D, Grünewald M, Agar DW (2006) Determination of the intrinsic reaction rate and active immobilised amount of an enzyme by means of a rigorous dynamic simulation. submitted to Biotechnology & Bioengineering

Lawrence PS, Grünewald M, Agar DW (2005) Spatial distribution of functionalities in an adsorptive reactor at a particle level. Catalysis Today 105:582-588

Lawrence SP, Dietrich W, Grünewald M, Agar DW (2003) Studies on the microstructuring of multifunctional catalysts. 3rd Int. Symp. on multifunctional reactors (ISMR3), Bath

Morbidelli M, Servida A, Varma A (1982) Optimal catalyst activity profiles in pellets 1. The case of negligible external mass transfer resistance. Ind. Eng. Chem. Fundam. 21:278-284

Reschetilowski W (2003) Katalysatormultifunktionalität- ein Weg der Prozessintensivierung. Wiss. Zeitung TU Dresden 52:121-126

Richrath M, Grünewald M, Agar DW (2005) Particle-scale Heat Removal in Fixed-bed Catalytic Reactors: Modelling and Optimisation of a Desorptive Cooling Process. ESCAPE-15, Barcelona

Schuit GCA, Gates BC (1983a) The winning catalysts are multifunctional (part 1). Chem. Tech. 09-1983

Schuit GCA, Gates BC (1983b) The winning catalysts are multifunctional (part 2). Chem. Tech. 11-1983

TA_Luft (1986) Technische Anweisung Luft, Nr. 2.5.1., Bundesrepublik Deutschland,

Wakao N, Funazkri LT (1978) Effect of fluid dispersion coefficients on particle-to-fluid mass transfer coefficients in packed beds. Chem. Eng. Sci. 33:1375-1384

Wernke R, Bergins C, Strauss K, Kahnis H, Grünewald M (2004) Flow in micro porous media. 4th Europ. Therm. Sci. Conf., Birmingham

Xiu GH, Rodrigues AE (2002) Sorption enhanced reaction process with reactive regeneration. Chem. Eng. Sci. 57:3893 - 3908

Xiu PL, Rodrigues AE (2003) New generalized strategy for improving sorption-enhanced reaction process. Chem. Eng. Sci. 58:3425-3437

Yang RT (1997) Gas separation by adsorption processes. Imperial College Press

Zirkwa I, Kahnis H, Grünewald M, Agar DW (2004) Intensification of Mass Transfer in Porous Catalyst Particles by Means of Pressure Oscillation. 4th Europ. Therm. Sci. Conf., Birmingham

Index

A

acetic acid, 96
achievable bandwidth, 309, 326, 327
acidic ion exchanger resin, 99
activity coefficient, 110
adsorbent fraction, 270
adsorption isotherm, 214, 270, 277
adsorptive catalyst, 343
adsorptive reactor, 149
ammonia, 169
annual costs, 66
Antoine equation, 109
approach to equilibrium, 352
asymptotic theory, 322
asynchronous switching, 262, 266
auto-catalytic, 103
average zone length, 262
axial concentration profile, 272
axial dispersion, 104, 112, 203
azeotrope, 301, 306

B

batch distillation, 116
big-M constraint, 65, 71
binary reaction, 192
biocatalyst, 98
bioreaction, 198
biphasic system, 245
Bodenstein number, 114, 206
Boudouard-reaction, 171
branch-and-bound, 67
breakthrough curve, 213
butadiene, 244, 247
butanol, 44
butyl acetate, 44, 97

by-passing, 112

C

CaO, 170
carbon monoxide, 169, 173
carbonylation, 98
catalyst leaching, 247
catalyst recovery, 101
catalytic activation, 101
catalytic dehydrogenation, 99
catalytic distillation, 95
catalytic hydrogenation, 99
catalytic membrane reactor, 103
cell model, 112
chemical equilibrium, 96
chromatographic column, 260
chromatographic parameter, 211
chromatographic reactor, 192, 226
chromatographic separation, 259
Claus process, 161, 344
closed-loop recycling, 193
cocurrent, 250
column internal, 98
combined parameter and state estimation scheme, 286
competitive Langmuir adsorption isotherm, 273
component space, 21, 30
computational fluid dynamics (CFD), 115
condition number, 309
continuous annular reactor, 193
continuous chromatographic process, 269
control performance specification, 324

G

gain crossover frequency, 316
gain crossover frequency range, 313
gain sign change, 308
gain validation, 321
gas chromatography, 117
General Rate Model, 269
generalised binary noise signal, 313
generalized reduced gradient, 67
global optimality, 68
glucose, 37, 231
gradient-based solver, 288

H

Hashimoto, 43, 218, 231, 259, 266,
    268, 271, 287
Hatta number, 105
HCN, 168
heat duty, 129
heat integration, 96
Height Equivalent to a Theoretical
    Stage (HETS), 105, 228
heuristic rule, 11
high-order controller, 335
hold-up, 97
homogeneous, 14, 241
hydroformylation, 98, 244
hydrogenation, 151
hydrotalcite, 98

I

identification, 282, 312, 334
impulse balance, 158
induced 2-norm, 320, 326
initialization strategy, 62, 68
in-situ extraction, 245
integral-squared error, 324
integrated heat storage, 340
integrated mass storage, 340
integrated second catalytic activity,
    340
internal accumulation, 25
isobutene, 27
isomerization, 198, 231

isotherm, 202

K

KATAPAK-S$^{®}$, 121

L

Langmuir-Hinshelwood-Hougen-
    Watson model (LHHW), 111
Le Chatelier's principle, 149
level of integration, 14
ligand, 98
Linear Driving Force (LDF), 157,
    203
linear system identification, 312
linear transfer function, 314
linearised dynamic model, 308
liquid-liquid phase separation, 120
lithium carbonate, 170
lithium zirconate, 170
logic constraint, 65
low-order controller, 328, 335

M

macromixing, 112
macrostructured adsorptive reactor,
    357
manipulated variable, 304
Mars-van Krevelen type mechanism,
    179
mass transfer correlation, 108
material balance line, 23
mathematical modelling, 269
maximize profit, 276
Maxwell-Stefan equation, 108, 157
157
mechanical stability, 100
MESH equation, 63
methanol, 27, 96, 173
methyl acetate, 11, 79, 96, 301
methyl formate, 168
methyl tertiary butyl ether, 11, 27,
    70, 97
$MgCO_3$, 169
$MgSO_4$, 176